Springer Series in Statistics

Springer Series in Statistics

Measures of Association for Cross-Classifications
Leo A. Goodman and **William H. Kruskal**
1979 / 146 pp. / 12 illus. / cloth
ISBN 0-387-**90443**-3

Statistical Decision Theory: Foundations, Concepts, and Methods
James O. Berger
1980 / 425 pp. / 20 illus. / cloth
ISBN 0-387-**90471**-9

Simultaneous Statistical Inference, Second Edition
Rupert G. Miller, Jr.
1981 / 299 pp. / 25 illus. / cloth
ISBN 0-387-**90548**-0

Point Processes and Queues: Martingale Dynamics
Pierre Brémaud
1981 / 354 pp. / 31 illus. / cloth
ISBN 0-387-**90536**-7

Non-negative Matrices and Markov Chains, Second Edition
E. Seneta
1981 / 279 pp. / cloth
ISBN 0-387-**90598**-7

Computing in Statistical Science through APL
Francis John Anscombe
1981 / 426 pp. / 70 illus. / cloth
ISBN 0-387-**90549**-9

Concepts of Nonparametric Theory
John W. Pratt and **Jean D. Gibbons**
1981 / 462 pp. / 23 illus. / cloth
ISBN 0-387-**90582**-0

Estimation of Dependences based on Empirical Data
Vladimir Vapnik
1982 / xvi, 399 pp. / 22 illus. / cloth
ISBN 0-387-**90733**-5

Applied Statistics: A Handbook of Techniques
Lothar Sachs
1982 / xxviii, 706 pp. / 59 illus. / cloth
ISBN 0-387-**90558**-8

H. Heyer

Theory of Statistical Experiments

Springer-Verlag
New York Heidelberg Berlin

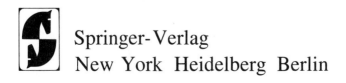

H. Heyer
Universität Tübingen
Mathematisches Institut
7400 Tübingen 1
Auf der Morgenstelle 10
West Germany

AMS Subject Classifications (1980): 62-02, 62-AXX, 62-BXX, 62-CXX

Library of Congress Cataloging in Publication Data

Heyer, Herbert.
Theory of statistical experiments.

(Springer series in statistics)
Rev. translation of: Mathematische Theorie statistischer Experimente. 1973.
Bibliography: p.
Includes indexes.
1. Mathematical statistics. I. Title. II. Series.
QA276.H49313 1982 519.5 82-19125

This is a new edition of a book, Mathematische Theorie statistischer Experimente, © 1973 by Springer-Verlag Berlin Heidelberg New York.

Printed and bound by R.R. Donnelley & Sons, Harrisonburg, VA.
Printed in the United States of America.

9 8 7 6 5 4 3 2 1

ISBN 0-387-**90785-8** Springer-Verlag New York Heidelberg Berlin
ISBN 3-540-**90785-8** Springer-Verlag Berlin Heidelberg New York

Preface

By a statistical experiment we mean the procedure of drawing a sample with the intention of making a decision. The sample values are to be regarded as the values of a random variable defined on some measurable space, and the decisions made are to be functions of this random variable.

Although the roots of this notion of statistical experiment extend back nearly two hundred years, the formal treatment, which involves a description of the possible decision procedures and a conscious attempt to control errors, is of much more recent origin. Building upon the work of R. A. Fisher, J. Neyman and E. S. Pearson formalized many decision problems associated with the testing of hypotheses. Later A. Wald gave the first completely general formulation of the problem of statistical experimentation and the associated decision theory. These achievements rested upon the fortunate fact that the foundations of probability had by then been laid bare, for it appears to be necessary that any such quantitative theory of statistics be based upon probability theory.

The present state of this theory has benefited greatly from contributions by D. Blackwell and L. LeCam whose fundamental articles expanded the mathematical theory of statistical experiments into the field of comparison of experiments. This will be the main motivation for the approach to the subject taken in this book.

The decision theory of Neyman and Wald was set up with an emphasis on those decision procedures which make the correct decisions. The quality of a decision procedure was described in probabilistic terms through its risk function, where this function describes the expected losses associated with the use of the procedure. An approach which supposes that the quality of a decision function should be based only upon its associated risk function clearly neglects further criteria which might well be important for applications, for example, the validity of the given model, the accuracy of the measurements, robustness, and computability. Despite this neglect the purely decision theoretic set up can be expanded to a stage at which other directions of applications become evident. As for examples we only mention two problems arising in the theory of Markov chains and in ergodic theory respectively where the comparison of risk functions yields answers to long standing questions. It was Bo Lindqvist who, resuming a problem of Feller's on diffusion processes in genetics, gave a decision theoretic solution to the problem of how fast a Markov chain forgets its initial state or how to measure the loss of information incurred by lumping states, and also achieved a decision theoretic characterization of weak ergodicity. Although we will not include the detailed analysis of these problems into this book, since they are lying a bit apart of the main topics of mathematical statistics, their actual importance outside mathematics has been a motivation for the general concept.

The first step of our approach will be the association of an experiment with a given decision problem. Next we shall introduce comparison relations with respect to decision problems. These relations are originally due to D. Blackwell and L. LeCam and now form the basis of the theory of comparison of statistical experiments. It turns out that the notion of an experiment chosen for this book relates to the statistical information it contains. This justifies the third step of the approach which associates with two experiments the amount of information lost, under the worst possible circumstances, by using the one experiment instead of the other.

In conclusion the concept of our presentation of the theory of statistical experiments will be the interplay of statistical decision theory with game and information theory. Along these lines we shall develop the basic ideas of non-asymptotic mathematical statistics.

The book has been conceived like its predecessor which appeared as a Hochschultext (in German) and served as a text book accompanying the standard graduate courses in mathematical statistics. At the same time it was intended to emphasize mathematical generality and rigor whenever the statistical background would support such a desire. This largely extended edition of the former Hochschultext appears in English in order to meet the penetrating demand for accessibility which has been articulated outside the German speaking professional community since the German version is on the market.

Any presentation of the basic notions of mathematical statistics should at least cover the elements of the field, treat the main problems of the theory without too much dependence on neighboring subjects and arrange the discussion around a unifying aspect. We made the attempt to fulfill these three requirements. The selection of the material discussed in the book is determined by the mathematical treatment of the basic facts from the theory of testing statistical hypotheses and estimation theory within the non-asymptotic set up. We start with an introduction to concave-convex games with emphasis on the fundamental game of statistics in which the statistician gambles against nature, proceed with the standard exposition of mathematical statistics including the existence of most powerful tests and minimal variance unbiased estimators, and turn to the theory of comparison of experiments where the various notions of sufficiency some of which have been discussed at an earlier stage, are incorporated into the general framework of LeCam's deficiency. Thus the comparison of experiments, considerably extended beyond the previous edition, in the direction of invariance, standard measures, Bayesian comparison, and extreme informativity, provides the unifying structure of this part of statistical decision theory.

Obviously the presentation of the theory relies on standard methods of functional analysis including measure and integration theory. In particular, we shall consistently apply the functional analytic properties of stochastic kernels. In order to facilitate the reading of the book we have decided to elaborate some auxiliaries and add them as appendices. Here the reader will also find references to the supplementary literature. Moreover a choice of notational conventions, classified by the subjects measure and integration and probability are designed to support a smooth handling of the text. The reader interested in the inference background of the theory is referred to the pioneering and still standard text books of D. A. S. Fraser, E. L. Lehmann, L. Schmetterer and H. Witting. For the decision theoretic approach to mathematical statistics one might also consult the books by J. O. Berger and Th. S. Ferguson. Functional analytic methods are the domineering tool in the monographs by J.-R. Barra and J.-P. Raoult.

The German edition of this book grew out of courses the author gave at the Universities of Erlangen-Nürnberg and Tübingen during the years 1969 to 1973. Naturally this primary edition had to be brought up to date, the more as the theory developed fast in a promising direction, opening new views and deeper insights into some of the major problems. Despite the obvious temptation we did not aim at the highest level of generality (suppressing for example the order theoretic approach and its implications to conical measures) but rather stayed sufficiently below in order to make the book enjoyable reading for all students working in the field.

Numerous friends and colleagues have helped me improving the presentation in this book by supplementing the material at various points, and have communicated errors and insufficiencies. I am grateful to all of them, stressing the names V. Baumann, E. Dettweiler, J. Fleming, W. Hazod, W. Hummitzsch, K. Huntzinger, H. Luschgy, D. Plachky, D. Schäfer, E. Siebert, T. P. Speed and E.-W. Zachow. Special thanks go to H. Zeuner who read the entire manuscript with much care and thought and provided many useful suggestions. Last but not least, I would like to extend my deep appreciation to L. LeCam and E. N. Torgersen for their profound contributions to the theory. Their work and the voices of friendly critics have encouraged me to present my book in expanded and reworked form to a wider public.

Tübingen, West Germany — Herbert Heyer

Spring, 1982

Contents

CHAPTER I

Games and Statistical Decisions

§1. TWO-PERSON ZERO SUM GAMES

We start with an introduction to the basic notions and properties of two-person zero sum games and their randomizations. Much emphasis is given to the interpretations of the formal definitions. A few standard examples indicate the route from the theory of games to statistical decision theory.

Definition 1.1. A *two-person zero sum game* is a triple $\Gamma = (A,B,M)$ where A and B are non empty sets and M denotes a mapping from $A \times B$ into $\overline{\mathbb{R}}$.

Remark 1.2. The sets A and B are interpreted as the *sets of strategies of two players* P_I and P_{II} resp. It will be assumed throughout the chapter that P_I and P_{II} play against each other and that they choose their strategies simultaneously, but independently. M is called the *pay-off function* of Γ. If P_I chooses $a \in A$ and P_{II} chooses $b \in B$, then P_{II} pays the amount $M(a,b)$ to P_I. Evidently the sum of gain and loss occuring in such a game Γ is 0.

Example 1.3. In the case of *roulette* P_I corresponds to the bank and P_{II} to the gambler. The set A can be identified with the set $\{0,1,\ldots,36\}$ of 37 equally probable strategies, the set B with a choice of combinations of possible outcomes at the roulette table, determined by the gambler by placing jetons. Within this framework M represents the loss of the gambler, which equals the difference between the gambled money and the amount payed out by the bank.

Example 1.4. Various *statistical problems* can be considered as two-person zero sum games. In such situations nature appears as player P_I

who chooses an element θ from a set Θ of parameters, and the statistician appears as player P_{II} who makes a decision $d \in D$ without knowledge of the choice of nature. Since P_{II} looses the amount $M(\theta,d)$ in favor of P_I (or P_I gains $M(\theta,d)$ at the cost of P_{II}), M is called the *loss function*. An example of a loss function occurring in estimating a real parameter θ is given by $M(\theta,d) := |\theta - d|^2$ for all $(\theta,d) \in \Theta \times D$.

In practical problems P_{II} has access to *observations* represented by random vectors X such that the strategies of P_{II} correspond to *decision procedures* $X \to \delta \circ X$ where $\delta \circ X$ denotes the decision to be made on the basis of the observation X. To complete the model, one introduces the pay-off as the *expected loss* which gives rise to the definition of the *risk function*.

Example 1.5. In the case of *finite games* $\Gamma = (A,B,M)$ it is assumed per definition that A and B are finite sets, i.e., of the form $\{a_1,\ldots,a_m\}$ and $\{b_1,\ldots,b_n\}$ resp. For all $i = 1,\ldots,m$; $j = 1,\ldots,n$ one defines $m_{ij} := M(a_i,b_j)$, and the pay-off function M can be considered as the *pay-off matrix* $(m_{ij}) \in \mathbb{M}(m \times n, \overline{\mathbb{R}})$. We note that, conversely, for every matrix $(m_{ij}) \in \mathbb{M}(m \times n, \overline{\mathbb{R}})$ there exists a finite game $\Gamma = (A,B,M)$ where A and B are finite sets as above and m_{ij} becomes the pay-off if P_I chooses i from $\{1,\ldots,m\}$ and P_{II} chooses j from $\{1,\ldots,n\}$.

Let $\Gamma = (A,B,M)$ denote an arbitrary (two-person zero sum) game. Since both players P_I and P_{II} tend to maximize their respective gains, they are interested in the mappings $b \to M(a_0,b)$ for all $a_0 \in A$ and $a \to M(a,b_0)$ for all $b_0 \in B$ respectively.

Definition 1.6. Let $a_1,a_2 \in A$. One says that a_1 *dominates* a_2, in symbols $a_1 > a_2$, if

$$M(a_1,b) \geq M(a_2,b)$$

for all $b \in B$. Given $A_1 \subset A$, $A_2 \subset A$, then A_1 is said to *dominate* A_2, in symbols $A_1 > A_2$, if for every $a_2 \in A_2$ there exists an $a_1 \in A_1$ such that $a_1 > a_2$. A set $A_1 \subset A$ with the property $A_1 > A$ is called *essentially complete*.

Definition 1.7. Let $b_1,b_2 \in B$. One says that b_1 *dominates* b_2, in symbols $b_1 > b_2$, if

$$M(a,b_1) \leq M(a,b_2)$$

for all $a \in A$. Given $B_1 \subset B$, $B_2 \subset B$, then B_1 is said to *dominate* B_2, in symbols $B_1 > B_2$, if for every $b_2 \in B_2$ there exists a $b_1 \in B_1$ such that $b_1 > b_2$.

Definition 1.8. For every $a \in A$ and every $b \in B$ we introduce the extended real numbers

$$M_I(a) := \inf_{b \in B} M(a,b)$$

and

$$M_{II}(b) := \sup_{a \in A} M(a,b)$$

resp.

Remark 1.9. If P_I chooses $a \in A$, P_I will certainly get an amount $\geq M_I(a)$, but not necessarily more. Thus $M_I(a)$ is a measure for the high quality of strategy a: $M_I(a)$ is the *minimum gain* of P_I if a has been chosen. M_I defines an order relation in the set A. Analogously, one interprets $M_{II}(b)$ as the *maximum loss* of P_{II} if $b \in B$ has been chosen. Again, M_{II} introduces an order relation in B.

Definition 1.10. Let

$$\underline{V}(\Gamma) := \sup_{a \in A} M_I(a)$$

and

$$\overline{V}(\Gamma) := \inf_{b \in B} M_{II}(b)$$

denote the *lower* and the *upper value* resp. of Γ. A strategy $a_0 \in A$ is called *maximin* if

$$M_I(a_0) = \underline{V}(\Gamma).$$

Analogously, a strategy $b_0 \in B$ is called *minimax* if

$$M_{II}(b_0) = \overline{V}(\Gamma).$$

For a given game Γ we shall often abbreviate $\underline{V} := \underline{V}(\Gamma)$ and $\overline{V} := \overline{V}(\Gamma)$.

If $|A| = \infty$, then $\sup_{a \in A} M_I(a)$ will not necessarily be attained in A. In this situation one is inclined to look for strategies $a \in A$ with the property that $M_I(a)$ is arbitrarily close to $\underline{V}(\Gamma)$. A corresponding statement holds for M_{II} and $\overline{V}(\Gamma)$.

Theorem 11. For all $a \in A$, $b \in B$ we have

$$M_I(a) \leq \underline{V}(\Gamma) \leq \overline{V}(\Gamma) \leq M_{II}(b).$$

Proof: Let $a' \in A$, $b' \in B$. Then

$$M_I(a') = \inf_{b \in B} M(a',b) \leq M(a',b') \leq \sup_{a \in A} M(a,b') = M_{II}(b').$$

It follows that

$$M_I(a') \leq \inf_{b' \in B} M_{II}(b') = \overline{V},$$

whence

$$\sup_{a' \in A} M_I(a') \leq \sup_{a' \in A} \inf_{b \in B} M_{II}(b) = \overline{V},$$

which implies $\underline{V}(\Gamma) \leq \overline{V}(\Gamma)$. □

Definition 1.12. Γ is said to *admit a value* if

$\underline{V}(\Gamma) = \overline{V}(\Gamma)$.

The extended real number

$V(\Gamma) := \underline{V}(\Gamma) = \overline{V}(\Gamma)$

is called the *value of* Γ.

Theorem 1.13. The following statements are equivalent:

(i) Γ admits a value, $a_0 \in A$ is maximin, and $b_0 \in B$ is minimax.

(ii) For all $a \in A$ and $b \in B$ one has

$$M(a_0,b) \geq M(a_0,b_0) \geq M(a,b_0).$$

Moreover, if one (and hence both) of the statements (i) and (ii) are fulfilled, then $V(\Gamma) = M(a_0,b_0)$.

Proof: 1. (ii) ⇒ (i). Obviously $\inf_{b \in B} M(a_0,b) \geq \sup_{a \in A} M(a,b_0)$ holds or equivalently

$$M_I(a_0) \geq M_{II}(b_0).$$

By Theorem 1.11 we get $M_I(a_0) = M_{II}(b_0)$ and $\underline{V}(\Gamma) = \overline{V}(\Gamma)$.

2. (i) ⇒ (ii). Let $\underline{V}(\Gamma) = \overline{V}(\Gamma) = V(\Gamma)$, and let a_0, b_0 be a maximin and a minimax strategy resp. Then

$$M(a_0,b) \geq M_I(a_0) = V = M_{II}(b_0) \geq M(a,b_0).$$

We choose $a := a_0$ and $b := b_0$, hence we get $V = M(a_0,b_0)$ and thus (ii). □

Example 1.14. We consider a finite game Γ with pay-off matrix $(m_{ij}) \in \mathbb{M}(m \times n, \mathbb{R})$. The inequality in (ii) of Theorem 1.13 shows that the value $V(\Gamma)$ of Γ is an element of the matrix (m_{ij}) which is minimal in its row and maximal in its column. (It determines a saddle point of (m_{ij}).) The numbers of the corresponding row and column define the maximin strategy of P_I and the minimax strategy of P_{II} resp.

Definition 1.15. Let $\Gamma = (A,B,M)$ be a game. A is called *concave* (with respect to Γ) if for all $a_1, a_2 \in A$ and $\theta \in [0,1]$ there exists an $a \in A$ such that

$$M(a,b) \geq (1 - \theta)M(a_1,b) + \theta M(a_2,b)$$

for all $b \in B$. If the above inequality is in fact an equality for all $b \in B$, then A is called *affine* (with respect to Γ).

Remark 1.16. In order to avoid the difficulty arising from an undefined convex combination in the above inequality we restrict for each $b \in B$ the function $M(\cdot,b)$ to attain at most one of the values $+\infty$ and $-\infty$.

Remark 1.17. For an *interpretation of concavity* we assume P_I to choose between $a_1 \in A$ and $a_2 \in A$. If P_I picks a_2 with probability θ, then

$$(1 - \theta)M(a_1,b) + \theta M(a_2,b)$$

is the expected gain under the condition that P_{II} picks $b \in B$.

Now let A be concave (with respect to Γ). Then P_I enjoys a strategy $a \in A$ which provides at least as large a gain as the above convex combination.

Lemma 1.18. Let A be concave, $a_1,\ldots,a_r \in A$, and $\theta_1,\ldots,\theta_r \in \mathbb{R}_+$ such that $\Sigma_{i=1}^{r} \theta_i = 1$. Then there exists an $a \in A$ satisfying

$$M(a,b) \geq \sum_{i=1}^{r} \theta_i M(a_i,b)$$

for all $b \in B$. If, moreover, A is affine, then the inequality turns into an equality.

The direct proof is left to the reader.

Definition 1.19. Let $\Gamma = (A,B,M)$ be a game. B is called *convex* (with respect to Γ) if for all $b_1, b_2 \in B$ and $\theta \in [0,1]$ there exists a $b \in B$ such that

$$M(a,b) \leq (1 - \theta)M(a,b_1) + \theta M(a,b_2)$$

holds for all $a \in A$. B is called *affine* if we have equality for all $a \in A$.

Remarks 1.16 and 1.17 and Lemma 1.18 can be rephrased in terms of convexity instead of concavity.

Definition 1.20. $\Gamma = (A,B,M)$ is said to be *concave-convex* if A is concave and B is convex (both with respect to Γ).

Example 1.21. Let A be a convex subset of $\mathbb{R}^p$ (in the traditional sense) and let $a \to M(a,b)$ be a concave function on A for all $b \in B$. Then A is concave with respect to Γ.

In fact, given $a_1, a_2 \in A$ and $\theta \in [0,1]$, one considers $a := (1 - \theta)a_1 + \theta a_2$. Then a satisfies the defining inequality of Definition 1.15.

An analogous statement is true in the case of convexity of B with respect to Γ.

But these geometric conditions implying concavity and convexity with respect to Γ are just sufficient, in general not necessary. In fact, consider a finite game Γ with pay-off matrix

$$(m_{ij}) := \begin{pmatrix} -2 & 0 \\ 0 & 1 \end{pmatrix}.$$

It is easily checked that Γ is concave-convex. Since 0 is a saddle point of (m_{ij}), we get $V(\Gamma) = 0$.

In the following we are going to introduce for a given game Γ its randomization and aim at showing that this randomization is always concave-convex with respect to itself.

For any set E the set of all probability measures on $(E, \mathfrak{P}(E))$ with finite support will be abbreviated by $\mathcal{M}_f^1(E)$.

Definition 1.22. Let $\Gamma = (A,B,M)$ be a game. We introduce the sets $A^* := \mathcal{M}_f^1(A)$, $B^* := \mathcal{M}_f^1(B)$ and the mapping $M^*: A^* \times B^* \to \overline{\mathbb{R}}$ defined by

$$M^*(a^*,b^*) := \sum_{\substack{a \in A \\ b \in B}} M(a,b)a^*(a)b^*(b)$$

for all $a^* \in A^*$, $b^* \in B^*$. The game $\Gamma^* := (A^*,B^*,M^*)$ is called a *randomization* of Γ. A^* and B^* are known as the sets of *mixed strategies* of Γ (in contrast to the sets A and B of *pure strategies* of Γ).

Discussion 1.23. We assume that player P_I^* of the randomized game Γ^* chooses strategy $a^* \in A^*$. In terms of the game Γ this means that player P_I uses strategy $a \in A$ with probability $a^*(a)$. A similar assumption is made concerning players P_{II}^* and P_{II}. From the definition of M^* we see that $M^*(a^*,b^*)$ is the expected gain of P_I.

Identifying the strategy $a \in A$ with the strategy $\varepsilon_a \in A^*$ we may consider A as a subset of A^*. Similarly one considers B as a subset of B^*. Clearly, $M^*(a,b) = M(a,b)$ for all $a \in A$, $b \in B$.

Theorem 1.24. The randomization $\Gamma^* = (A^*,B^*,M^*)$ of the game $\Gamma = (A,B,M)$ is concave-convex. Moreover, A^* and B^* are affine with respect to Γ^*.

Proof: We restrict ourselves to showing that A^* is affine with respect to Γ^*.

Let $a^{1*},a^{2*} \in A^*$ and $\theta \in [0,1]$. We define

$$a^*(a) := (1 - \theta)a^{1*}(a) + \theta a^{2*}(a)$$

for all $a \in A$. Evidently $a^* \in A^*$.

Let $b^* \in B^*$. Then

$$\begin{aligned} M^*(a^*,b^*) &= \sum_{a,b} M(a,b)a^*(a)b^*(b) \\ &= \sum_{a,b} M(a,b)[(1 - \theta)a^{1*}(a) + \theta a^{2*}(a)]b^*(b) \\ &= (1 - \theta)M^*(a^{1*},b^*) + \theta M^*(a^{2*},b^*), \end{aligned}$$

thus A^* is affine with respect to Γ^*. □

Theorem 1.25. Given the games Γ and Γ^* one has

(i) $M_I^*(a^*) = \inf_{b \in B} M^*(a^*,b)$

(ii) $M_{II}^*(b^*) = \sup_{a \in A} M^*(a,b^*)$.

Proof: For all $a^* \in A^*$, $b^* \in B^*$ we get

$$\begin{aligned} M^*(a^*,b^*) &= \sum_b \left[\sum_a M(a,b)a^*(a)\right]b^*(b) \\ &= \sum_b M^*(a^*,b)b^*(b) \\ &\geq \inf_b M^*(a^*,b), \end{aligned}$$

whence $M_I^*(a^*) \geq \inf_b M^*(a^*,b)$. On the other hand, $B \subset B^*$ implies

$$M_I^*(a^*) = \inf_{b^*} M^*(a^*,b^*) \leq \inf_{b} M^*(a^*,b).$$

This shows (i). The proof of (ii) runs similarly. □

Corollary 1.26. (i) $M_I^*(a) = M_I(a)$ for all $a \in A$.
(ii) $M_{II}^*(b) = M_{II}(b)$ for all $b \in B$.

The proof follows from the equality $M^*(a,b) = M(a,b)$ valid for all $a \in A$, $b \in B$.

Corollary 1.27. $\underline{V}(\Gamma) \leq \underline{V}(\Gamma^*) \leq \overline{V}(\Gamma^*) \leq \overline{V}(\Gamma)$.

Proof: Corollary 1.26 together with $A \subset A^*$ implies

$$\underline{V}(\Gamma^*) = \sup_{a^*} M_I^*(a^*) \geq \sup_{a} M_I^*(a) = \sup_{a} M_I(a) = \underline{V}(\Gamma)$$

and analogously, $\overline{V}(\Gamma^*) \leq \overline{V}(\Gamma)$. □

Corollary 1.28. If Γ admits the value $V(\Gamma)$, then Γ^* admits the value $V(\Gamma^*)$, and one has

$$V(\Gamma^*) = V(\Gamma).$$

The proof is an immediate consequence of Corollary 1.27. □

Remark 1.29. It is sufficient to consider a game Γ from one of the player's, say player P_I's point of view. In fact, given $\Gamma = (A,B,M)$ we consider the game $\tilde{\Gamma} := (B,A,\tilde{M})$ where

$$\tilde{M}(b,a) = -M(a,b)$$

for all $a \in A$, $b \in B$. In this case

$$\underline{V}(\tilde{\Gamma}) = -\overline{V}(\Gamma)$$

and

$$\overline{V}(\tilde{\Gamma}) = -V(\Gamma).$$

Thus $\tilde{\Gamma}$ admits a value $V(\tilde{\Gamma})$ iff Γ admits the value $V(\Gamma)$, and $\tilde{\tilde{\Gamma}} = \Gamma$.

§2. CONCAVE-CONVEX GAMES AND OPTIMALITY

The first part of this section will be devoted to establishing sufficient conditions for a game $\Gamma = (A,B,M)$ to admit a value $V(\Gamma) = \underline{V}(\Gamma) = \overline{V}(\Gamma)$.

For every $\tau \in \mathbb{R}$ and $b \in B$ the symbol $[M \geq \tau]_b$ denotes the set

$$\{a \in A: M(a,b) \geq \tau\}.$$

Clearly $[M \geq \tau]_b \subset A$.

Theorem 2.1. For every game $\Gamma = (A,B,M)$ the following conditions are equivalent:

(i) $\underline{V}(\Gamma) = \overline{V}(\Gamma)$.

(ii) For all $\tau \in \mathbb{R}$, $\tau < \overline{V}(\Gamma)$ we have

$$\bigcap_{b \in B} [M \geq \tau]_b = \emptyset.$$

Proof: 1. (ii) ⇒ (i). Let $\tau \in \mathbb{R}$, $\tau < \overline{V}(\Gamma)$. Then, by assumption there exists an $a \in A$ satisfying $M(a,b) \geq \tau$ for all $b \in B$. This implies

$$M_I(a) = \inf_b M(a,b) \geq \tau.$$

Since $\underline{V}(\Gamma) = \sup_a M_I(a)$, we obtain $\underline{V}(\Gamma) \geq \tau$. Now, we choose τ sufficiently close to $\overline{V}(\Gamma)$ and get $\underline{V}(\Gamma) \geq \overline{V}(\Gamma)$, which implies $\underline{V}(\Gamma) = \overline{V}(\Gamma)$.

2. (i) ⇒ (ii). Let $\tau \in \mathbb{R}$, $\tau < \overline{V}(\Gamma)$. Then

$$\tau < \overline{V}(\Gamma) = \underline{V}(\Gamma) = \sup_{a \in A} M_I(a).$$

It follows that there exists an $a \in A$ such that

$$\tau < M_I(a) = \inf_{b' \in B} M(a,b') \leq M(a,b)$$

holds for all $b \in B$. This implies $a \in [M \geq \tau]_b$ for all $b \in B$, whence $\bigcap_b [M \geq \tau]_b \neq \emptyset$. □

Theorem 2.2. Let $\Gamma = (A,B,M)$ be a concave-convex game such that $M < \infty$. Furthermore, let $b_1,\dots,b_m \in B$ and assume that for every $i = 1,\dots,m$ the condition $M_{II}(b_i) = \infty$ implies $M(a,b_i) > -\infty$ for all $a \in A$. Then for every $\tau < \overline{V}(\Gamma)$ we have

$$\bigcap_{i=1}^{m} [M \geq \tau]_{b_i} \neq \emptyset.$$

Before we go into the proof of the theorem proper we establish some *auxiliaries*. First of all we introduce the set

$$S: = \{(M(a,b_1),\dots,M(a,b_m)): a \in A\}.$$

Clearly $S \subset [-\infty,\infty[^m$. Next we set $H: = [\tau,\infty[^m$ and assume that

$$\bigcap_{i=1}^{m} [M \geq \tau]_{b_i} = \emptyset.$$

Then $S \cap H = \emptyset$, and

$$S \subset T: = \{y \in [-\infty,\infty[^m: y \leq x \text{ for some } x \in S\}.$$

Under these assumptions we obtain the following two lemmas.

Lemma 2.3. For all $y^1,\ldots,y^n \in T$ and $\xi_1,\ldots,\xi_n \geq 0$ such that $\Sigma_{i=1}^n \xi_i = 1$ we have $\Sigma_{i=1}^n \xi_i y^i \in T$.

Proof: For every $i = 1,\ldots,n$ let $y^i \leq x^i \in S$. Then

$$\sum_{i=1}^n \xi_i y^i \leq \sum_{i=1}^n \xi_i x^i.$$

Putting $x_j^i := M(a_i,b_j)$ for all $i = 1,\ldots,n;\ j = 1,\ldots,m$ we get

$$\sum_{i=1}^n \xi_i x_j^i = \sum_{i=1}^n \xi_i M(a_i,b_j).$$

Since A is assumed to be concave with respect to Γ, there exists an $a \in A$ satisfying

$$\sum_{i=1}^n \xi_i M(a_i,b_j) \leq M(a,b_j) = z_j$$

for all $j = 1,\ldots,m$. This implies $\Sigma_{i=1}^n \xi_i x^i \leq z \in S$ and hence the assertion. □

Lemma 2.4. The set $T' :\neq T \cap \mathbb{R}^m$ is a non-empty convex subset of $\mathbb{R}^m$ such that $T' \cap H = \emptyset$.

Proof: 1. We show $T' \neq \emptyset$. Suppose that $T' := T \cap \mathbb{R}^m = \emptyset$, i.e., that every vector $\in S$ admits at least one component $= -\infty$. Then $\Sigma_{i=1}^m M(a,b_i) = -\infty$ for all $a \in A$. Since B is convex with respect to Γ, there exists a $\tilde{b} \in B$ satisfying

$$M(a,\tilde{b}) \leq \frac{1}{m} \sum_{i=1}^m M(a,b_i) = -\infty$$

for all $a \in A$. It follows that $M_{II}(\tilde{b}) = -\infty$, hence $\overline{V}(\Gamma) = -\infty$ which serves as a contradiction of $\tau < \overline{V}(\Gamma)$. Therefore $T' \neq \emptyset$.

2. T' is convex, since by Lemma 2.3, T is convex. The statement $T' \cap H = \emptyset$ follows from the definition of T together with the fact that $S \cap H = \emptyset$. □

Proof of the Theorem: As above we assume that $\bigcap_{i=1}^m [M \geq \tau]_{b_i} = \emptyset$ holds and deduce a contradiction. From Lemma 2.4 it follows that $H := [\tau,\infty[^m$ and T' are disjoint convex subsets of $\mathbb{R}^m$. By the Hahn-Banach separation theorem there exist $\ell_1,\ldots,\ell_m \in \mathbb{R}$ with at least one $\ell_i \neq 0$ $(i = 1,\ldots,m)$ such that

$$\sum_{i=1}^{m} \ell_i x_i \geq \sum_{i=1}^{m} \ell_i y_i \quad \text{holds for all} \quad x \in H,\ y \in T'.$$

Since H is not bounded from above, $\ell_i \geq 0$ for all $i \in \{1,\ldots,m\}$, and without loss of generality we may assume that $\Sigma_{i=1}^{m} \ell_i = 1$. Since $(\tau,\ldots,\tau) \in H$ we get

$$\sum_{i=1}^{m} \ell_i y_i \leq \tau \quad \text{for all} \quad y \in T'.$$

Suppose that this inequality also holds for all $y \in S$. Then there exists a $b \in B$ satisfying

$$M(a,b) \leq \sum_{i=1}^{m} \ell_i M(a,b_i) \leq \tau$$

for all $a \in A$. It follows that $M_{II}(b) \leq \tau$, thus $\overline{V} \leq \tau$ which is a contradiction. Consequently there is a $\tilde{y} \in S \setminus T'$ such that $\Sigma_{i=1}^{m} \ell_i \tilde{y}_i > \tau$. But $\tilde{y} \in S \setminus T'$ implies $\tilde{y}_i = -\infty$ for some i. We define

$$I := \{i: \tilde{y}_i > -\infty \quad \text{for all} \quad \tilde{y} \in S \setminus T': \sum_{i=1}^{m} \ell_i \tilde{y}_i > \tau\}.$$

Therefore $I \neq \{1,\ldots,m\}$ and $\ell_i = 0$ for all $i \notin I$. Let $p \in]0,1[$ and put $p_i : \ell_i p$ for all $i \in I$. Then $\Sigma_{i \in I} p_i = p$. Define

$$p_i := \frac{1-p}{m - |I|} \quad \text{for all} \quad i \notin I.$$

Then $\Sigma_{i \notin I} p_i = 1 - p$, whence $\Sigma_{i=1}^{m} p_i = 1$, and $p_i \geq 0$ for all $i \in \{1,\ldots,m\}$. Since B is convex with respect to Γ, there exists a $b_p \in B$ such that for all $a \in A$ we have

$$\begin{aligned} M(a,b_p) &\leq \sum_{i=1}^{m} p_i M(a,b_i) \\ &= p \sum_{i \in I} \ell_i M(a,b_i) + \frac{1-p}{m-|I|} \sum_{i \notin I} M(a,b_i) \\ &= p \sum_{i=1}^{m} \ell_i M(a,b_i) + \frac{1-p}{m-|I|} \sum_{i \notin I} M(a,b_i). \end{aligned}$$

If $\Sigma_{i=1}^{m} \ell_i M(a,b_i) \leq \tau$, then

$$\begin{aligned} M(a,b_p) &\leq p\tau + \frac{1-p}{m-|I|} \sum_{i \notin I} M(a,b_i) \\ &\leq p\tau + \frac{1-p}{m-|I|} \sum_{i \notin I} M_{II}(b_i). \end{aligned}$$

If, on the other side, $\Sigma_{i=1}^{m} \ell_i M(a,b_i) > \tau$, then there exists an i satisfying $M(a,b_i) = -\infty$, i.e., $i \notin I$. In this case $M(a,b_p) = -\infty$. Thus, for all $a \in A$,

$$M(a,b_p) \leq p\tau + \frac{1-p}{m-|I|} \sum_{i \notin I} M_{II}(b_i),$$

which implies

$$\tau < \overline{V} \leq M_{II}(b_p) \leq p\tau + \frac{1-p}{m-|I|} \sum_{i \notin I} M_{II}(b_i). \qquad (*)$$

For every $i \notin I$ there is by definition of I a vector $\tilde{y} := (M(\tilde{a},b_1), \ldots, M(\tilde{a},b_m))$ such that $M(\tilde{a},b_i) = -\infty$ which by hypothesis implies $M_{II}(b_i) < \infty$. Hence $\Sigma_{i \notin I} M_{II}(b_i) < \infty$.

Now we infer from (*) with $p \to 1$ the inequalities $\tau < \overline{V} \leq \tau$ which provides the desired contradiction. □

Theorem 2.5. Let $\Gamma = (A,B,M)$ be a concave-convex game such that $|M| < \infty$. We assume

(a) There exists a sequence $(b_n)_{n \geq 1}$ in B satisfying

$$\inf_b M(a,b) = \inf_i M(a,b_i)$$

for all $a \in A$.

(b) For every sequence $(a_n)_{n \geq 1}$ in A there exists an $a \in A$ with the property

$$\underline{\lim}_n M(a_n,b) \leq M(a,b)$$

for all $b \in B$.

Then Γ admits a value.

Proof: Let $m \geq 1$ and $\tau < \overline{V}(\Gamma)$. By Theorem 2.2 we have

$$\bigcap_{i=1}^{m} [M \geq \tau]_{b_i} \neq \emptyset \quad \text{if} \quad \tau < \overline{V}(\Gamma).$$

We choose $a_m \in \bigcap_{i=1}^{m} [M \geq \tau]_{b_i}$. Assumption (b) implies that there exists an $a \in A$ such that

$$\underline{\lim}_m M(a_m,b) \leq M(a,b) \quad \text{for all} \quad b \in B.$$

By Theorem 2.1 it remains to be shown that $M(a,b) \geq \tau$ for all $b \in B$.

Taking assumption (a) into account it suffices to show that

$M(a,b_i) \geq \tau$ for all $i \geq 1$. But for all $m \geq i$ we have $M(a_m,b_i) \geq \tau$ which implies

$$M(a,b_i) \geq \overline{\lim_m} M(a_m,b_i) \geq \tau. \quad \square$$

Remark 2.6. If $A \subset \mathbb{R}^k$, $B \subset \mathbb{R}^m$ are given as in Example 1.21, if A is compact and M separately continuous on $A \times B$, then the hypotheses of the theorem are satisfied and Γ admits a value.

Theorem 2.7. Let $\Gamma = (A,B,M)$ be a concave-convex game with $M < \infty$. We assume that on A there exists a topology $\mathcal{T}$ such that

(a) A is $\mathcal{T}$-compact, and

(b) $a \to M(a,b)$ is upper semicontinuous with respect to $\mathcal{T}$, on A for all $b \in B$.

Then Γ admits a value, and P_I has a maximin strategy.

Proof: Let $\tau < \overline{V}$. Since $a \to M(a,b)$ is upper semicontinuous on A and A is compact, $[M \geq \tau]_b$ is a compact subset of A for all $b \in B$. The family $([M \geq \tau]_b)_{b \in B}$ possesses by Theorem 2.2 the finite intersection property and therefore $\bigcap_{b \in B} [M \geq \tau]_b \neq \emptyset$. Thus, by Theorem 2.1, Γ admits a value.

Moreover, $M_I(a) = \inf_b M(a,b)$, whence M_I is upper semicontinuous. It follows that there exists an $a_0 \in A$ satisfying

$$M_I(a_0) = \sup_a M_I(a) = \underline{V}(\Gamma).$$

But this shows that a_0 is a maximin strategy. $\square$

In order to reformulate the hypotheses of Theorem 2.7 we add a few topological

Properties 2.8. We consider the coarsest topology $\mathcal{T}$ on A for which all mappings $a \to M(a,b)(b \in B)$ are upper semicontinuous (on A).

2.8.1. $\mathcal{T}$ is generated by the system

$$\{O_{b,t}: b \in B, t \in \mathbb{R}\}$$

of open sets $O_{b,t} := \{a \in A: M(a,b) < t\}$.

The following two statements can be proved as exercises.

2.8.2. A net (a_α) in A $\mathcal{T}$-converges to $a \in A$ iff

$$\overline{\lim_\alpha} M(a_\alpha,b) \leq M(a,b) \quad \text{for all } b \in B.$$

2.8.3. A is $\mathscr{T}$-compact iff for every net (a_α) in A there exists an $a \in A$ satisfying

$$\underline{\lim}_{\alpha} M(a_\alpha, b) \leq M(a,b) \quad \text{for all } b \in B.$$

With the preceding properties in mind Theorem 2.7 reads as follows:

Theorem 2.9. Let $\Gamma = (A,B,M)$ be a concave-convex game with $M < \infty$. We assume that for every net (a_α) in A there exists an $a \in A$ satisfying

$$\underline{\lim}_{\alpha} M(a_\alpha, b) \leq M(a,b) \quad \text{for all } b \in B.$$

Then Γ admits a value, and P_I has a maximin strategy.

Theorem 2.10. Let $\Gamma = (A,B,M)$ be a game with $A := \{a_1,\ldots,a_m\}$, and let $|M| < \infty$. Then the randomization $\Gamma^* = (A^*,B^*,M^*)$ of Γ admits a value.

Proof: First of all we identify $a^* \in A^*$ with $(\theta_1,\ldots,\theta_m) \in \mathbb{R}^m$ where $\theta_i \geq 0$ for all $i = 1,\ldots,m$, and $\Sigma_{i=1}^m \theta_i = 1$, such that $a^*(a_i) = \theta_i$ for all $i = 1,\ldots,m$.

Let $\mathscr{T}(\mathbb{R}^m)$ denote the natural topology of $\mathbb{R}^m$. Clearly, A^* is compact with respect to the restriction to A^* of $\mathscr{T}(\mathbb{R}^m)$. But we have

$$M^*(a^*,b^*) = \sum_{a,b} M(a,b)a^*(a)b^*(b)$$

for all $a^* \in A^*$, $b^* \in B^*$. Then for fixed $b^* \in B^*$, $M^*(a^*,b^*)$ is a linear combination of $\theta_1,\ldots,\theta_m$. Therefore $a^* \to M^*(a^*,b^*)$ is continuous on A^* for all $b^* \in B^*$. But Γ^* is concave-convex by Theorem 1.24, and Theorem 2.7 implies the assertion. □

Now we are ready to introduce optimal strategies. Let $\Gamma = (A,B,M)$ be a given game.

Definition 2.11. Let $a_0 \in A$, $b \in B$, and let $\varepsilon \geq 0$. a_0 is said to be *ε-optimal for* b, in symbols $a_0 >_\varepsilon b$, if $M(a_0,b) \geq M(a,b) - \varepsilon$ for all $a \in A$. For $\varepsilon = 0$ we obtain the notion of *optimality;* the corresponding order relation will be obtained by $>$ without any subscript.

Lemma 2.12. Let $\Gamma = (A,B,M)$ be a game admitting a value $V(\Gamma)$ with $|V(\Gamma)| < \infty$. Moreover, let $a_0 \in A$ be a maximin strategy for P_I, i.e., $M_I(a_0) = V(\Gamma)$. Then for every $\varepsilon > 0$ there exists a $b_\varepsilon \in B$ satisfying

$$a_0 >_\varepsilon b_\varepsilon.$$

Proof: We have

$$V(\Gamma) = \overline{V}(\Gamma) = \inf_{b \in B} M_{II}(b).$$

Let $\varepsilon > 0$. Then there exists a $b_\varepsilon \in B$ such that

$$M_{II}(b_\varepsilon) - \varepsilon \leq V(\Gamma).$$

But $M(a_0,b_\varepsilon) \geq M_I(a_0) = V(\Gamma)$. Therefore

$$M(a_0,b_\varepsilon) \geq M_{II}(b_\varepsilon) - \varepsilon \geq M(a,b_\varepsilon) - \varepsilon$$

for all $a \in A$. □

Given a game $\Gamma = (A,B,M)$ we denote by $\tilde{A}$ the set of all strategies of player P_I that are ε-optimal for a strategy $b_\varepsilon \in B$ for all $\varepsilon > 0$.

Theorem 2.13. Let $\Gamma = (A,B,M)$ be a concave-convex game with $|M| < \infty$. We assume that

(a) there exists a topology $\mathscr{T}$ on A such that A is $\mathscr{T}$-compact and $M(\cdot,b)$ is upper semicontinuous with respect to $\mathscr{T}$ for all $b \in B$, and that

(b) B is affine with respect to Γ.

Then $\tilde{A} > A$, i.e., $\tilde{A}$ is essentially complete.

Proof: Let $\hat{a} \in A$. We define the game $\hat{\Gamma} := (A,B,\hat{M})$ with $\hat{M}(a,b) := M(a,b) - M(\hat{a},b)$ for all $a \in A$, $b \in B$. Clearly, A is concave and B is affine with respect to $\hat{\Gamma}$. Thus $\hat{\Gamma}$ is concave-convex. But then, by Theorem 2.7, $\hat{\Gamma}$ admits a value $\hat{V} := V(\hat{\Gamma})$, and P_I has a maximin strategy $\tilde{a} \in A$. From $\hat{M}_I(\hat{a}) = 0$ we conclude $\hat{V} \geq 0$. Moreover,

$$\hat{V} = \hat{M}_I(\tilde{a}) = \inf_b \hat{M}(\tilde{a},b) < \infty,$$

whence $0 \leq \hat{V} < \infty$. It is easily seen that $\tilde{a} > \hat{a}$. In fact, $\hat{V} = \hat{M}_I(\tilde{a}) \geq 0$ implies $\hat{M}(\tilde{a},b) \geq 0$ for all $b \in B$, thus $M(\tilde{a},b) \geq M(\hat{a},b)$ for all $b \in B$. It remains to be shown that $\tilde{a} \in \tilde{A}$. For this let $\varepsilon > 0$. By the above lemma there exists a $b_\varepsilon \in B$ such that

$$\hat{M}(\tilde{a},b_\varepsilon) \geq \hat{M}(a,b_\varepsilon) - \varepsilon \quad \text{for all } a \in A.$$

Since $|M| < \infty$ we may add $M(\hat{a},b_\varepsilon)$ to both sides of this inequality and get

$$\hat{M}(\tilde{a},b_\varepsilon) + M(\hat{a},b_\varepsilon) \geq \hat{M}(a,b_\varepsilon) + M(\hat{a},b_\varepsilon) - \varepsilon.$$

From this follows

$M(\tilde{a},b_\varepsilon) \geq M(a,b_\varepsilon) - \varepsilon$ for all $a \in A$,

i.e., $\tilde{a} \in \tilde{A}$. □

Definition 2.14. $a_0 \in A$ is called *admissible* if for all $a \in A$

$M(a,b) \geq M(a_0,b)$ for all $b \in B$ implies
$M(a,b) = M(a_0,b)$ for all $b \in B$.

Corollary 2.15. This corollary is to the above theorem. Every admissible strategy $a_0 \in A$ belongs to $\tilde{A}$.

Proof: Theorem 2.13 implies the existence of an $\tilde{a} \in \tilde{A}$ such that for all $b \in B$ we have

$M(\tilde{a},b) \geq M(a_0,b)$.

Since a_0 is admissible, this implies

$M(\tilde{a},b) = M(a_0,b)$ for all $b \in B$,

i.e., $a_0 \in \tilde{A}$. □

Remark 2.16. The assumption of concave-convexity in Theorem 2.13 cannot be dropped without replacement. To see this, take $\Gamma = (A,B,M)$ with $A: = [-1,1]$, $B: = \{-1,1\}$ and M defined by $M(a,b): = ab$ for all $a \in A$, $b \in B$. First of all one notes that B is not convex and hence not affine with respect to Γ. Next one shows that $\tilde{A} = \{-1,1\} = B$. Finally, $\tilde{A} \nsucc A$, since $a = 0$ is not dominated by any strategy in $\tilde{A}$.

§3. BASIC PRINCIPLES OF STATISTICAL DECISION THEORY

In order to introduce the fundamental notions of statistical decision theory we first clarify some terminology from statistical inference.

Let $(\Omega_1,\mathbf{A}_1)$ and $(\Omega,\mathbf{A})$ be two measurable spaces and let X be a measurable mapping from $(\Omega_1,\mathbf{A}_1)$ into $(\Omega,\mathbf{A})$. The observation upon which any statistical decision is based can be interpreted as the image $X(\omega_1)$ under such an X of an unobserved $\omega_1 \in \Omega_1$. In this case X is called the *sample variable* and $\omega: = X(\omega_1)$ the *sample* corresponding to the sample variable X. The measurable space $(\Omega,\mathbf{A})$ is then said to be the *sample space* and so it is this which is the space of all possible samples.

In the following X will denote a random vector $(X_1,\ldots,X_n)$ with real-valued components $X_1,\ldots,X_n$ on $(\Omega_1,\mathbf{A}_1)$, whence ω will be a point $(x_1,\ldots,x_n)$ of $\mathbb{R}^n$. We thus specialize $(\Omega,\mathbf{A})$ to be a measurable space $(A,\mathfrak{B}(A))$, where A denotes a Borel subset of $\mathbb{R}^n$ and $\mathfrak{B}(A)$ the trace on A of the Borel σ-algebra $\mathbb{B}^n := \mathbb{B}(\mathbb{R}^n)$ of $\mathbb{R}^n$. In this situation x_k is the sample corresponding to the sample variable X_k for each $k = 1,\ldots,n$ and $x = (x_1,\ldots,x_n)$ is referred to as a *sample of size* n. Under the assumption that the sample variables $X_1,\ldots,X_n$ are mutually stochastically independent and possess the same distribution, we shall speak of a *random sample of size* n. Every measurable mapping T from the sample space into a second measurable space $(\Omega',\mathbb{A}')$ is called a *statistic*. If in particular $\Omega' := \mathbb{R}$ and $\mathbb{A}' := \mathbb{B}(\mathbb{R})$, then T is said to be a *real-valued statistic* on $(\Omega,\mathbb{A})$. Any statistic T can be looked on as a measurable function of the sample variable X, i.e., as a measurable mapping $T \circ X$ from $(\Omega_1,\mathbb{A}_1)$ into $(\Omega',\mathbb{A}')$.

Definition 3.1. A (statistical) *experiment* (or model) is a triple $\mathsf{X} = (\Omega,\mathbb{A},\mathscr{P})$, where $(\Omega,\mathbb{A})$ is a measurable space and $\mathscr{P}$ is a nonempty family of measures in $\mathscr{M}^1(\Omega,\mathbb{A})$.

$(\Omega,\mathbb{A})$ is called the *sample space* (or basic space) *of the experiment* X.

For this section we shall restrict the discussion to *parametrized experiments* $\mathsf{X} = (\Omega,\mathbb{A},\mathscr{P})$, where $\mathscr{P}$ is a parametrized family $(P_i)_{i\in I}$ with *parameter set* I.

Remark 3.2. Once the statistician has started to study a phenomenon by establishing a model of the form $\mathsf{X} = (\Omega,\mathbb{A},(P_i)_{i\in I})$ he can interpret Ω as the set of measurements, $\mathbb{A}$ as the set of assertions on the measurements, and $(P_i)_{i\in I}$ as the family of all possible distributions.

The most important classes of experiments which come up in any mathematical theory of statistical experiments and take an important place in the theory of statistical decisions are the classes of testing experiments and estimation experiments.

Having established a model X the statistician in a next step performs an experiment and makes his decisions on the basis of his observations. Decisions will be statements about the "true" parameter $i \in I$ of the model X.

Standard Examples 3.3.

3.3.1. *Testing.* One considers a (null-)hypothesis of the form $H: i \in I_0$, where $I_0 \subset I$. Two decisions are possible: One can reject or accept H.

3.3.2. *Estimation.* One tries to estimate a real-valued function g of the "true" parameter $i \in I$. In this case the set of all possible decisions is a subset of $\mathbb{R}$.

Any decision procedure for an experiment which for every sample of size n specifies a decision, yields the definition of a decision function as a mapping from the sample space into the set of all possible decisions.

More precisely we proceed as follows:

Definition 3.4. A *decision space* is a measurable space $(D,\mathcal{D})$ which serves as the space of all possible decisions for the given experiment. The elements of D are called *decisions*.

If D is finite, $\mathcal{D}$ will be taken as $\mathfrak{P}(D)$.

Definition 3.5. Let $X = (\Omega,\mathfrak{A},(P_i)_{i\in I})$ be an experiment and let $(D,\mathcal{D})$ be a decision space. A *decision function* corresponding to X and $(D,\mathcal{D})$ is a Markov kernel δ from $(\Omega,\mathfrak{A})$ to $(D,\mathcal{D})$.

The totality of decision functions corresponding to an experiment X (and a fixed decision space $(D,\mathcal{D})$) will be abbreviated by $\mathcal{D}(X)$. By definition, $\mathcal{D}(X) = \mathrm{Stoch}((\Omega,\mathfrak{A}),(D,\mathcal{D}))$.

Remark 3.6. A decision function $\delta \in \mathcal{D}(X)$ defines for each $\omega \in \Omega$ a measure $\delta(\omega,\cdot)$ in $\mathcal{M}^1(D,\mathcal{D})$. The statistician chooses a decision according to this measure $\delta(\omega,\cdot)$. Such decision functions are called *randomized* in contrast to those decision functions which for each $\omega \in \Omega$ fix a decision deterministically. The latter are called *non-randomized* decision functions.

If $\delta \in \mathcal{D}(X)$ is a randomized decision function we suppose that it is used as follows: When a sample $\omega = X(\omega_1)$ arises, we perform the random experiment $(D,\mathcal{D},\delta(\omega,\cdot))$ and obtain a decision $d \in D$. In other words $\delta(\omega,A)$ is the probability that the decision arrived at belongs to $A \in \mathcal{D}$ when ω has been observed.

Definition 3.7. $\delta \in \mathcal{D}(X)$ is called a *non-randomized decision function* if there exists a mapping $\psi\colon \Omega \to D$ satisfying

$$\delta(\omega,A) = (1_A \circ \psi)(\omega) = 1_A(\psi(\omega))$$

$$= \begin{cases} 1 & \text{if } \psi(\omega) \in A \\ 0 & \text{otherwise} \end{cases},$$

whenever $\omega \in \Omega$, $A \in \mathcal{D}$.

Remark 3.8. We note that $\delta(\omega,\cdot)$ is the probability measure on $(D,\mathcal{D})$ assigning mass 1 to the set $\{\psi(\omega)\} \subset D$ provided $\{\psi(\omega)\} \in \mathcal{D}$. This restriction is the reason for choosing the σ-algebra $\mathcal{D}$ so that it contains all one-point subsets. Since

$$\delta(\omega,A) = 1_{\psi^{-1}(A)}(\omega) \quad \text{for all} \quad \omega \in \Omega,\ A \in \mathcal{D},$$

we have $\psi^{-1}(A) \in \mathfrak{A}$ for all $A \in \mathcal{D}$, and hence ψ is a measurable mapping.

For any given $\omega \in \Omega$ the element $\psi(\omega)$ is interpreted as the decision to take when $\omega \in \Omega$ has been observed.

Example 3.9. *k-decisions* (for $k \geq 1$). Let $D: = \{1,\ldots,k\}$ and $\mathcal{D}: = \mathfrak{P}(D)$. In this case, any $\delta \in \mathcal{D}(X)$ is determined by its values

$$\delta(\omega,d): = \delta(\omega,\{d\})$$

for all $\omega \in \Omega$, $d \in D$. Obviously, $\Sigma_{d=1}^{k} \delta(\omega,d) = 1$ for every $\omega \in \Omega$. If δ is a nonrandomized decision function, then for every $\omega \in \Omega$, there is a decision $d_\omega \in D$ such that $\delta(\omega,d_\omega) = 1$.

Decision spaces $(D,\mathcal{D})$ with $D: = \{1,\ldots,k\}$ are called *k-decision spaces*.

A subexample of the preceding one is

Example 3.10. *Testing statistical hypotheses.* We resume the set up of Example 3.3.1. Let I_0 be a subset of I. We are going to test the *hypothesis* $H = H_0: i \in I_0$ against the *alternative* $H_1: i \in I_1: = I \setminus I_0$. Here we assume $I_0, I_1 \neq \emptyset$, $I_0 \cup I_1 = I$ and $I_0 \cap I_1 = \emptyset$. In this situation we have $D = \{0,1\}$, where 0 corresponds to *accepting* H and 1 to *rejecting* H, and $\mathcal{D}: = \mathfrak{P}(D)$. The resulting decision function δ is defined by the numbers

$$\begin{cases} \delta(\omega,1) = P\,(\text{rejecting } H \mid \omega \text{ has been observed}) =: \phi(\omega) \\ \delta(\omega,0) = 1 - \phi(\omega), \end{cases}$$

whenever $\omega \in \Omega$.

In this case δ is called a *randomized test* of the hypothesis H_0 against (versus) the alternative H_1. δ is non-randomized if δ attains only the values 0 and 1. Under this assumption the sets

$$W: = [\phi = 1] = \{\omega \in \Omega : \phi(\omega) = 1\}$$

and

$$U: = [\phi = 0]$$

are called the *rejection region* and the *acceptance region* resp. *of the test* δ for H_0 versus H_1.

Example 3.11. *Estimating parameters.* Let I be a subset of $\mathbb{R}$. We are going to proceed with the discussion of Example 3.3.2 and aim at estimating the "true" parameter $i \in I$. In this case D is chosen as a Borel subset of $\mathbb{R}$, and $\mathcal{D}: = \mathfrak{B}(D)$. Any decision function δ determines for each observation $\omega \in \Omega$ a measure $\delta(\omega,\cdot) \in \mathcal{M}^1(D,\mathcal{D})$. Non-randomized decision functions are given by measurable functions $\psi: \Omega \to \mathbb{R}$. They are called *estimators* (for the parameter i).

For the following we shall give ourselves an experiment $X = (\Omega,\mathbf{A},(P_i)_{i\in I})$, a decision space $(D,\mathcal{D})$ and a decision function $\delta \in \mathcal{D}(X)$.

Since δ is a Markov kernel, $\delta P_i \in \mathcal{M}^1(D,\mathcal{D})$ for every $i \in I$, and δP_i can be interpreted as the expected decision when the "true" parameter is $i \in I$.

Definition 3.12. The mapping $OC_\delta: I \times \mathcal{D} \to \mathbb{R}$ given by

$$OC_\delta(i,A) := \delta P_i(A)$$

for all $(i,A) \in I \times \mathcal{D}$ is said to be the *operational characteristic* of δ.

Standard Examples 3.13.

3.13.1. *Testing*. In the case of Example 3.10 we have

$$OC_\delta(i,1) = \int \delta(\omega,1)P_i(d\omega) = \int \phi(\omega)P_i(d\omega)$$

for all $i \in I$. The mapping $i \to OC_\delta(i,1)$ on I is called the *power function* of the test δ.

If δ is non-randomized, then $OC_\delta(i,1) = P_i(W)$ for all $i \in I$, where W denotes the rejection region of δ.

3.13.2. *Estimation*. In the non-randomized case of Example 3.11 we have

$$OC_\delta(i,A) = P_i[\psi \in A]$$

for all $(i,A) \in I \times \mathcal{D}$. The mapping $A \to OC_\delta(i,A)$ on $\mathcal{D}$ is the probability distribution of ψ under the measure P_i (for $i \in I$).

Remark 3.14. The basic model of statistical decision theory can be viewed as a two-person null sum game: Player P_I is "nature" with I as its set of strategies, player P_{II} is the statistician with $\mathcal{D}(X)$ as his set of strategies. The corresponding pay-off function is called the risk function and will be constructed from a loss function as follows.

Definition 3.15. A *loss function* corresponding to the experiment X and decision space $(D,\mathcal{D})$ is a mapping $V: I \times D \to \mathbb{R}$ with the property that for every $i \in I$ the mapping $V_i: d \to V(i,d)$ is measurable.

By $\mathcal{V}$ we denote the totality of all loss functions corresponding to X and $(D,\mathcal{D})$.

Remark 3.16. For given $i \in I$ and $d \in D$ the number $V(i,d)$ can be regarded as the loss which has to be dealt with when d had been chosen and i is the "true" parameter. For d to become the correct decision for the parameter i, we should have $V(i,d) = 0$.

Definition 3.17. For every $\delta \in \mathcal{D}(X)$ and $V \in \mathcal{V}$ we introduce the *risk function* $R_\delta^V: I \to \mathbb{R}$ by

$$R_\delta^V(i) := \int V(i,t) OC_\delta(i,dt) = \iint V(i,t)\delta(\omega,dt)P_i(d\omega)$$

for all $i \in I$, whenever the defining integral exists.

Standard Examples 3.18.

3.18.1. *Testing.* Given the loss function V defined by

$$V(i,0) := \begin{cases} 0 & \text{if } i \in I_0 \\ a > 0 & \text{if } i \in I \setminus I_0 \end{cases} \quad \text{and}$$

$$V(i,1) := \begin{cases} b > 0 & \text{if } i \in I_0 \\ 0 & \text{if } i \in I \setminus I_0, \end{cases}$$

we obtain for all $i \in I_0$

$$R_\delta^V(i) = b\ OC_\delta(i,1) = b \int \phi dP_i,$$

and for all $i \in I_1 := I \setminus I_0$

$$R_\delta^V(i) = a\ OC_\delta(i,0) = a\left(1 - \int \phi dP_i\right).$$

3.18.2. *Estimation.* One often uses the loss function V defined by (the quadratic losses)

$$V(i,d) := C(i-d)^2 \quad \text{for all } (i,d) \in I \times D,$$

where C denotes a given constant.

Now we consider the two-person null sum game $(I, \mathcal{D}(X), r)$, where I denotes the parameter set of our experiment X, $\mathcal{D}(X)$ is the set of randomized decision functions corresponding to X and $(D, \mathcal{D})$, and r is the corresponding risk function defined for a given loss function $V \in \mathcal{V}$ as a mapping $I \times \mathcal{D}(X) \to \mathbb{R}$ by

$$r(i,\delta) := R_\delta^V(i)$$

for all $(i,\delta) \in I \times \mathcal{D}(X)$.

The aim of player P_{II}, i.e., the statistician, will be to search for strategies, i.e., decision functions, that are "best" in the sense of minimal risk.

We quote two of the most applicable *principles* attached to the following two notions.

Definition 3.19. $\delta_0 \in \mathscr{D}(X)$ is called *minimax* if

$$\sup_{i\in I} r(i,\delta_0) = \inf_{\delta\in\mathscr{D}(X)} \sup_{i\in I} r(i,\delta).$$

Definition 3.20. Let $(I,\mathfrak{I})$ be a measurable space. Given a decision function $\delta \in \mathscr{D}(X)$ and an *apriori distribution* $\Lambda \in \mathscr{M}^1(I,\mathfrak{I})$ we introduce the corresponding *Bayes risk* as the integral

$$r(\delta|\Lambda) := \int r(i,\delta)\Lambda(di)$$

if it exists.

$\delta_0 \in \mathscr{D}(X)$ is called *Bayesian* with respect to $\Lambda \in \mathscr{M}^1(I,\mathfrak{I})$ if

$$r(\delta_0|\Lambda) = \inf_{\delta\in\mathscr{D}(X)} r(\delta|\Lambda).$$

Remark 3.21. It should be noted that $\delta_0 \in \mathscr{D}(X)$ is minimax iff δ_0 is a minimax strategy for the game $(I,\mathscr{D}(X),r)$ in the sense of Definition 1.10, and $\delta_0 \in \mathscr{D}(X)$ is Bayesian with respect to $\Lambda \in \mathscr{M}^1(I,\mathfrak{I})$ iff δ_0 is optimal for Λ provided the underlying game is $(\mathscr{D}(X),\mathscr{M}^1(I,\mathfrak{I}), -r(\cdot|\cdot))$.

For experiments of the form $X = (\Omega,\mathfrak{A},(P_i)_{i\in I})$ with the same parameter set I *comparison relations* in the sense of decision theory can be introduced. These relations form the basis of the theory of comparison of statistical experiments. At this point we just present the main definition and illustrate an application.

Let X and $Y := (\Omega',\mathfrak{A}',(Q_i)_{i\in I})$ be two experiments with the same parameter set I, and let ε be a real function $I \to \mathbb{R}_+$.

Definition 3.22. Let $k \geq 1$. X is said to be *more informative than* Y *at level* ε (or ε-deficient with respect to Y) *for* k-*decision problems,* in symbols $X \underset{\varepsilon}{\overset{k}{>}} Y$, if for the decision space $(D,\mathscr{D})$ with $D := \{1,\ldots,k\}$ and $\mathscr{D} := \mathfrak{P}(D)$, for all bounded loss functions $V \in \mathscr{V}$ and for every $\sigma \in \mathscr{D}(Y)$ there exists a $\rho \in \mathscr{D}(X)$ such that

$$R_\rho^V \leq R_\sigma^V + \varepsilon||V||$$

holds, where

$$||V||: = \sup_{i \in I} \sup_{d \in D} |V(i,d)| = \sup_{i \in I} ||V_i||.$$

For $k = 2$ we obtain informativity at level ε for *testing problems*.

X is said to be *more informative than* Y *at level* ε (or ε-deficient with respect to Y), in symbols $X >_\varepsilon Y$, if $X >_\varepsilon^k Y$ for all $k \geq 1$.

Discussion 3.23. The comparison relation of informativity at level ε originates from the fact, that the statistician performs his observations corresponding to one of the models X or Y, but not to both of them. The question arises which of the models should favorably be chosen.

If $X >_0 Y$, then independently of any decision function $\sigma \in \mathscr{D}(Y)$ there exists a decision function $\rho \in \mathscr{D}(X)$ such that the risk of ρ is smaller than the risk of σ. In this case X is to be preferred.

If $X >_\varepsilon Y$ and $\varepsilon(i)$ is small for every $i \in I$, then to every decision function $\sigma \in \mathscr{D}(Y)$ one can find a decision function $\rho \in \mathscr{D}(X)$ which is "almost" as good as σ. In this situation only small amounts of information are lost if X is observed in place of Y.

Clearly, ε deserves to be called a *tolerance function* for X (and Y).

Example 3.24. We wish to describe an experiment leading to a test of the independence of two characteristics A and B present in a large population. The relative frequencies $p: = P(A)$ and $\pi: = P(B)$ are supposed known, and without loss of generality we may suppose that $0 \leq p \leq \pi \leq \frac{1}{2}$. The unknown joint probability is denoted by $\rho = P(A \cap B)$, and we wish to test the null hypothesis H_0: $\rho = \pi p$ against the alternative H_1: $\rho \neq \pi p$, where $0 \leq \rho \leq p$.

A reasonable experiment would be one in which a random sample of individuals is drawn, and then the individuals are cross-classified according as they possess the characteristics A and B, or not. The test would then be based upon a probabilistic model for this classification procedure (a 2×2-*contingency table*).

Initially we recognize at least five ways of doing the sampling. For example, the entire population could be sampled at random; again those individuals possessing characteristic A could be sampled; those possessing B, and so on. If we randomly sampled the population of individuals possessing characteristic A, the number also possessing B has a binomial distribution with probability $P(B|A) = \frac{\pi}{p}$, equalling π under H_0 and belonging to $[0,1] \setminus \{\pi\}$ under H_1. Similarly for the other sampling procedures.

By virtue of the assumption $0 \leq p \leq \pi \leq \frac{1}{2}$ it can be shown that for the four sampling procedures (sampling A, $\bar{A}$, B, $\bar{B}$ individuals) the first, i.e., the one involving sampling the set of individuals with the rarest characteristic, is best. By this we mean that for *any* test of H_0 versus H_1 associated with any of the last three sampling procedures, there is a test associated with the first sampling procedure which is "uniformly more powerful". A detailed solution of the problem will follow in §21.

CHAPTER II

Sufficient σ-Algebras and Statistics

§4. GENERALITIES

A notion from the theory of probability which proves to be of basic interest in mathematical statistics, is the conditional probability of a measure with respect to a σ-algebra. In particular we are interested in such σ-algebras, for which there are versions of the corresponding conditional probability that are independent of the individual measures involved. To make this idea precise we present

Definition 4.1. Let $(\Omega,\mathfrak{A},\mathscr{P})$ be an experiment and $\mathfrak{S} \subset \mathfrak{A}$ a sub-σ-algebra of $\mathfrak{A}$. $\mathfrak{S}$ is called *sufficient* (for $\mathscr{P}$ or $(\Omega,\mathfrak{A},\mathscr{P})$) if for every $A \in \mathfrak{A}$ there exists a function $Q_A \in \mathfrak{M}_{(1)}(\Omega,\mathfrak{S})$ satisfying

$$E_P^{\mathfrak{S}}(1_A) = Q_A \ [P] \quad \text{for all} \quad P \in \mathscr{P}. \tag{S}$$

In order to reformulate this definition in various ways we insert a few *notational conventions*.

A σ-ring $\mathfrak{R}$ of subsets of a set Ω is said to be a *system of null sets for a σ-algebra* $\mathfrak{A}$ in Ω if $R \in \mathfrak{R}$ and $A \in \mathfrak{A}$ implies $R \cap A \in \mathfrak{R}$.

If $\mathfrak{R}$ is a system of null sets for $\mathfrak{A}$, then on $\mathfrak{M}(\Omega,\mathfrak{A})$ we define an equivalence relation $\underset{\mathfrak{R}}{\sim}$ by

$$f \underset{\mathfrak{R}}{\sim} g :\iff [f \neq g] \in \mathfrak{R}.$$

This equivalence relation is compatible with addition, multiplication and the passage to limits of sequences of functions.

Examples of systems of null sets for $\mathfrak{A}$ are

(1) $\mathfrak{R}_\mu := \{A \in \mathfrak{A}: \mu(A) = 0\}$ for $\mu \in \mathscr{M}_+(\Omega,\mathfrak{A})$.

(2) $\mathfrak{R}_{\mathscr{M}} := \bigcap_{\mu \in \mathscr{M}} \mathfrak{R}_\mu$ for $\mathscr{M} \subset \mathscr{M}_+(\Omega,\mathfrak{A})$.

Given a system $\mathfrak{R}$ of null sets for a σ-algebra $\mathfrak{A}$ in Ω, one shows easily that the σ-algebra $\tilde{\mathfrak{A}}$ generated by $\mathfrak{A}$ and $\mathfrak{R}$ is equal to the system $\{A \triangle R: A \in \mathfrak{A}, R \in \mathfrak{R}\}$.

For any system $\mathfrak{R}$ of null sets for $\mathfrak{A}$, $f \in \mathfrak{M}(\Omega,\mathfrak{A})$ and $M \subset \mathfrak{M}(\Omega,\mathfrak{A})$ we introduce

$$[f]_{\mathfrak{R}} := \{g \in \mathfrak{M}(\Omega,\mathfrak{A}), f \underset{\mathfrak{R}}{\sim} g\}$$

and

$$[M]_{\mathfrak{R}} := \bigcup_{f \in M} [f]_{\mathfrak{R}}.$$

For two subsets M_1 and M_2 of $\mathfrak{M}(\Omega,\mathfrak{A})$ we write

$M_1 = M_2[\mathfrak{R}]$ instead of $[M_1]_{\mathfrak{R}} = [M_2]_{\mathfrak{R}}$,

$M_1 = M_2[\mu]$ instead of $M_1 = M_2[\mathfrak{R}_\mu]$ if $\mu \in \mathscr{M}_+(\Omega,\mathfrak{A})$,

and

$M_1 = M_2[\mathscr{M}]$ in place of $M_1 = M_2[\mathfrak{R}_{\mathscr{M}}]$.

Now let $(\Omega,\mathfrak{A},\mathscr{P})$ be any experiment and $\mathfrak{S}$ a sub-σ-algebra of $\mathfrak{A}$. If for every $f \in \mathfrak{M}(\Omega,\mathfrak{A})$ there exists a function $Q_f \in \mathfrak{M}(\Omega,\mathfrak{S})$ satisfying $E_P^{\mathfrak{S}}(f) = Q_f[P]$ for all $P \in \mathscr{P}$, then we shall also write $E_{\mathscr{P}}^{\mathfrak{S}}(f) = Q_f[\mathscr{P}]$.

Moreover, let T be a statistic $(\Omega,\mathfrak{A}) \to (\Omega',\mathfrak{A}')$. If for every $f \in \mathfrak{M}(\Omega,\mathfrak{A})$ there is a function $Q_f' \in \mathfrak{M}(\Omega',\mathfrak{A}')$ such that $E_P^T(f) = Q_f'$ holds $[T(P)]$ for all $P \in \mathscr{P}$, then we shall write $E_{\mathscr{P}}^T(f) = Q_f'[T(\mathscr{P})]$.

We finally put $\mathscr{P}^{\mathfrak{S}}(A) := E_{\mathscr{P}}^{\mathfrak{S}}(1_A)$ and $\mathscr{P}^T(A) := E_{\mathscr{P}}^T(1_A)$ for every $A \in \mathfrak{A}$.

<u>Theorem 4.2</u>. Let $(\Omega,\mathfrak{A},\mathscr{P})$ be an experiment and $\mathfrak{S} \subset \mathfrak{A}$ a sub-σ-algebra. The following statements are equivalent:

(i') $\mathfrak{S}$ is sufficient (for $\mathscr{P}$).

(i'') For every $A \in \mathfrak{A}$ there exists $Q_A \in \mathfrak{M}(\Omega,\mathfrak{S})$ with $E_{\mathscr{P}}^{\mathfrak{S}}(1_A) = Q_A[\mathscr{P}]$.

(i''') For all $A \in \mathfrak{A}$ we have $\bigcap_{P \in \mathscr{P}} E_P^{\mathfrak{S}}(1_A) \neq \emptyset$.

(ii') For every $f \in \mathfrak{M}_+(\Omega,\mathfrak{A})$ there exists $Q_f \in \mathfrak{M}_+(\Omega,\mathfrak{S})$ satisfying $E_{\mathscr{P}}^{\mathfrak{S}}(f) = Q_f[\mathscr{P}]$.

(ii'') For every $f \in \mathfrak{M}_+(\Omega,\mathfrak{A})$ there exists $Q_f \in \mathfrak{M}(\Omega,\mathfrak{S})$ with $E_{\mathscr{P}}^{\mathfrak{S}}(f) = Q_f[\mathscr{P}]$.

(ii''') For all $f \in \mathfrak{M}_+(\Omega,\mathfrak{A})$ we have $\bigcap_{P \in \mathscr{P}} E_P^{\mathfrak{S}}(f) \neq \emptyset$.

(iii') For every $f \in \bigcap_{P\in\mathscr{P}} \mathscr{L}^1(\Omega,\mathfrak{A},P)$ there exists a

$$Q_f \in \bigcap_{P\in\mathscr{P}} \mathscr{L}^1(\Omega,\mathfrak{S},P_{\mathfrak{S}}) \text{ satisfying } E_{\mathscr{P}}^{\mathfrak{S}}(f) = Q_f[\mathscr{P}].$$

(iii") For all $f \in \bigcap_{P\in\mathscr{P}} \mathscr{L}^1(\Omega,\mathfrak{A},P)$ we have $\bigcap_{P\in\mathscr{P}} E_P^{\mathfrak{S}}(f) \neq \emptyset$.

Proof: First of all we note that the equivalences within the three groups of properties are clear. Moreover, the implications (iii") ⇒ (i''') and (ii''') ⇒ (i''') are obvious.

1. We prove (i') ⇒ (ii'): Let (i') be satisfied and let

$$K = \{f \in \mathfrak{M}_+(\Omega,\mathfrak{A}): \text{ There exists } Q_f \in \mathfrak{M}_+(\Omega,\mathfrak{S}) \text{ satisfying } E_{\mathscr{P}}^{\mathfrak{S}}(f) = Q_f\}.$$

Plainly K is a convex cone having the following additional properties:

(α) For all $A \in \mathfrak{A}$ the indicator $1_A \in K$.

(β) If $(f_n)_{n\in\mathbb{N}}$ is a sequence in K with $f_n \uparrow f \in \mathfrak{M}(\Omega,\mathfrak{A})$, then $f \in K$.

(This is just the monotone convergence theorem for conditional expectations.)

Thus $K = \mathfrak{M}_+(\Omega,\mathfrak{A})$.

2. (i') ⇒ (iii'): Let (i') be satisfied. We put

$$L := \{f \in \bigcap_{P\in\mathscr{P}} \mathscr{L}^1(\Omega,\mathfrak{A},P): \text{ There exists } Q_f \in \bigcap_{P\in\mathscr{P}} \mathscr{L}^1(\Omega,\mathfrak{S},P_{\mathfrak{S}}) \text{ satisfying } E_{\mathscr{P}}^{\mathfrak{S}}(f) = Q_f\}.$$

Then L is a linear space possessing the following properties:

(α) $1_A \in L$ for all $A \in \mathfrak{A}$.

(β) If $(f_n)_{n\in\mathbb{N}}$ is a sequence in L with $f_n \to f \in \mathfrak{M}(\Omega,\mathfrak{A})$, and if there exists $g \in \bigcap_{P\in\mathscr{P}} \mathscr{L}^1(\Omega,\mathfrak{A},P)$ with $|f_n| \leq g$ for all $n \in \mathbb{N}$, then $f \in L$.

(This is Lebesgue's dominated convergence theorem for conditional expectations.)

Thus $L = \bigcap_{P\in\mathscr{P}} \mathscr{L}^1(\Omega,\mathfrak{A},P)$. □

We now turn to sufficient statistics. The relationship of these to sufficient σ-algebras will be discussed later.

Definition 4.3. Let $(\Omega,\mathfrak{A},\mathscr{P})$ be an experiment and $T: (\Omega,\mathfrak{A}) \to (\Omega',\mathfrak{A}')$ a statistic. T is said to be *sufficient* if for every $A \in \mathfrak{A}$ there exists a function $Q'_A \in \mathfrak{M}_{(1)}(\Omega',\mathfrak{A}')$ satisfying

$$E_{\mathscr{P}}^{T}(1_A) = Q'_A[T(\mathscr{P})]. \tag{S'}$$

Theorem 4.4. Let $(\Omega,\mathfrak{A},\mathscr{P})$ be an experiment and $T: (\Omega,\mathfrak{A}) \to (\Omega',\mathfrak{A}')$ a statistic. The following statements are equivalent:

(i) T is sufficient.

(ii) $\mathfrak{A}(T) := T^{-1}(\mathfrak{A}')$ is sufficient.

Proof: 1. Let T be sufficient and $A \in \mathbf{A}$. We choose a function $Q'_A \in \mathfrak{M}_{(1)}(\Omega',\mathbf{A}')$ with $E^T_{\mathscr{P}}(1_A) = Q'_A[T(\mathscr{P})]$ and then put $Q_A := Q'_A \circ T$. If $S := T^{-1}(A')$ (for $A' \in \mathbf{A}'$) is a set of $\mathbf{A}(T)$, then for every $P \in \mathscr{P}$ we have

$$\int_S Q_A \, dP = \int_{T^{-1}(A')} Q'_A \circ T \, dP = \int_{A'} Q'_A \, dT(P)$$
$$= \int_{A'} E^T_P(1_A) \, dT(P) = \int_{T^{-1}(A')} 1_A \, dP = \int_S 1_A \, dP,$$

whence $Q_A = E^{\mathbf{A}(T)}_{\mathscr{P}}(1_A)[\mathscr{P}]$.

2. Conversely, let $\mathbf{A}(T)$ be sufficient and take $A \in \mathbf{A}$. By definition there exists a function $Q_A \in \mathfrak{M}_{(1)}(\Omega,\mathbf{A}(T))$ with $E^{\mathbf{A}(T)}_{\mathscr{P}}(1_A) = Q_A[\mathscr{P}]$. Since Q_A is $\mathbf{A}(T)$-measurable, there exists $Q'_A \in \mathfrak{M}_{(1)}(\Omega',\mathbf{A}')$ satisfying $Q_A = Q'_A \circ T$. But for all $A' \in \mathbf{A}'$ we then have the chain of equalities

$$\int_{A'} Q'_A \, dT(P) = \int_{T^{-1}(A')} Q'_A \circ T \, dP$$
$$= \int_{T^{-1}(A')} Q_A \, dP = \int_{T^{-1}(A')} 1_A \, dP,$$

thus $Q'_A = E^T_{\mathscr{P}}(1_A)[T(\mathscr{P})]$. □

Theorem 4.5. Let $(\Omega,\mathbf{A},\mathscr{P})$ be an experiment, $T: (\Omega,\mathbf{A}) \to (\Omega',\mathbf{A}')$ a statistic, $\mathbf{S}' \subset \mathbf{A}'$ a sub-σ-algebra and $\mathbf{S} := T^{-1}(\mathbf{S}')$.

(i) If $\mathbf{S}$ is sufficient for $\mathscr{P}$, then $\mathbf{S}'$ is sufficient for $T(\mathscr{P})$.

(ii) If T is sufficient and $\mathbf{S}'$ is sufficient for $T(\mathscr{P})$, then $\mathbf{S}$ is sufficient for $\mathscr{P}$.

Proof: (i) Let $\mathbf{S}$ be sufficient for $\mathscr{P}$ and $A' \in \mathbf{A}'$. Then there exists a function $Q \in \mathfrak{M}_{(1)}(\Omega,\mathbf{S})$ with $E^{\mathbf{S}}_{\mathscr{P}}(1_{T^{-1}(A')}) = Q[\mathscr{P}]$. By $\mathbf{S} = T^{-1}(\mathbf{S}')$ there is a function $Q_{A'} \in \mathfrak{M}_{(1)}(\Omega',\mathbf{S}')$ satisfying $Q = Q_{A'} \circ T$. But then we obtain for all $S' \in \mathbf{S}'$ and $P \in \mathscr{P}$

$$\int_{S'} Q_{A'} \, dT(P) = \int_{T^{-1}(S')} Q_{A'} \circ T \, dP = \int_{T^{-1}(S')} Q \, dP$$
$$= P(T^{-1}(S') \cap T^{-1}(A')) = (T(P))(S' \cap A')$$
$$= \int_{S'} 1_{A'} \, dT(P),$$

whence $Q_{A'} = E^{\mathbf{S}'}_{T(\mathscr{P})}(1_{A'})[T(\mathscr{P})]$.

(ii) Let T be a sufficient statistic and $\mathbf{S}' \subset \mathbf{A}'$ a sub-σ-algebra of $\mathbf{A}'$ which is sufficient for $T(\mathscr{P})$. Then for every $A \in \mathbf{A}$ there exists a function $q'_A \in \mathfrak{M}_{(1)}(\Omega',\mathfrak{A}')$ with $E^T_{\mathscr{P}}(1_A) = q'_A[T(\mathscr{P})]$. Moreover, by Theorem 4.2 there exists a function $Q'_A \in \mathfrak{M}_{(1)}(\Omega',\mathfrak{A}')$ satisfying $E^{\mathfrak{S}'}_{T(\mathscr{P})}(q'_A) = Q'_A[T(\mathscr{P})]$.

Putting $Q_A := Q'_A \circ T$ we obtain for all $S := T^{-1}(S') \in \mathfrak{S}$ with $S' \in \mathfrak{S}'$ and for all $P \in \mathscr{P}$

$$\int_S Q_A\, dP = \int_{T^{-1}(S')} Q'_A \circ T\, dP = \int_{S'} Q'_A dT(P) = \int_{S'} q'_A\, dT(P)$$
$$= \int_{T^{-1}(S')} 1_A\, dP = \int_S 1_A\, dP,$$

whence $Q_A = E^{\mathfrak{S}}_{\mathscr{P}}(1_A)[\mathscr{P}]$. □

Example 4.6. Let $(\Omega,\mathfrak{A},P)$ be an arbitrary experiment. Then

(a) $\mathfrak{A}$ is always sufficient.

(b) $\{\emptyset,\Omega\}$ is sufficient if and only if $\mathscr{P}$ is a one element set.

Concerning the proof of (a) one just observes that for every $A \in \mathfrak{A}$ the indicator 1_A equals $E^{\mathfrak{A}}_P(1_A)[P]$ for all $P \in \mathscr{P}$. Now let $\{\emptyset,\Omega\}$ be sufficient and P_0, $P \in \mathscr{P}$. For every $A \in \mathfrak{A}$ there exists a function $Q_A \in \mathfrak{M}_{(1)}(\Omega,\{\emptyset,\Omega\})$, i.e., a function of the form $Q_A = q_A 1_\Omega$ with $q_A \in [0,1]$, satisfying $Q_A = E_P^{\{\emptyset,\Omega\}}(1_A) = E_{P_0}^{\{\emptyset,\Omega\}}(1_A)$. Therefore for all $A \in \mathfrak{A}$ one has $P(A) = \int_\Omega Q_A\, dP = q_A = \int_\Omega Q_A\, dP_0 = P_0(A)$ which implies $P = P_0$ or $\mathscr{P} = \{P_0\}$.

Example 4.7. Let $(\Omega,\mathfrak{A},\mathscr{P})$ be an experiment with $\Omega := \{0,1\}^n$, $\mathfrak{A} := \mathfrak{P}(\Omega)$ and $\mathscr{P} := \{(p\varepsilon_0 + (1-p)\varepsilon_1)^{\otimes n} : p \in]0,1[\}$ and $T: (\Omega,\mathfrak{A}) \to (\mathbb{Z},\mathfrak{P}(\mathbb{Z}))$ be defined by

$$T(x) := \sum_{k=1}^n x_k \quad \text{for all} \quad x := (x_1,\ldots,x_n) \in \Omega.$$

Then T is sufficient.

Indeed, for $A \in \mathfrak{A}$, $i \in \{0,1,\ldots,n\}$ and $p \in]0,1[$ we have

$$[T(1_A(p\varepsilon_0 + (1-p)\varepsilon_1)^{\otimes n})](\{i\})$$
$$= [1_A(p\varepsilon_0 + (1-p)\varepsilon_1)^{\otimes n}]\Big(\{(x_1,\ldots,x_n) \in \Omega : \sum_{k=1}^n x_k = i\}\Big)$$
$$= \mathrm{card}\Big(A \in \{(x_1,\ldots,x_n) \in \Omega : \sum_{k=1}^n x_k = i\}\Big)p^{n-i}(1-p)^i.$$

In particular,

$$T((p\varepsilon_0 + (1-p)\varepsilon_1)^{\otimes n})(\{i\}) = \binom{n}{i}p^{n-i}(1-p)^i.$$

We now define the function $Q'_A \in \mathfrak{M}_{(1)}(\mathbb{Z},\mathfrak{P}(\mathbb{Z}))$ by

$$Q'_A(i): = \begin{cases} \frac{1}{\binom{n}{i}} \operatorname{card}(A \cap \{(x_1,\ldots,x_n) \in \Omega: \sum_{k=1}^{n} x_k = i\}) & \text{if } i \in \{0,1,\ldots,n\}, \\ 0 & \text{if } i \notin \{0,1,\ldots,n\}. \end{cases}$$

It is easily checked that $E_P^T(1_A) = Q'_A$ holds for all $P \in \mathscr{P}$.

Example 4.8. Let $(\Omega',\mathfrak{A}')$ be a measurable space and $(\Omega,\mathfrak{A}): = (\Omega'^n,\mathfrak{A}'^{\otimes n})$ for some $n \in \mathbb{N}$. The points of Ω will be denoted by $\omega: = (\omega'_1,\ldots,\omega'_n)$.

Let Σ_n be the set of all permutations of the set $\{1,\ldots,n\}$. For every $\pi \in \Sigma_n$ one defines a mapping $T_\pi: (\Omega,\mathfrak{A}) \to (\Omega,\mathfrak{A})$ by $T_\pi(\omega'_1,\ldots,\omega'_n): = (\omega'_{\pi(1)},\ldots,\omega'_{\pi(n)})$ for all $(\omega'_1,\ldots,\omega'_n) \in \Omega$. Furthermore, let us introduce the set $\mathscr{P}$ and the σ-algebra $\mathfrak{S}$ where $\mathscr{P}: = \{P'^{\otimes n}: P' \in \mathscr{M}^1(\Omega',\mathfrak{A}')\}$ and $\mathfrak{S}: = \{A \in \mathfrak{A}: T_\pi^{-1}(A) = A$ for all $\pi \in \Sigma_n\}$. Then $\mathfrak{S}$ is sufficient for $\mathscr{P}$.

To see this take $A \in \mathfrak{A}$ and define the function $Q_A \in \mathfrak{M}_{(1)}(\Omega,\mathfrak{A})$ by

$$Q_A(\omega): = \frac{1}{n!} \sum_{\pi\in\Sigma_n} 1_A(\omega'_{\pi(1)},\ldots,\omega'_{\pi(n)})$$

for all $\omega: = (\omega'_1,\ldots,\omega'_n) \in \Omega$.

It is easily verified that Q_A satisfies $E_{\mathscr{P}}^{\mathfrak{S}}(1_A) = Q_A[\mathscr{P}]$.

Remark 4.9. The last two examples indicate that sufficient statistics in some sense preserve the information contained in the underlying experiment. A detailed treatment of this aspect of the notion will follow in Chapter VII, where sufficiency will be reinterpreted within the theory of comparison of experiments.

§5. PROPERTIES OF THE SYSTEM OF ALL SUFFICIENT σ-ALGEBRAS

It is the aim of this section to study operations on sufficient σ-algebras which preserve the sufficiency. For any measurable space $(\Omega,\mathfrak{A})$ and a family $\mathscr{M} \subset \mathscr{M}_+(\Omega,\mathfrak{A})$ we define the system

$$\mathfrak{N}_{\mathscr{M}}: = \{A \in \mathfrak{A}: \text{ Either } A \in \mathfrak{R}_{\mathscr{M}} \text{ or } CA \in \mathfrak{R}_{\mathscr{M}}\}.$$

Clearly $\mathfrak{N}_{\mathscr{M}}$ is a sub-σ-algebra of $\mathfrak{A}$.

Now let $(\Omega,\mathfrak{A},\mathscr{P})$ be an experiment. The sub-σ-algebra $\mathfrak{N}_{\mathscr{P}}$ specific to the family $\mathscr{P}$ of the experiment plays an important technical role in the theory of sufficiency.

Definition 5.1. Given two sub-σ-algebras $\mathfrak{S}_1,\mathfrak{S}_2$ of $\mathfrak{A}$ we say that $\mathfrak{S}_1$ is *$\mathscr{P}$-contained* in $\mathfrak{S}_2$, in symbols $\mathfrak{S}_1 \subset \mathfrak{S}_2[\mathscr{P}]$, if $\mathfrak{S}_1 \vee \mathfrak{N}_{\mathscr{P}} \subset \mathfrak{S}_2 \vee \mathfrak{N}_{\mathscr{P}}$, and that $\mathfrak{S}_1$ is *$\mathcal{P}$-equivalent* to $\mathfrak{S}_2$, in symbols $\mathfrak{S}_1 \sim \mathfrak{S}_2[\mathcal{P}]$, if $\mathfrak{S}_1 \subset \mathfrak{S}_2[\mathcal{P}]$ and $\mathfrak{S}_2 \subset \mathfrak{S}_1[\mathcal{P}]$.

Theorem 5.2. Let $(\Omega,\mathfrak{A},\mathcal{P})$ be an experiment. Then

(i) For two sub-σ-algebras $\mathfrak{S}_1,\mathfrak{S}_2$ of $\mathfrak{A}$ the following conditions are equivalent:

(i') For every $S_1 \in \mathfrak{S}_1$ there is an $S_2 \in \mathfrak{S}_2$ with $P(S_1 \Delta S_2) = 0$ for all $P \in \mathcal{P}$.

(i") For every $f_1 \in \mathfrak{M}(\Omega,\mathfrak{S}_1)$ there exists $f_2 \in \mathfrak{M}(\Omega,\mathfrak{S}_2)$ with $P[f_1 \neq f_2] = 0$ for all $P \in \mathcal{P}$.

(i"') $\mathfrak{S}_1 \subset \mathfrak{S}_2[\mathcal{P}]$.

(ii) If $\mathfrak{S}_1,\mathfrak{S}_2$ are two sub-σ-algebras of $\mathfrak{A}$ with $\mathfrak{S}_1 \sim \mathfrak{S}_2[\mathcal{P}]$, then $\mathfrak{S}_1$ is sufficient if and only if $\mathfrak{S}_2$ is sufficient (for $\mathcal{P}$).

Proof: Since we have the obvious implication (i) ⇒ (ii), it suffices to show the equivalences in (i).

1. (i') ⇒ (i"). Let (i') be satisfied and let us define a class L by

$$\{f \in \mathfrak{M}(\Omega,\mathfrak{S}_1)\colon \text{There is } g \in \mathfrak{M}(\Omega,\mathfrak{S}_2) \text{ with } P[f \neq g] = 0 \text{ for all } P \in \mathcal{P}\}.$$

The class L is clearly a linear space and since $P(S_1 \Delta S_2) = P[1_{S_1} \neq 1_{S_2}] = 0$ for $S \in \mathfrak{S}_1$, $S_2 \in \mathfrak{S}_2$ and $P \in \mathcal{P}$, L contains the indicator functions of all sets in $\mathfrak{S}_1$. It is also closed under the passage to limits of sequences and so we conclude that $L = \mathfrak{M}(\Omega,\mathfrak{S}_1)$.

2. (i") ⇒ (i"'). Take any $S_1 \in \mathfrak{S}_1$. By (i") we can choose a function $f_2 \in \mathfrak{M}(\Omega,\mathfrak{S}_2)$ with $P[1_{S_1} \neq f_2] = 0$ for all $P \in \mathcal{P}$. From $[1_{S_1} \neq f_2] \in \mathfrak{N}_{\mathcal{P}}$ and $S_1 \cap [1_{S_1} \neq f_2] \in \mathfrak{N}_{\mathcal{P}}$ we obtain

$$\begin{aligned} 1_{S_1} &= 1_{S_1} 1_{[1_{S_1}=f_2]} + 1_{S_1} 1_{[1_{S_1} \neq f_2]} \\ &= f_2 1_{[1_{S_1}=f_2]} + 1_{S_1 \cap [1_{S_1} \neq f_2]} \in \mathfrak{M}(\Omega,\mathfrak{S}_2 \vee \mathfrak{N}_{\mathcal{P}}), \end{aligned}$$

whence $\mathfrak{S}_1 \subset \mathfrak{S}_2 \vee \mathfrak{N}_{\mathcal{P}}$ and thus $\mathfrak{S}_1 \vee \mathfrak{N}_{\mathcal{P}} \subset \mathfrak{S}_2 \vee \mathfrak{N}_{\mathcal{P}}$.

3. (i"') ⇒ (i'). This follows from the fact that for any system $\mathfrak{R}$ of null sets for the σ-algebra $\mathfrak{A}$, the σ-algebra $\tilde{\mathfrak{A}}$ generated by $\mathfrak{A}$ and $\mathfrak{R}$ equals $\{A \Delta R\colon A \in \mathfrak{A}, R \in \mathfrak{R}\}$. □

The following results concern permanence properties of the system of all sub-σ-algebras of $\mathfrak{A}$ sufficient for a fixed family $\mathcal{P}$ of probability measures on $\mathfrak{A}$. These properties include transitivity, passage to the limit of isotone or antitone sequences, and the formation of intersections.

Theorem 5.3. Let $(\Omega,\mathfrak{A},\mathcal{P})$ be an experiment, $\mathfrak{S}$ and $\mathfrak{A}_0$ sub-σ-algebras of $\mathfrak{A}$ with $\mathfrak{S} \subset \mathfrak{A}_0$. If $\mathfrak{A}_0$ is sufficient for $\mathcal{P}$ and $\mathfrak{S}$ is sufficient for $\{P_{\mathfrak{A}_0} : P \in \mathcal{P}\}$, then $\mathfrak{S}$ is sufficient for $\mathcal{P}$.

Proof: Let the assumptions of the theorem be satisfied and take $A \in \mathfrak{A}$. We first choose a function $Q_A^0 \in \mathfrak{M}_{(1)}(\Omega,\mathfrak{A}_0)$ such that for all $A_0 \in \mathfrak{A}_0$ and all $P \in \mathcal{P}$ we have

$$\int_{A_0} Q_A^0 \, dP = P(A \cap A_0).$$

Then we apply Theorem 4.2 to this function Q_A^0 obtaining a function $Q_A \in \mathfrak{M}_{(1)}(\Omega,\mathfrak{S})$ satisfying

$$\int_S Q_A dP_{\mathfrak{A}_0} = \int_S Q_A^0 dP_{\mathfrak{A}_0} \quad \text{for all } S \in \mathfrak{S} \text{ and } P_{\mathfrak{A}_0} \text{ with } P \in \mathcal{P}.$$

Thus for every $A \in \mathfrak{A}$ there is a function $Q_A \in \mathfrak{M}_{(1)}(\Omega,\mathfrak{A})$ such that for all $S \in \mathfrak{S}$ we have

$$\int_S Q_A dP = \int_S Q_A dP_{\mathfrak{A}_0} = \int_S Q_A^0 dP_{\mathfrak{A}_0} = \int_S Q_A^0 dP = P(A \cap S),$$

that is, $\mathfrak{S}$ is sufficient for $\mathcal{P}$. □

Theorem 5.4. Let $(\Omega,\mathfrak{A},\mathcal{P})$ be an experiment and $(\mathfrak{S}_n)_{n\in\mathbb{N}}$ a sequence of sufficient sub-σ-algebras of $\mathfrak{A}$. If $(\mathfrak{S}_n)_{n\in\mathbb{N}}$ is isotone or antitone (in the natural ordering of inclusion), then $\bigvee_{n\in\mathbb{N}} \mathfrak{S}_n$ or $\bigcap_{n\in\mathbb{N}} \mathfrak{S}_n$ is sufficient (for $\mathcal{P}$), respectively.

Proof: Both statements are consequences of the martingale convergence theorem for ascending or descending families of σ-algebras, respectively. We carry out the proof for the first mentioned case only. Let $A \in \mathfrak{A}$ be given and for every $n \in \mathbb{N}$ take a function $Q_A^n \in \mathfrak{M}_{(1)}(\Omega,\mathfrak{S}_n)$ such that $E_P^{\mathfrak{S}_n}(1_A) = Q_A^n$ holds $[P]$ for all $P \in \mathcal{P}$. Since $(\mathfrak{S}_n)_{n\in\mathbb{N}}$ is isotone, for $n,m \in \mathbb{N}$ with $n < m$ and all $P \in \mathcal{P}$ we have

$$E_P^{\mathfrak{S}_n}(Q_A^m) = E_P^{\mathfrak{S}_n}(E_P^{\mathfrak{S}_m}(1_A)) = E_P^{\mathfrak{S}_n}(1_A) = Q_A^n \; [P].$$

Now $0 \leq Q_A^n \leq 1$ for all $n \in \mathbb{N}$ and so the sequence $\{Q_A^n : n \in \mathbb{N}\}$ forms a nonnegative martingale with respect to $\{\mathfrak{S}_n : n \in \mathbb{N}\}$ for each $P \in \mathcal{P}$.

Let

$$K := \{\omega \in \Omega : \underline{\lim}_{n\geq 1} Q_A^n(\omega) = \overline{\lim}_{n\geq 1} Q_A^n(\omega)\} \quad \text{and} \quad Q_A := 1_K \cdot \lim_{n\to\infty} Q_A^n.$$

The martingale convergence theorem for isotone sequences of σ-algebras then yields the relations $P(K) = 1$ and

$$Q_A = E_P^{\bigvee_{n \in \mathbb{N}} \mathfrak{S}_n}(1_A)\,[P] \quad \text{for all} \quad P \in \mathcal{P},$$

from which the sufficiency of $\bigvee_{n \in \mathbb{N}} \mathfrak{S}_n$ for $\mathcal{P}$ follows. □

Theorem 5.5. Let $(\Omega, \mathfrak{A}, \mathcal{P})$ be an experiment and let $\mathfrak{S}_1, \mathfrak{S}_2$ be sufficient sub-σ-algebras of $\mathfrak{A}$. If either $\mathfrak{N}_{\mathcal{P}} \subset \mathfrak{S}_1$ or $\mathfrak{N}_{\mathcal{P}} \subset \mathfrak{S}_2$, then $\mathfrak{S}_1 \cap \mathfrak{S}_2$ is also sufficient (for $\mathcal{P}$).

Remark 5.6. In order to understand the hypothesis of the theorem we point out that in general

$$(\mathfrak{S}_1 \cap \mathfrak{S}_2) \vee \mathfrak{N}_{\mathcal{P}} \neq (\mathfrak{S}_1 \vee \mathfrak{N}_{\mathcal{P}}) \cap (\mathfrak{S}_2 \vee \mathfrak{N}_{\mathcal{P}}).$$

Any condition which gives equality instead of inequality in this relationship implies the assertion of the theorem.

Proof of Theorem 5.5: 1. Let $A \in \mathfrak{A}$ and $f_0 := 1_A$. Since $\mathfrak{S}_1$ and $\mathfrak{S}_2$ are assumed to be sufficient for $\mathcal{P}$, we can construct recursively $\mathfrak{S}_1$-measurable functions f_{2k+1} and $\mathfrak{S}_2$-measurable functions f_{2k+2} satisfying $E_{\mathcal{P}}^{\mathfrak{S}_1}(f_{2k}) = f_{2k+1}\,[\mathcal{P}]$ and $E_{\mathcal{P}}^{\mathfrak{S}_2}(f_{2k+1}) = f_{2k+2}\,[\mathcal{P}]$ for all $k \geq 1$.

2. For a fixed $P \in \mathcal{P}$ we consider the linear operator $T_P := E_P^{\mathfrak{S}_1} E_P^{\mathfrak{S}_2}$ on $L^2(\Omega, \mathfrak{A}, P)$. We have $||T_P||_2 \leq 1$, and by the L^2-ergodic theorem there exists a projection operator Π_P with the properties

$$L_2 - \lim_{n \to \infty} \frac{1}{n} \sum_{k=0}^{n-1} T_P^k f = \Pi_P f \quad \text{for all} \quad f \in L^2(\Omega, \mathfrak{A}, P),$$

and

$$\Pi_P = T_P \circ \Pi_P.$$

Since $||\Pi_P||_2 \leq 1$, Π_P is an orthogonal projection.

Now we shall show that $\Pi_P = E_P^{\mathfrak{S}_P}$ with $\mathfrak{S}_P := (\mathfrak{S}_1 \vee \mathfrak{N}_P) \cap (\mathfrak{S}_2 \vee \mathfrak{N}_P)$. For this it suffices to show that $\Pi_P f$ is $\mathfrak{S}_P$-measurable for all $f \in L^1(\Omega, \mathfrak{A}, P)$ and that $\Pi_P f = f\,[P]$ holds for all $\mathfrak{S}_P$-measurable $f \in L^1(\Omega, \mathfrak{A}, P)$.

Since $T_P^k f$ is $\mathfrak{S}_2$-measurable for all $k \geq 1$, also

$$L_2 - \lim_{n \to \infty} \frac{1}{n} \sum_{k=1}^{n} T_P^k f = L_2 - \lim_{n \to \infty} \frac{1}{n} \sum_{k=0}^{n-1} T_P^k f = \Pi_P f$$

is $\mathfrak{S}_2$-measurable. Moreover, we have

$$||\Pi_P f||_2 \geq ||E_P^{\mathfrak{S}_1}\Pi_P f||_2 \geq ||E_P^{\mathfrak{S}_2}E_P^{\mathfrak{S}_1}\Pi_P f||_2$$
$$= ||T_P\Pi_P f||_2 = ||\Pi_P f||_2 .$$

$E_P^{\mathfrak{S}_1}$ is an orthogonal projection, whence $E_P^{\mathfrak{S}_1}(\Pi_P f) = \Pi_P f[P]$. Therefore $\Pi_P f$ is $\mathfrak{S}_1 \vee \mathfrak{S}_P$-measurable and thus also $\mathfrak{S}_P$-measurable. For any $\mathfrak{S}_P$-measurable $f \in L^1(\Omega,\mathfrak{A},P)$ we have

$$T_P f = E_P^{\mathfrak{S}_2}E_P^{\mathfrak{S}_1} f = E_P^{\mathfrak{S}_2} f = f[P],$$

whence $\Pi_P f = f[P]$.

3. By the Dunford-Schwartz ergodic theorem for positive contractions on $L^1(\Omega,\mathfrak{A},P)$ and $L^\infty(\Omega,\mathfrak{A},P)$ we obtain

$$\lim_{n\to\infty} \frac{1}{n}\sum_{k=0}^{n-1} T_P^k f = E_P^{\mathfrak{S}_P}(f)[P].$$

4. We now reconsider the functions f_k defined in 1., and observe that $f_{2k} = T_P^k(1_A)[P]$ for all $P \in \mathscr{P}$. We put

$$g(\omega) := \begin{cases} \lim\limits_{n\to\infty} \frac{1}{n}\sum\limits_{k=0}^{n-1} f_{2k}(\omega) & \text{if the limit exists} \\ 0 & \text{otherwise.} \end{cases}$$

From 3. follows that $g = E_P^{\mathfrak{S}_P}(1_A)[P]$ for all $P \in \mathscr{P}$. Analogously one shows that for the function h defined by

$$h(\omega) := \begin{cases} \lim\limits_{n\to\infty} \frac{1}{n}\sum\limits_{k=0}^{n-1} f_{2k+1}(\omega) & \text{if the limit exists} \\ 0 & \text{otherwise,} \end{cases}$$

one has

$$h = E_P^{\mathfrak{S}_P}(f_1) = E_P^{\mathfrak{S}_P}(E_P^{\mathfrak{S}_1 \vee \mathfrak{N}_P}(1_A)) = E_P^{\mathfrak{S}_P}(1_A)[P]$$

for all $P \in \mathscr{P}$. Consequently, $h = g[\mathscr{P}]$.

5. Since $\mathfrak{S}_1$ is sufficient, there exists a $\mathfrak{S}_1$-measurable function g' on Ω with $E_P^{\mathfrak{S}_1}(g) = g'[\mathscr{P}]$. Lebesgue's dominated convergence theorem implies for all $P \in \mathscr{P}$

$$g' = E_P^{\mathfrak{S}_1}(g) = \lim_{n\to\infty} \frac{1}{n}\sum_{k=0}^{n-1} E_P^{\mathfrak{S}_1}(f_{2k})$$
$$= \lim_{n\to\infty} \frac{1}{n}\sum_{k=0}^{n-1} f_{2k+1} = h = g[P].$$

Hence g is $\mathfrak{S}_1 \vee \mathfrak{N}_{\mathscr{P}}$-measurable. Similarly one shows that g is $\mathfrak{S}_2 \vee \mathfrak{N}_{\mathscr{P}}$-measurable and thus that g is measurable with respect to $\mathfrak{S} := (\mathfrak{S}_1 \vee \mathfrak{N}_{\mathscr{P}}) \cap (\mathfrak{S}_2 \vee \mathfrak{N}_{\mathscr{P}}) = (\mathfrak{S}_1 \cap \mathfrak{S}_2) \vee \mathfrak{N}_{\mathscr{P}}$. Therefore there exists an $\mathfrak{S}_1 \cap \mathfrak{S}_2$-measurable function g'' with $g = g''[\mathscr{P}]$. It remains to show that $E_P^{\mathfrak{S}_1 \cap \mathfrak{S}_2}(1_A) = g''[P]$ for all $P \in \mathscr{P}$, which yields the sufficiency of $\mathfrak{S}_1 \cap \mathfrak{S}_2$ for $\mathscr{P}$.

In fact, for all $S \in \mathfrak{S}_1 \cap \mathfrak{S}_2$ we have

$$\int_S 1_A \, dP = \int_S E_P^{\mathfrak{S}_1}(1_A)\,dP = \int_S f_1 \, dP$$
$$= \int_S f_2 \, dP = \int_S \frac{1}{n} \sum_{k=0}^{n-1} f_{2k}\,dP,$$

whence by the Lebesgue dominated convergence theorem,

$$\int_S 1_A \, dP = \int_S \left(\lim_{n\to\infty} \frac{1}{n} \sum_{k=0}^{n-1} f_{2k}\right) dP = \int_S g \, dP = \int_S g'' \, dP,$$

which had to be proved. □

§6. COMPLETENESS AND MINIMAL SUFFICIENCY

When we introduced the notion of sufficiency, we noted that the trivial σ-algebra $\{\emptyset,\Omega\}$ in Ω is sufficient for the experiment $(\Omega,\mathfrak{A},\mathscr{P})$ if and only if $\mathscr{P}$ is a one element set. Thus for a family $\mathscr{P}$ of at least two measures in $\mathscr{M}^1(\Omega,\mathfrak{A})$ the smallest sufficient σ-algebra, if it exists is different from $\{\emptyset,\Omega\}$. A property of the trivial sub-σ-algebra $\{\emptyset,\Omega\}$ of the σ-algebra $\mathfrak{A}$ of an experiment $(\Omega,\mathfrak{A},\{P\})$ is that any function which is P - a.e. $\{\emptyset,\Omega\}$-measurable (i.e., any P - a.e. constant function) $f \in \mathfrak{M}(\Omega,\mathfrak{A})$ satisfying $\int f\,dP = 0$ equals 0 P - a.e. This property will be axiomatized in this section. We shall arrive at the notion of a $\mathscr{P}$-complete σ-algebra. Moreover, minimality properties of sufficient σ-algebras and statistics will be studied, the intention being to describe in detail the close relationship between completeness and minimal sufficiency for general experiments.

<u>Definition 6.1</u>. Let $(\Omega,\mathfrak{A},\mathscr{P})$ be an experiment, $\mathfrak{S} \subset \mathfrak{A}$ a sub-σ-algebra and $p \in [1,\infty[$.

(a) $\mathfrak{S}$ is called p-*complete* (for $\mathscr{P}$ or $(\Omega,\mathfrak{A},\mathscr{P})$) if for every $\mathfrak{S}$-measurable function $f \in \bigcap_{P\in\mathscr{P}} \mathscr{L}^p(\Omega,\mathfrak{A},P)$ the statement $E_P(f) = 0$ for all $P \in \mathscr{P}$ implies $f = 0[\mathscr{P}]$. A 1-complete sub-σ-algebra $\mathfrak{S}$ of $\mathfrak{A}$ is called *complete*.

(b) $\mathfrak{S}$ is said to be *boundedly complete* (for $\mathscr{P}$ or $(\Omega,\mathfrak{A},\mathscr{P})$) if for every $f \in \mathfrak{M}^b(\Omega,\mathfrak{S})$ with $E_P(f) = 0$ for all $P \in \mathscr{P}$ we have $f = 0[\mathscr{P}]$.

Definition 6.2. A statistic $T: (\Omega,\mathfrak{A}) \to (\Omega',\mathfrak{A}')$ is called *complete (boundedly complete)* if the σ-algebra $\mathfrak{A}(T)$ is complete (boundedly complete

We illustrate the notions of complete and boundedly complete σ-algebras and statistics by a theorem and a few particular cases.

Theorem 6.3. Let $(\Omega,\mathfrak{A},\mathscr{P})$ be an experiment and $\mathfrak{S} \subset \mathfrak{A}$ a sub-σ-algebr

(i) $\mathfrak{S}$ is p-complete for $p \in [1,\infty[$ (boundedly complete) if and only if for some statistic $T: (\Omega,\mathfrak{A}) \to (\Omega',\mathfrak{A}')$ with $\mathfrak{S} = \mathfrak{A}(T)$ the σ-algebra $\mathfrak{A}'$ is p-complete (boundedly complete) for $T(\mathscr{P}) = \{T(P): P \in \mathscr{P}\}$.

(ii) Let $\mathfrak{S}$ be complete for $\mathscr{P}$ and $\mathscr{P}'$ a subset of $\mathscr{M}^1(\Omega,\mathfrak{A})$ with $\mathscr{P}' \supset \mathscr{P}$ such that $\mathfrak{N}_{\mathscr{P}} = \mathfrak{N}_{\mathscr{P}'}$. Then $\mathfrak{S}$ is complete for $\mathscr{P}'$.

Proof: (i) Let $\mathfrak{S}$ be p-complete for $\mathscr{P}$ and T be a statistic $(\Omega,\mathfrak{A}) \to (\Omega',\mathfrak{A}')$ with $\mathfrak{S} = \mathfrak{A}(T)$. Moreover, let $f' \in \bigcap_{P\in\mathscr{P}} \mathscr{L}^p(\Omega',\mathfrak{A}',T(P))$ satisfy $\int f'\, dT(P) = 0$ for all $P \in \mathscr{P}$. By the transformation theorem for integrals we obtain

$$\int f' \circ T\, dP = \int f'\, dT(P) = 0 \quad \text{for all} \quad P \in \mathscr{P}.$$

On the other hand, one has $f' \circ T \in \bigcap_{P\in\mathscr{P}} \mathscr{L}^p(\Omega,\mathfrak{A},P)$ and that $f' \circ T$ is measurable with respect to $\mathfrak{A}(T) = \mathfrak{S}$. The p-completeness of $\mathfrak{S}$ implies that $P[f' \circ T \neq 0] = 0$ for all $P \in \mathscr{P}$, whence $T(P)[f' \neq 0] = 0$ for all $P \in \mathscr{P}$. In other words, $\mathfrak{A}'$ is p-complete for $T(\mathscr{P})$.

Suppose that conversely, the σ-algebra $\mathfrak{A}'$ is p-complete for $T(\mathscr{P})$ where T is some statistic $(\Omega,\mathfrak{A}) \to (\Omega',\mathfrak{A}')$ with $\mathfrak{S} = \mathfrak{A}(T)$, and let $f \in \bigcap_{P\in\mathscr{P}} \mathscr{L}^p(\Omega,\mathfrak{S},P)$ with $E_P(f) = 0$ for all $P \in \mathscr{P}$. Then there exists an $\mathfrak{A}'$-measurable function f' on Ω' with $f = f' \circ T$. It is clear that $f' \in \bigcap_{P\in\mathscr{P}} \mathscr{L}^p(\Omega',\mathfrak{A}',T(P))$ and $E_{T(P)}(f') = 0$ for all $P \in \mathscr{P}$. Hence by the p-completeness of $\mathfrak{A}'$ we get $f' = 0[T(\mathscr{P})]$, i.e. $f = 0[\mathscr{P}]$. Consequently $\mathfrak{S}$ is p-complete for $\mathscr{P}$.

Both implications hold also for bounded completeness, since $f' \in \mathfrak{M}^b(\Omega',\mathfrak{A}')$ implies $f' \circ T \in \mathfrak{M}^b(\Omega,T^{-1}(\mathfrak{A}'))$.

(ii) Let $f \in \mathfrak{M}(\Omega,\mathfrak{S}) \cap \bigcap_{P'\in\mathscr{P}'} \mathscr{L}^1(\Omega,\mathfrak{A},P')$ be such that $E_{P'}(f) = 0$ for all $P' \in \mathscr{P}'$. Then clearly $E_P(f) = 0$ for all $P \in \mathscr{P}$. Since $\mathfrak{S}$ is complete for $\mathscr{P}$, this implies $P[f \neq 0] = 0$ for all $P \in \mathscr{P}$. But by assumption we conclude $P'[f \neq 0] = 0$ for all $P' \in \mathscr{P}'$. This, however, shows the completeness of $\mathfrak{S}$ for $\mathscr{P}'$. □

Example 6.4. Let $\Omega := \{1,2,3\}$, $\mathfrak{A} := \mathfrak{P}(\Omega)$, $\mathfrak{S} := \{\{1\}\{2,3\},\emptyset,\Omega\}$, $\mathscr{P}_1 := \{\varepsilon_3\}$, $\mathscr{P}_2 := \{\frac{1}{2}(\varepsilon_1 + \varepsilon_2)\}$ and $\mathscr{P} := \mathscr{P}_1 \cup \mathscr{P}_2$.

Then $\mathfrak{S}$ is complete for $\mathscr{P}$ and $\mathscr{P}_1$, but not for $\mathscr{P}_2$, and $\mathfrak{A}$ is complete for $\mathscr{P}_1$, but not for $\mathscr{P}_2$.

Example 6.5. Let $(\Omega,\mathfrak{A})$ be an arbitrary measurable space and $\mathscr{P} := \{\varepsilon_\omega : \omega \in \Omega\}$. Then every sub-$\sigma$-algebra $\mathfrak{S}$ of $\mathfrak{A}$ is complete for $\mathscr{P}$.

Example 6.6. Let $(\Omega,\mathfrak{A},\mathscr{P})$ be an experiment with $\Omega = \mathbb{R}^n$, $\mathfrak{A} := \mathfrak{B}^n$ and $\mathscr{P} := \{\nu_{a,1}^{\otimes} : a \in \mathbb{R}\}$. Moreover let $T := \overline{X}$ with

$$\overline{X}(x_1,\dots,x_n) := \frac{1}{n}\sum_{k=1}^{n} x_k \quad \text{for all } (x_1,\dots,x_n) \in \mathbb{R}^n.$$

Then T is complete for $\mathscr{P}$. Indeed, $T(\nu_{a,1}^{\otimes n}) = \overline{X}(\nu_{a,1}^{\otimes n}) = \nu_{a,\frac{1}{n}}$.

Now let $f \in \bigcap_{a\in\mathbb{R}} \mathscr{L}^1(\mathbb{R},\mathfrak{B},\nu_{a,\frac{1}{n}})$ satisfy

$$0 = \int f \, dT(\nu_{a,1}^{\otimes n}) = \int f(y) e^{-\frac{n}{2}(y^2+a^2)} e^{yna} dy = e^{-\frac{n}{2}a^2} \int f(y) e^{-\frac{n}{2}y^2} e^{yna} dy$$

for all $a \in \mathbb{R}$. Then by the uniqueness theorem for Laplace transforms we obtain $f(y)e^{-\frac{n}{2}y^2} = 0$ for λ - a.a. $y \in \mathbb{R}$, hence $f = 0[\lambda]$ which implies $f = 0\,[\nu_{a,\frac{1}{n}}]$ for all $a \in \mathbb{R}$, or equivalently, $f = 0[T(\nu_{a,1}^{\otimes n})]$ for all $a \in \mathbb{R}$. But this is simply the completeness of T for $\mathscr{P}$.

Example 6.7. There exists a σ-algebra $\mathfrak{A}$ in a set Ω, which is boundedly complete, but not complete for a set $\mathscr{P}$ of measures in $\mathscr{M}^1(\Omega,\mathfrak{A})$.

In fact: Let $\Omega := \{-1,0,1,2,3,\dots\}$, $\mathfrak{A} := \mathfrak{P}(\Omega)$ and

$$\mathscr{P} := \left\{\frac{1}{2-\alpha}\left(\varepsilon_{-1} + (1-\alpha)^2 \sum_{n\geq 0} \alpha^n \varepsilon_n : \alpha \in \,]0,1]\right\}.\right.$$

Then $\mathfrak{A}$ serves as the desired example.

Let $f \in \mathfrak{M}^b(\Omega,\mathfrak{A})$ with $E_P(f) = 0$ for all $P \in \mathscr{P}$, i.e., with $f(-1) + f(0) + (f(1) - 2f(0))\alpha + \Sigma_{n\geq 2} (f(n) - 2f(n-1) + f(n-2))\alpha^n = 0$ for all $\alpha \in \,]0,1[$.

From the identity theorem for power series we conclude that all coefficients vanish, i.e., that $f(0) = -f(-1)$, $f(1) = -2f(-1)$, $f(2) = 2f(1) - f(0) = -3f(-1),\dots,$ finally by induction, that $f(n) = -(n+1)f(-1)$ $(n \geq 0)$ holds.

Since f was assumed to be bounded, we obtain $f \equiv 0$. If $f \in \mathbb{R}^\Omega$ is not bounded, then the above conclusion breaks down: Simply define f by

$f(-1): = -1$ and $f(n): = n + 1$ for all $n \geq 0$. Then the previous computation yields $E_p(f) = 0$ for all $P \in \mathscr{P}$, whereas f does not vanish
identically. Thus $\mathbb{A}$ is not complete for $\mathscr{P}$.

Remark 6.8. Example 6.7 can be slightly extended: The σ-algebra
$\mathbb{A}$ introduced above fails to be p-complete for all $p \in [1,\infty[$ and $\mathscr{P}$.
The proof is based on the fact that the function f of the example lies
in $\bigcap\limits_{P \in \mathscr{P}} \mathscr{L}^p(\Omega,\mathbb{A},P)$ for all $p \in [1,\infty[$, this following from the quotient
test for infinite series.

We now introduce the notions of minimal sufficient σ-algebras and
statistics.

Definition 6.9. Let $(\Omega,\mathbb{A},\mathscr{P})$ be an experiment and $\mathfrak{S}$ a sufficient
sub-σ-algebra of $\mathbb{A}$.

$\mathfrak{S}$ is called a *minimal sufficient* *σ-algebra* if for any other suffici-
ent σ-algebra $\mathfrak{T}$ in $\mathbb{A}$ we have $\mathfrak{S} \vee \mathfrak{N}_{\mathscr{P}} \subset \mathfrak{T} \vee \mathfrak{N}_{\mathscr{P}}$.

Remark 6.10. Clearly minimal sufficient σ-algebras are $\mathscr{P}$-equi-
valent.

In contrast with the definition of a minimal sufficient sub-σ-algebra
we define the minimal sufficient statistic as follows.

Definition 6.11. Let $(\Omega,\mathbb{A},\mathscr{P})$ be an experiment and T: $(\Omega,\mathbb{A})$ →
$(\Omega',\mathbb{A}')$ any statistic.

T is said to be *minimal sufficient* if for every sufficient statistic
V: $(\Omega,\mathbb{A})$ → $(\Omega'',\mathbb{A}'')$, there exists a (not necessarily measurable) mapping
S: $\Omega'' \to \Omega'$ satisfying $T = S \circ V[\mathscr{P}]$.

Remark 6.12. In the examples 7 and 8 of §9 we shall show, that for
a statistic T the σ-algebra $\mathbb{A}(T)$ can be minimal sufficient without
T being minimal sufficient, and that T can be minimal sufficient without $\mathbb{A}(T)$ being minimal sufficient.

We are discussing minimal sufficiency as it has been introduced in
the literature, but realize that this minimality is defined in a fairly
general sense and not necessarily with respect to an order relation.

First properties on minimal sufficient statistics are contained in
the following

Theorem 6.13. Let $(\Omega,\mathbb{A},\mathscr{P})$ be an experiment.

(i) If $\mathscr{P}' \subset \mathscr{P}$ is such that $\mathfrak{N}_{\mathscr{P}} = \mathfrak{N}_{\mathscr{P}'}$ holds and if T is a
      statistic $(\Omega,\mathbb{A})$ → $(\Omega',\mathbb{A}')$ sufficient for $\mathscr{P}$ and minimal
      sufficient for $\mathscr{P}'$, then T is minimal sufficient for $\mathscr{P}$.

(ii) Let $(\Omega',\mathbf{A}')$ and $(\Omega'',\mathbf{A}'')$ be two measurable spaces and ϕ an isomorphism $(\Omega',\mathbf{A}') \to (\Omega'',\mathbf{A}'')$ (of measurable spaces). Then a statistic $T: (\Omega,\mathbf{A}) \to (\Omega',\mathbf{A}')$ is minimal sufficient if and only if the statistic $\phi \circ T: (\Omega,\mathbf{A}) \to (\Omega'',\mathbf{A}'')$ is minimal sufficient.

Proof: (i) Let $V: (\Omega,\mathbf{A}) \to (\Omega'',\mathbf{A}'')$ be another statistic sufficient for $\mathcal{P}$. Then V is also sufficient for $\mathcal{P}'$ and so there exists a mapping $S: \Omega'' \to \Omega'$ satisfying $P[T \neq S \circ V] = 0$ for all $P \in \mathcal{P}'$. It then follows from our assumptions that $P[T \neq S \circ V] = 0$ for all $P \in \mathcal{P}$. But this implies the minimal sufficiency of T for $\mathcal{P}$.

(ii) Let T be minimal sufficient and let V be a sufficient statistic $(\Omega,\mathbf{A}) \to (\Omega''',\mathbf{A}''')$. Then there exists a mapping $S: \Omega''' \to \Omega'$ with $T = S \circ V[\mathcal{P}]$. We then note that the mapping $\phi \circ S: \Omega''' \to \Omega'$ satisfies $\phi \circ T = (\phi \circ S) \circ V[\mathcal{P}]$, proving the assertion in one direction. On the other hand let $\phi \circ T$ be minimal sufficient and suppose that the mapping $V: (\Omega,\mathbf{A}) \to (\Omega''',\mathbf{A}''')$ satisfies $\phi \circ T = S \circ V[\mathcal{P}]$. Then the mapping $\phi^{-1} \circ S: \Omega''' \to \Omega'$ satisfies $T = \phi^{-1} \circ \phi \circ T = (\phi^{-1} \circ S) \circ V[\mathcal{P}]$. □

The following result is important for it gives a complete description of the relationship between the minimal sufficiency of a statistic T and that of its generated σ-algebra $\mathbf{A}(T)$.

Theorem 6.14. Let $(\Omega,\mathbf{A},\mathcal{P})$ be an experiment admitting the following property:

(P) For every function $f \in \mathfrak{M}(\Omega,\mathbf{A})$ and every set $M \subset \mathbb{R}$ with $f^{-1}(M) \in \mathbf{A}$ there exist sets $B_1, B_2 \in \mathfrak{B}$ with $B_1 \subset M \subset B_2$ such that $f^{-1}(B_2 \setminus B_1) \in \mathfrak{R}_{\mathcal{P}}$.

Then a statistic $T: (\Omega,\mathbf{A}) \to (\mathbb{R}^n,\mathfrak{B}^n)$ is minimal sufficient if and only if $\mathbf{A}(T)$ is a minimal sufficient sub-σ-algebra of $\mathbf{A}$.

Proof: The proof is based upon some results concerning Borel isomorphisms and induced σ-algebras, and so we begin by stating these separately.

1. For every $n \in \mathbb{N} \cup \{\infty\}$ the measurable spaces $(\mathbb{R},\mathfrak{B})^{\otimes n}$, $(\overline{\mathbb{R}},\overline{\mathfrak{B}})^{\otimes n}$ and $([0,1],\mathfrak{B}([0,1]))^{\otimes n}$ are isomorphic to $(\mathbb{R},\mathfrak{B})$.

2. For every measurable space $(\Omega,\mathbf{A})$ and every separable sub-σ-algebra $\mathfrak{S}$ of $\mathbf{A}$ there exists an $f \in \mathfrak{M}^b(\Omega,\mathbf{A})$ with $\mathfrak{S} = \mathbf{A}(f)$.

To see this let $\mathfrak{E} := \{A_k: k \in \mathbb{N}\}$ be a finite or countably infinite system of generators for $\mathfrak{S}$. We define a mapping $\psi: (\Omega,\mathbf{A}) \to (\mathbb{R},\mathfrak{B})^{\otimes \mathbb{N}}$ by $pr_k \circ \psi := 1_{A_k}$ for all $k \in \mathbb{N}$, where pr_k denotes the kth coordinate mapping $(\mathbb{R},\mathfrak{B})^{\mathbb{N}} \to (\mathbb{R},\mathfrak{B})$ for all $k \in \mathbb{N}$.

Plainly $A_k = \psi^{-1}[pr_k = 1] \in \mathfrak{A}(\psi)$ for $A_k \in \mathfrak{S}$ $(k \in \mathbb{N})$, whence $\mathfrak{S} \subset \mathfrak{A}(\psi)$. On the other hand the mappings $pr_k \circ \psi = 1_{A_k}$ are $\mathfrak{S}$-measurable for all $k \in \mathbb{N}$. This implies that ψ is $\mathfrak{S}$-measurable, i.e., $\mathfrak{A}(\psi) \subset \mathfrak{S}$. Thus we have shown that $\mathfrak{A}(\psi) = \mathfrak{S}$. Composing ψ with an isomorphism $\phi: (\mathbb{R},\mathfrak{B})^{\otimes \mathbb{N}} \to (\mathbb{R},\mathfrak{B})$ gives us $\mathfrak{S} = \mathfrak{A}(f)$ where $f := \phi \circ \psi$.

3. Let $(\Omega,\mathfrak{A})$ be a measurable space and $f,g \in \mathfrak{M}(\Omega,\mathfrak{A})$. It is known that $\mathfrak{A}(f) \subset \mathfrak{A}(g)$ if and only if there exists a measurable mapping $h: (\mathbb{R},\mathfrak{B}) \to (\mathbb{R},\mathfrak{B})$ with $f = h \circ g$.

This statement follows readily from the factorization theorem for measurable mappings.

We turn now to the main part of the proof of the theorem.

4. By Theorem 6.13(ii) we may assume without loss of generality that $n = 1$. Let $T: (\Omega,\mathfrak{A}) \to (\mathbb{R},\mathfrak{B})$ be minimal sufficient and suppose $\mathfrak{S}$ to be a sub-σ-algebra of $\mathfrak{A}$ which is sufficient for $\mathscr{P}$. In order to show $\mathfrak{A}(T) \vee \mathfrak{N}_{\mathscr{P}} \subset \mathfrak{S} \vee \mathfrak{N}_{\mathscr{P}}$ we consider the σ-algebra $\mathfrak{C} := (\mathfrak{A}(T) \vee \mathfrak{N}_{\mathscr{P}}) \cap (\mathfrak{S} \vee \mathfrak{N}_{\mathscr{P}})$. By Theorem 5.4 $\mathfrak{C}$ is sufficient for $\mathscr{P}$. Let $\{A_k: k \in \mathbb{N}\}$ be a countable system of generators of $\mathfrak{A}(T)$. Then for every $k \in \mathbb{N}$ we can choose a function $Q_{A_k} \in \mathfrak{M}_{(1)}(\Omega,\mathfrak{C})$ with $E_P^{\mathfrak{C}}(1_{A_k}) = Q_{A_k}[P]$ for all $P \in \mathscr{P}$. Let the σ-algebra generated by the set $\{Q_{A_k}: k \in \mathbb{N}\}$ be denoted by $\mathfrak{D}$.

By 2. there are functions $f,g \in \mathfrak{M}^b(\Omega,\mathfrak{A})$ with $\mathfrak{A}(f) = \mathfrak{D}$ and $\mathfrak{A}(g) = \mathfrak{D} \vee \mathfrak{A}(T)$. From 3. we infer the existence of a function $h: (\mathbb{R},\mathfrak{B}) \to (\mathbb{R},\mathfrak{B})$ satisfying $h \circ g = f$. Since $\mathfrak{A}(T) \subset \mathfrak{A}(g) \subset \mathfrak{A}(T) \vee \mathfrak{N}_{\mathscr{P}}$, we get without loss of generality $P[g \neq T] = 0$ for all $P \in \mathscr{P}$, hence $f = h \circ T[\mathscr{P}]$. On the other side there is a mapping $S: \mathbb{R} \to \mathbb{R}$ with $T = S \circ f[\mathscr{P}]$, since f is sufficient and T is minimal sufficient. Let $N_1, N_2 \in \mathfrak{N}_{\mathscr{P}}$ satisfy $P(N_1) = P(N_2) = 0$ for all $P \in \mathscr{P}$ and $[f \neq h \circ T] \subset N_1$, $[T \neq S \circ f] \subset N_2$ respectively. Then we have for all $B \in \mathfrak{B}$

$$\begin{aligned} T^{-1}(B) &\,\Delta\, (f^{-1}(S^{-1}(B)) \cap CN_2) \\ &= (T^{-1}(B) \cap f^{-1}(S^{-1}(CB))) \cup (T^{-1}(B) \cap N_2) \\ &\quad \cup (T^{-1}(CB) \cap CN_2 \cap f^{-1}(S^{-1}(B))) \subset N_2 \end{aligned}$$

and

$$\begin{aligned} f^{-1}(B) &\,\Delta\, T^{-1}(h^{-1}(B)) \\ &= (f^{-1}(B) \cap T^{-1}(h^{-1}(CB))) \cup (f^{-1}(CB) \cap T^{-1}(h^{-1}(B))) \subset N_1, \end{aligned}$$

i.e.,

$$f^{-1}(\mathfrak{B}) \vee \mathfrak{N}_{\mathscr{P}} = T^{-1}(\mathfrak{B}) \vee \mathfrak{N}_{\mathscr{P}},$$

which implies

$$\mathfrak{C} \subset T^{-1}(\mathfrak{B}) \vee \mathfrak{N}_{\mathscr{P}} = f^{-1}(\mathfrak{B}) \vee \mathfrak{N}_{\mathscr{P}} = \mathfrak{D} \vee \mathfrak{N}_{\mathscr{P}} \subset \mathfrak{C} \vee \mathfrak{N}_{\mathscr{P}} = \mathfrak{C},$$

whence $T^{-1}(\mathfrak{B}) \vee \mathfrak{N}_{\mathscr{P}} = \mathfrak{C} \subset \mathfrak{S} \vee \mathfrak{N}_{\mathscr{P}}$, and this shows the minimal sufficiency of the σ-algebra $\mathfrak{A}(T) = T^{-1}(\mathfrak{B})$.

4. Conversely, let $\mathfrak{A}(T)$ be a minimal sufficient sub-σ-algebra of $\mathfrak{A}$ and let $V: (\Omega,\mathfrak{A}) \to (\Omega',\mathfrak{A}')$ be an arbitrary sufficient statistic. By assumption we have $\mathfrak{A}(T) \subset \mathfrak{A}(V)[\mathscr{P}]$, and so there exists an $\mathfrak{A}(V)$-measurable real function T_0 on Ω satisfying $T_0 = T[\mathscr{P}]$. This function can be factorized in the form $T_0 = S \circ V$ and so we have $T = S \circ V[\mathscr{P}]$. But this shows the minimal sufficiency of T.

Theorem 6.15. Let $(\Omega,\mathfrak{A},\mathscr{P})$ be an experiment and $\mathfrak{S}$ a sub-σ-algebra of $\mathfrak{A}$, which is sufficient and boundedly complete for $\mathscr{P}$. Then $\mathfrak{S}$ is minimal sufficient for $\mathscr{P}$.

Proof: Let $\mathfrak{S}_1$ be any sufficient sub-σ-algebra of $\mathfrak{A}$. We shall show that $\mathfrak{S} \subset \mathfrak{S}_1[\mathscr{P}]$ holds. Choose $S \in \mathfrak{S}$. Since $\mathfrak{S}$ and $\mathfrak{S}_1$ are sufficient σ-algebras, there exist functions $Q_S^{(1)} \in \mathfrak{M}_{(1)}(\Omega,\mathfrak{S}_1)$ and $\overline{Q}_S \in \mathfrak{M}_{(1)}(\Omega,\mathfrak{S})$ with $E_P^{\mathfrak{S}_1}(1_S) = Q_S^{(1)}[P]$ and $E_P^{\mathfrak{S}}(E_P^{\mathfrak{S}_1}(1_S)) = E_P^{\mathfrak{S}}(Q_S^{(1)}) = \overline{Q}_S\,[P]$ respectively for all $P \in \mathscr{P}$. But then for every $P \in \mathscr{P}$ we have $\int \overline{Q}_S\, dP = \int E_P^{\mathfrak{S}}(E_P^{\mathfrak{S}_1}(1_S))dP = \int 1_S\, dP$ and so, by the bounded completeness of $\mathfrak{S}$ we conclude that $\overline{Q}_S = 1_S[P]$. Since conditional expectations are orthogonal projections, we have

$$\int 1_S\, dP \geq \int [Q_S^{(1)}]^2\, dP \geq \int [\overline{Q}_S]^2\, dP = \int 1_S\, dP,$$

whence $S \,\Delta\, [Q_S^{(1)} = 1] \in \mathfrak{N}_{\mathscr{P}}$. This completes the proof. □

CHAPTER III

Sufficiency under Additional Assumptions

§7. SUFFICIENCY IN THE SEPARABLE CASE

In this section we pose the question of how far we can simplify or strengthen certain results concerning sufficiency if the given experiment $(\Omega,\mathfrak{A},\mathscr{P})$ admits a separable σ-algebra $\mathfrak{A}$.

We recall that in general sub-σ-algebras of separable σ-algebras are not separable. For example, the Borel σ-algebra $\mathfrak{B}$ of $\mathbb{R}$ is separably generated, but the sub-σ-algebra $\mathfrak{S}$ of $\mathfrak{B}$ containing all subsets which are either countable or have countable complements is not separably generated.

Let $(\Omega,\mathfrak{A},P)$ be a probability space with a separable σ-algebra $\mathfrak{A}$. Then every sub-σ-algebra of $\mathfrak{A}$ is at least P-equivalent to a separably generated σ-algebra. The example $(\mathbb{R},\mathfrak{B},\{\varepsilon_x: x \in \mathbb{R}\})$ shows, however, that even this weak version of separability is in general not inherited by sub-σ-algebras.

Theorem 7.1. Let $(\Omega,\mathfrak{A},\mathscr{P})$ be an experiment and let $\mathfrak{S}$ be a sub-σ-algebra of $\mathfrak{A}$ which is sufficient for $\mathscr{P}$. Then for any separable sub-σ-algebra $\mathfrak{C}$ of $\mathfrak{A}$ the σ-algebra $\mathfrak{S} \vee \mathfrak{C}$ is sufficient for $\mathscr{P}$.

The proof of the theorem is based on a lemma which contains a straightforward generalization of the well-known formula on conditional expectations with respect to a finite σ-algebra.

Lemma 7.2. Let $(\Omega,\mathfrak{A},P)$ be a probability space and let $\mathfrak{C}$ be the sub-σ-algebra of $\mathfrak{A}$ generated by an $\mathfrak{A}$-measurable partition $\{C_1,\ldots,C_n\}$ of Ω. Then for every sub-σ-algebra $\mathfrak{S}$ of $\mathfrak{A}$ and for each $f \in L^1(\Omega,\mathfrak{A},P)$ we have

$$E_P^{\mathfrak{S}\vee\mathfrak{C}}(f) = \sum_{k=1}^{n} 1_{C_k} \frac{E_P^{\mathfrak{S}}(f\cdot 1_{C_k})}{E_P^{\mathfrak{S}}(1_{C_k})} \; [P]. \qquad (*)$$

Proof: 1. First of all we note that for all $C \in \mathfrak{A}$, $C \subset [E_P^{\mathfrak{S}}(1_C) > 0][P]$. Indeed, for each $C \in \mathfrak{A}$ we have

$$\begin{aligned} P(C \cap [E_P^{\mathfrak{S}}(1_C) > 0]) &= \int_{[E_P^{\mathfrak{S}}(1_C)>0]} 1_C \, dP \\ &= \int_{[E_P^{\mathfrak{S}}(1_C)>0]} E_P^{\mathfrak{S}}(1_C)\,dP = \int_\Omega E_P^{\mathfrak{S}}(1_C)\,dP \\ &= P(C). \end{aligned}$$

Thus we have shown that the right side of (*) is defined P - a.e.

2. It suffices to prove the asserted formula for all $f \in L_+^1(\Omega,\mathfrak{A},P)$ and for any $\cap$-stable system generating the σ-algebra $\mathfrak{S} \vee \mathfrak{C}$. But for every $S \in \mathfrak{S}$ and $1 \leq i \leq n$ we get

$$\begin{aligned} \int_{S\cap C_i} \sum_{k=1}^{n} 1_{C_k} \frac{E_P^{\mathfrak{S}}(f\cdot 1_{C_k})}{E_P^{\mathfrak{S}}(1_{C_k})}\,dP &= \int_S 1_{C_i} \frac{E_P^{\mathfrak{S}}(f\cdot 1_{C_i})}{E_P^{\mathfrak{S}}(1_{C_i})}\,dP \\ &= \int_S E_P^{\mathfrak{S}}\left(1_{C_i} \frac{E_P^{\mathfrak{S}}(f\cdot 1_{C_i})}{E_P^{\mathfrak{S}}(1_{C_i})}\right)dP = \int_S E_P^{\mathfrak{S}}(1_{C_i}) \frac{E_P^{\mathfrak{S}}(f\cdot 1_{C_i})}{E_P^{\mathfrak{S}}(1_{C_i})}\,dP \\ &= \int_S E_P^{\mathfrak{S}}(f\cdot 1_{C_i})\,dP = \int_S f\cdot 1_{C_i}\,dP = \int_{S\cap C_i} f\,dP \\ &= \int_{S\cap C_i} E_P^{\mathfrak{S}\vee\mathfrak{C}}(f)\,dP. \qquad \square \end{aligned}$$

Proof of Theorem 7.1: 1. Let $\mathfrak{C}$ be a finite σ-algebra. Then there exists a finite $\mathfrak{A}$-measurable partition $\{C_1,\ldots,C_n\}$ of Ω generating $\mathfrak{C}$. For every $A \in \mathfrak{A}$ we put

$$Q_A^{\mathfrak{S}\vee\mathfrak{C}} := \sum_{k=1}^{n} 1_{C_k} \frac{\mathscr{P}^{\mathfrak{S}}(A \cap C_k)}{\mathscr{P}^{\mathfrak{S}}(C_k)} .$$

$Q_A^{\mathfrak{S}\vee\mathfrak{C}}$ is an $\mathfrak{S}\vee\mathfrak{C}$-measurable function on Ω, for which by the lemma

$$Q_A^{\mathfrak{S}\vee\mathfrak{C}} = P^{\mathfrak{S}\vee\mathfrak{C}}(A)\,[P]$$

for every $P \in \mathscr{P}$. Therefore $\mathscr{P}^{\mathfrak{S}\vee\mathfrak{C}}(A)$ exists and

$$\mathscr{P}^{\mathfrak{S}\vee\mathfrak{C}}(A) = Q_A^{\mathfrak{S}\vee\mathfrak{C}}\,[\mathscr{P}].$$

which shows the sufficiency of $\mathfrak{S} \vee \mathfrak{C}$ for $\mathscr{P}$ in the case of finite $\mathfrak{C}$.

2. Now let $\mathfrak{C}$ be a σ-algebra with a countably infinite generator $\{C_1, C_2, \ldots\}$. For every $n \geq 1$ let $\mathfrak{C}_n := \mathfrak{A}(\{C_1, \ldots, C_n\})$. By 1. $\mathfrak{S} \vee \mathfrak{C}_n$ is sufficient for $\mathscr{P}$ for every $n \geq 1$. Moreover, $\mathfrak{C}_n \uparrow \mathfrak{C}$, whence $\mathfrak{S} \vee \mathfrak{C}_n \uparrow \mathfrak{S} \vee \mathfrak{C}$. But then Theorem 5.4 implies that $\mathfrak{S} \vee \mathfrak{C}$ is sufficient for $\mathscr{P}$. □

<u>Corollary 7.3</u>. Let $\mathfrak{S}$ be a separable sub-σ-algebra of $\mathfrak{A}$ that contains a σ-algebra $\mathfrak{D}$ sufficient for $\mathscr{P}$. Then $\mathfrak{S}$ is sufficient for $\mathscr{P}$.

<u>Proof</u>: Since $\mathfrak{S} = \mathfrak{S} \vee \mathfrak{D}$, $\mathfrak{S}$ is sufficient by the theorem. □

<u>Theorem 7.4</u>. Let $(\Omega, \mathfrak{A}, \mathscr{P})$ be an experiment admitting a separable σ-algebra $\mathfrak{A}$ and let $\mathfrak{S}$ be a sub-σ-algebra of $\mathfrak{A}$ which is sufficient for $\mathscr{P}$. Then there exists a separable sub-σ-algebra $\mathfrak{T}$ of $\mathfrak{A}$ which is sufficient for $\mathscr{P}$ and satisfies $\mathfrak{T} \subset \mathfrak{S} \subset \mathfrak{T} \vee \mathfrak{N}_{\mathscr{P}}$.

<u>Proof</u>: By assumption we pick a countable $\cap$-stable generator $\mathfrak{E}$ of $\mathfrak{A}$ and for each $E \in \mathfrak{E}$ we choose a function $\mathscr{P}^{\mathfrak{S}}(E) \in \mathfrak{M}_{(1)}(\Omega, \mathfrak{S})$ such that

$$P^{\mathfrak{S}}(E) = \mathscr{P}^{\mathfrak{S}}(E)[\mathscr{P}].$$

Let $\mathfrak{T} := \mathfrak{A}(\{\mathscr{P}^{\mathfrak{S}}(E) : E \in \mathfrak{E}\})$. Then $\mathfrak{T}$ is countably generated, and $\mathfrak{T} \subset \mathfrak{S}$.

In order to show that $\mathfrak{T}$ is sufficient for $\mathscr{P}$ and that $\mathfrak{S} \subset \mathfrak{T} \vee \mathfrak{N}_{\mathscr{P}}$ we consider the system

$$\mathfrak{D} := \{A \in \mathfrak{A}: \text{ There exists a } Q_A \in \mathfrak{M}_{(1)}(\Omega, \mathfrak{T}) \text{ such that } \mathscr{P}^{\mathfrak{S}}(A) = Q_A[\mathscr{P}]\}.$$

Clearly, $\mathfrak{D}$ is a Dynkin system containing $\mathfrak{E}$. Thus $\mathfrak{D}$ contains the Dynkin system $\mathfrak{D}(\mathfrak{E})$ which equals $\mathfrak{A}$, since $\mathfrak{E}$ is $\cap$-stable. In particular, for every $S \in \mathfrak{S}$ there exists a $Q_S \in \mathfrak{M}_{(1)}(\Omega, \mathfrak{T})$ satisfying $1_S = Q_S[\mathscr{P}]$, i.e., $[1_S \neq Q_S] \in \mathfrak{N}_{\mathscr{P}}$, whence $S \in \mathfrak{T} \vee \mathfrak{N}_{\mathscr{P}}$. □

<u>Corollary 7.5</u>. If $\mathfrak{N}_{\mathscr{P}} = \{\emptyset, \Omega\}$, every sufficient sub-$\sigma$-algebra of the separable σ-algebra $\mathfrak{A}$ is separable.

<u>Proof</u>: Clear. □

<u>Corollary 7.6</u>. Let $(\mathfrak{S}_n)_{n \in \mathbb{N}}$ be a sequence of sufficient sub-σ-algebras of the separable σ-algebra $\mathfrak{A}$. Then $\bigvee_{n \in \mathbb{N}} \mathfrak{S}_n$ is sufficient.

<u>Proof</u>: From the theorem we get for each $n \in \mathbb{N}$ a separable sufficient σ-algebra $\mathfrak{T}_n$ satisfying $\mathfrak{T}_n \subset \mathfrak{S}_n \subset \mathfrak{T}_n \vee \mathfrak{N}_{\mathscr{P}}$. Clearly

$$\bigvee_{n \in \mathbb{N}} \mathfrak{T}_n \subset \bigvee_{n \in \mathbb{N}} \mathfrak{S}_n \subset \left(\bigvee_{n \in \mathbb{N}} \mathfrak{T}_n\right) \vee \mathfrak{N}_{\mathscr{P}}.$$

By Theorem 7.1, for every $k \in \mathbb{N}$ the σ-algebra $\bigvee_{n=1}^{k} \mathfrak{T}_n$ is sufficient for $\mathscr{P}$, whence $\bigvee_{n \in \mathbb{N}} \mathfrak{T}_n = \bigvee_{k \in \mathbb{N}} (\bigvee_{n=1}^{k} \mathfrak{T}_k)$ is sufficient for $\mathscr{P}$, since it is the limit of an increasing sequence of σ-algebras sufficient for $\mathscr{P}$. From $\bigvee_{n \in \mathbb{N}} \mathfrak{S}_n = \bigvee_{n \in \mathbb{N}} \mathfrak{T}_n[\mathscr{P}]$ and Theorem 5.2(ii) we conclude that $\bigvee_{n \in \mathbb{N}} \mathfrak{S}_n$ is sufficient for $\mathscr{P}$. □

§8. SUFFICIENCY IN THE DOMINATED CASE

In the following we shall discuss the special class of dominated experiments. The central results of the section will be the characterizations of sufficient σ-algebras and statistics given by Halmos-Savage and Neyman. Moreover we shall generalize the notion of sufficiency to the notion of pairwise sufficiency and introduce the most applicable class of dominated experiments.

Definition 8.1. Let $(\Omega, \mathfrak{A})$ be a measurable space and $\mathscr{M}_1, \mathscr{M}_2$ be two subsets of $\mathscr{M}_+^{\sigma}(\Omega, \mathfrak{A})$.

$\mathscr{M}_1$ is said to be *dominated by* $\mathscr{M}_2$ ($\mathscr{M}_2$-dominated), in symbols $\mathscr{M}_1 \ll \mathscr{M}_2$, if $\mathfrak{R}_{\mathscr{M}_2} \subset \mathfrak{R}_{\mathscr{M}_1}$.

$\mathscr{M}_1$ and $\mathscr{M}_2$ are called *equivalent*, in symbols $\mathscr{M}_1 \sim \mathscr{M}_2$, if $\mathscr{M}_1 \ll \mathscr{M}_2$ and $\mathscr{M}_2 \ll \mathscr{M}_1$.

In the special case of a one-point set $\mathscr{M}_2$ equal to $\{\mu\}$ the first defining relation will be written $\mathscr{M}_1 \ll \mu$ and we will say that $\mathscr{M}_1$ is μ-dominated with dominating (σ-finite) measure μ.

Lemma 8.2. Let $(\Omega, \mathfrak{A})$ be a measurable space. For every $\mu \in \mathscr{M}_+^{\sigma}(\Omega, \mathfrak{A})$ with $\mu \neq 0$ there exist a measure $P \in \mathscr{M}^1(\Omega, \mathfrak{A})$ and a strictly positive function $f \in \mathfrak{M}(\Omega, \mathfrak{A})$ such that $\mu = f \cdot P$ holds. In particular we have $\mu \sim P$.

Proof: Let $\mathfrak{Z} = \{Z_k : k \in \mathbb{N}\}$ be an $\mathfrak{A}$-measurable partition of Ω with $0 < \mu(Z_k) < \infty$ for all $k \in \mathbb{N}$. Putting

$$g := \sum_{k \geq 1} \frac{1}{2^k \mu(Z_k)} 1_{Z_k}$$

we get a function $g > 0$ in $\mathfrak{M}(\Omega, \mathfrak{A})$ such that $g \cdot \mu = P$ is a probability measure on $(\Omega, \mathfrak{A})$. Defining $f := \frac{1}{g}$ we achieve the desired representation $\mu = f \cdot P$ of μ. □

This lemma shows that in dealing with dominated experiments it suffices to consider probability measures. On the other hand, the general framework of arbitrary σ-finite measures does not involve any additional problem. So one can easily expose part of the theory for measures in $\mathcal{M}_+^\sigma(\Omega,\mathbf{A})$ rather than $\mathcal{M}^1(\Omega,\mathbf{A})$.

For any subset $\mathcal{M}$ of $\mathcal{M}_+^\sigma(\Omega,\mathbf{A})$ we denote by

$$\operatorname{conv}_\sigma \mathcal{M} := \{\mu \in \mathcal{M}_+^\sigma(\Omega,\mathbf{A}): \mu = \sum_{n\geq 1} c_n \nu_n \text{ where } c_n > 0 \text{ for all } n \geq 1, \sum_{n\geq 1} c_n = 1, \nu_n \in \mathcal{M} \text{ for all } n \geq 1\}$$

the σ-convex hull of $\mathcal{M}$.

Theorem 8.3. Let $\mathcal{M}$ be a subset of $\mathcal{M}^1(\Omega,\mathbf{A})$ and μ an element of $\mathcal{M}_+^\sigma(\Omega,\mathbf{A})$ such that $\mathcal{M} \ll \mu$ holds. Then there exists a measure $\mu_0 \in \operatorname{conv}_\sigma \mathcal{M}$ satisfying $\mathcal{M} \sim \mu_0$. In particular there exists a countable subset $\mathcal{M}'$ of $\mathcal{M}$ satisfying $\mathcal{M}' \sim \mathcal{M}$.

Proof: By the lemma we may suppose that $\mu \in \mathcal{M}^1(\Omega,\mathbf{A})$. The Radon-Nikodym theorem implies that for each $\nu \in \mathcal{M}$ there exists an $f_\nu \in \mathcal{L}_+^1(\Omega,\mathbf{A},\mu)$ satisfying $\nu = f_\nu \cdot \mu$. Now we define the system

$$\mathcal{H} := \{H \subset \mathcal{M}: |H| < \infty\}$$

and for every $H \in \mathcal{H}$, $g_H := 1 \wedge \sup_{\nu\in H} f_\nu$. But $0 \leq g_H \leq g_{H'} \leq 1$ for all $H,H' \in \mathcal{H}$ with $H \subset H'$, thus $g := \lim_{H\in\mathcal{H}} g_H$ exists, and by Beppo Levi's theorem we get $g = \lim_{H\in\mathcal{H}} g_H$ in the norm of $L^1(\Omega,\mathbf{A},\mu)$. It follows that there is an isotone sequence $(H_n)_{n\geq 1}$ in $\mathcal{H}$ such that $g = \lim_{n\to\infty} g_{H_n}$ in the norm of $L^1(\Omega,\mathbf{A},\mu)$. Obviously, the set $\mathcal{M}' := \bigcup_{n\geq 1} H_n$ is countable, and $\mathcal{M}' \ll \mathcal{M}$. But given $A \in \mathbf{A}$, $\nu(A) = 0$ for all $\nu \in \mathcal{M}'$ implies

$$\int_A g_{H_n}\, d\mu \leq \sum_{\nu\in H_n} \int_A f_\nu\, d\mu = \sum_{\nu\in H_n} \nu(A) = 0$$

for all $n \geq 1$, therefore

$$\int_A (1 \wedge f_\nu)\, d\mu = \int_A g_{\{\nu\}}\, d\mu \leq \int_A g\, d\mu = \lim_{n\to\infty} \int_A g_{H_n}\, d\mu = 0$$

for all $\nu \in \mathcal{M}$, whence $\nu(A) = 0$ for all $\nu \in \mathcal{M}$, and $\mathcal{M}' \sim \mathcal{M}$.

Since the set $\mathcal{M}'$ is countable, it is of the form $\{\mu_n: n \in \mathbb{N}\} \subset \mathcal{M}$. Thus $\mathcal{M} \sim \mu_0 := \Sigma_{n\geq 1} \frac{1}{2^n} \mu_n$ proves the assertion. □

Theorem 8.4. (P. R. Halmos, L. J. Savage). Let $(\Omega,\mathbf{A},\mathcal{P})$ be a dominated experiment with dominating measure $P_0 \in \operatorname{conv}_\sigma \mathcal{P}$ and let $\mathbf{S}$ be a sub-σ-algebra of $\mathbf{A}$. The following statements are equivalent:

(i) $\mathbf{S}$ is sufficient for $\mathcal{P}$.

(ii) For every $P \in \mathscr{P}$ there exists a function $f_P \in \mathfrak{M}(\Omega,\mathfrak{S})$ satisfying $P = f_P \cdot P_0$.

Proof: 1. (ii) ⇒ (i). For every $P \in \mathscr{P}$ we assume given an $\mathfrak{S}$-measurable P_0-density f_P. Moreover, we define for each $A \in \mathfrak{A}$ the function $Q_A := E_{P_0}^{\mathfrak{S}}(1_A)$. Then for all $S \in \mathfrak{S}$ and $P \in \mathscr{P}$ we obtain

$$\begin{aligned}\int_S Q_A \, dP &= \int_S Q_A f_P \, dP_0 = \int_S E_{P_0}^{\mathfrak{S}}(1_A) f_P \, dP_0 \\ &= \int_S E_{P_0}^{\mathfrak{S}}(1_A f_P) dP_0 = \int_S 1_A f_P \, dP_0 = \int_S 1_A \, dP \\ &= \int_S E_P^{\mathfrak{S}}(1_A) dP,\end{aligned}$$

whence $Q_A = E_{\mathscr{P}}^{\mathfrak{S}}(1_A)[\mathscr{P}]$, which implies the sufficiency of $\mathfrak{S}$ for $\mathscr{P}$.

2. (i) ⇒ (ii). Let $\mathfrak{S}$ be a sufficient sub-σ-algebra of $\mathfrak{A}$, $P \in \mathscr{P}$ and g_P a P_0-density of P. For every $A \in \mathfrak{A}$ there exists an $\mathfrak{S}$-measurable function Q_A on Ω such that

$$Q_A = E_{\mathscr{P}}^{\mathfrak{S}}(1_A)[\mathscr{P}].$$

Furthermore we choose $P_0 := \Sigma_{n \geq 1} c_n P_n \in \operatorname{conv}_\sigma \mathscr{P}$. Then for all $A \in \mathfrak{A}$ we obtain $Q_A = E_{P_0}^{\mathfrak{S}}(1_A)[P_0]$, and

$$\begin{aligned}\int_A E_{P_0}^{\mathfrak{S}}(g_P) dP_0 &= \int 1_A E_{P_0}^{\mathfrak{S}}(g_P) dP_0 = \int E_{P_0}^{\mathfrak{S}}(1_A E_{P_0}^{\mathfrak{S}}(g_P)) dP_0 \\ &= \int E_{P_0}^{\mathfrak{S}}(1_A) E_{P_0}^{\mathfrak{S}}(g_P) dP_0 \\ &= \int Q_A E_{P_0}^{\mathfrak{S}}(g_P) dP_0 = \int E_{P_0}^{\mathfrak{S}}(Q_A g_P) dP_0 \\ &= \int Q_A g_P \, dP_0 = \int Q_A \, dP = \int E_P^{\mathfrak{S}}(1_A) dP = P(A),\end{aligned}$$

i.e., $f_P := E_{P_0}^{\mathfrak{S}}(g_P)$ is the desired $\mathfrak{S}$-measurable P_0-density of P. □

Corollary 8.5. Let $\mathfrak{S}$ be a sufficient sub-σ-algebra of $\mathfrak{A}$ and let $\mathfrak{S}_1$ be another sub-σ-algebra of $\mathfrak{A}$ with $\mathfrak{S}_1 \supset \mathfrak{S}$. Then $\mathfrak{S}_1$ is sufficient.

Proof: By Theorem 8.3 there exists a measure $P_0 \in \operatorname{conv}_\sigma \mathscr{P}$ such that $\mathscr{P} \sim P_0$ holds. From the theorem we deduce that every $P \in \mathscr{P}$ admits an $\mathfrak{S}$-measurable P_0-density f_P. Since by assumption f_P is also an $\mathfrak{S}_1$-measurable P_0-density, another application of the theorem yields the result. □

Corollary 8.6. Let T be a statistic $(\Omega,\mathbf{A}) \to (\Omega',\mathbf{A}')$. T is sufficient if and only if for all $P \in \mathscr{P}$ there exists an $\mathbf{A}'$-measurable function g'_P on Ω' satisfying $P = (g'_P \circ T) \cdot P_0$.

Proof: 1. If T is sufficient, then by the theorem for every $P \in \mathscr{P}$ there exists a $T^{-1}(\mathbf{A}')$-measurable P_0-density such that $P = g_P \cdot P_0$. But g_P can be factorized via an $\mathbf{A}'$-measurable function g'_P on Ω' satisfying $g_P = g'_P \circ T$, which yields the assertion.

2. The converse follows from the fact that every function $g'_P \circ T$ where g'_P is $\mathbf{A}'$-measurable, is clearly $T^{-1}(\mathbf{A}')$-measurable. □

Theorem 8.7. (Neyman Criterion). Let $(\Omega,\mathbf{A},\mathscr{P})$ be an experiment which is μ-dominated by a measure $\mu \in \mathscr{M}_+^{\sigma}(\Omega,\mathbf{A})$.

(i) A sub-σ-algebra $\mathbf{S}$ of $\mathbf{A}$ is sufficient for $\mathscr{P}$ if and only if there exists an $h \in \mathfrak{M}_+(\Omega,\mathbf{A})$ and for every $P \in \mathscr{P}$ there is an $f_P \in \mathfrak{M}_+(\Omega,\mathbf{S})$ satisfying $P = f_P h\cdot\mu$.

(ii) A statistic $T: (\Omega,\mathbf{A}) \to (\Omega',\mathbf{A}')$ is sufficient for $\mathscr{P}$ if and only if there exists an $h \in \mathfrak{M}_+(\Omega,\mathbf{A})$ and for every $P \in \mathscr{P}$ there exists a $g_P \in \mathfrak{M}_+(\Omega',\mathbf{A}')$ such that $P = (g_P \circ T)h\cdot\mu$.

Proof: Since (i) $\Rightarrow$ (ii) is easily deduced from the factorization theorem for measurable mappings, it suffices to show (i). First of all we choose $P_0 = \Sigma_{n\geq 1} c_n P_n \in \mathrm{conv}_\sigma \mathscr{P}$ with $P_0 \sim \mathscr{P}$.

If $\mathbf{S}$ is sufficient for $\mathscr{P}$, then by Theorem 8.4 for each $P \in \mathscr{P}$ there exists $f_P \in \mathfrak{M}(\Omega,\mathbf{S})$ satisfying $P = f_P \cdot P_0$. This implies the representation $P = f_P h\cdot\mu$ where $h := dP_0/d\mu$.

If, conversely, every $P \in \mathscr{P}$ is of the form $P = f_P h\cdot\mu$ with $h \in \mathfrak{M}_+(\Omega,\mathbf{A})$ and $f_P \in \mathfrak{M}_+(\Omega,\mathbf{S})$, then in particular $P_0 = (\Sigma_{n\geq 1} c_n f_{P_n})h\cdot\mu$. Consequently

$$f_P P_0 = f_P\Big(\sum_{n\geq 1} c_n f_{P_n}\Big)h \cdot \mu = \Big(\sum_{n\geq 1} c_n f_{P_n}\Big) \cdot P,$$

i.e., every $P \in \mathscr{P}$ admits an $\mathbf{S}$-measurable P_0-density

$$\frac{f_P}{\sum_{n\geq 1} c_n f_{P_n}}.$$

Thus, by Theorem 8.4, $\mathbf{S}$ is sufficient for $\mathscr{P}$. □

Theorem 8.8. For every dominated experiment $(\Omega,\mathbf{A},\mathscr{P})$ there exists a sub-σ-algebra of $\mathbf{A}$ which is minimal sufficient for $\mathscr{P}$.

Proof: Theorem 8.3 provides us with a measure $P_0 \in \mathrm{conv}_\sigma \mathscr{P}$ satisfying $\mathscr{P} \sim P_0$. For every $P \in \mathscr{P}$ there exists an $f_P \in \mathfrak{M}(\Omega,\mathfrak{A})$ such that $P = f_P \cdot P_0$. Define $\mathfrak{S}_0 := \mathfrak{A}(\{f_P : P \in \mathscr{P}\})$. Then by Theorem 8.4, $\mathfrak{S}_0$ is sufficient for $\mathscr{P}$.

Let $\mathfrak{S}$ be a further sub-σ-algebra of $\mathfrak{A}$ which is sufficient for $\mathscr{P}$. Again by Theorem 8.4 for every $P \in \mathscr{P}$ there exists a function $g_P \in \mathfrak{M}(\Omega,\mathfrak{S})$ such that $P = g_P \cdot P_0$. Obviously $\mathfrak{S} \supset \mathfrak{A}(\{g_P : P \in \mathscr{P}\})$. But since $g_P = f_P [P_0]$, we obtain $\mathfrak{S}_0 \subset \mathfrak{S}[\mathscr{P}]$ which implies that $\mathfrak{S}_0$ is minimal sufficient for $\mathscr{P}$. □

Example 8.9. Let $(\Omega,\mathfrak{A},\mathscr{P})$ be an experiment with $\Omega := \mathbb{R}^n$, $\mathfrak{A} := \mathfrak{B}^n$ and $\mathscr{P} := \{\nu_{a,1}^{\otimes n} : a \in \mathbb{R}\}$ where for each $a \in \mathbb{R}$, $\nu_{a,1}^{\otimes n} = n_a \cdot \lambda^n$, n_a being defined by

$$n_a(x) := \left(\frac{1}{2\pi}\right)^{\frac{n}{2}} \exp\left(-\frac{1}{2}\sum_{k=1}^{n} (x_k - a)^2\right)$$

for all $x = (x_1,\ldots,x_n) \in \mathbb{R}^n$.

Since $\mathscr{P} \ll \lambda^n$, the hypotheses of Theorem 8.7 are satisfied. Define a statistic $T: (\mathbb{R}^n,\mathfrak{B}^n) \to (\mathbb{R},\mathfrak{B})$ by

$$T(x) := \bar{x} := \frac{1}{n}\sum_{k=1}^{n} x_k \quad \text{for all} \quad x = (x_1,\ldots,x_n) \in \mathbb{R}^n.$$

We shall show that T is sufficent for $\mathscr{P}$.

First of all we note that

$$\sum_{k=1}^{n} (x_k - a)^2 = \sum_{k=1}^{n} x_k^2 - 2\,a\,n\,\bar{x} + n\,a^2 = \sum_{k=1}^{n} (x_k - \bar{x})^2 + n(\bar{x} - a)^2,$$

whence

$$n_a(x) = \left(\frac{1}{2\pi}\right)^{\frac{n}{2}} \exp\left(-\frac{n}{2}(\bar{x} - a)^2\right)\exp\left(-\frac{1}{2}\sum_{k=1}^{n} (x_k - \bar{x})^2\right)$$

for all $x = (x_1,\ldots,x_n) \in \mathbb{R}^n$.

Putting

$$g_a(\xi) := \left(\frac{1}{2\pi}\right)^{\frac{n}{2}} \exp\left(-\frac{n}{2}(\xi - a)^2\right) \quad \text{for all} \quad \xi \in \mathbb{R}$$

and

$$h(x) := \exp\left(-\frac{1}{2}\sum_{k=1}^{n} (x_k - \bar{x})^2\right) \quad \text{for all} \quad x \in \mathbb{R}^n$$

we obtain non-negative Borel measurable functions g_a and h on $\mathbb{R}$ and $\mathbb{R}^n$ respectively satisfying $\nu_{a,1}^{\otimes n} = (g_a \circ T)h \cdot \lambda^n$. Theorem 8.7(ii) yields that T is sufficient for $\mathscr{P}$.

More general examples of dominated experiments admitting sufficient statistics are provided by the exponential families.

Definition 8.10. $(\Omega,\mathbf{A},\mathcal{P})$ is called an *exponential experiment* if there exist a measure $\mu \in \mathcal{M}_+^\sigma(\Omega,\mathbf{A})$ and a function $\tilde{f} \in \mathbb{R}^{\mathcal{P}} \otimes \mathbf{M}(\Omega,\mathbf{A})$ satisfying

$$P = [\exp \tilde{f}(P,\cdot)]\cdot\mu \quad \text{for all} \quad P \in \mathcal{P}. \tag{EX}$$

Clearly $(\Omega,\mathbf{A},\mathcal{P})$ is an exponential experiment if and only if there are functions $\zeta_1,\ldots,\zeta_n \in \mathbb{R}^{\mathcal{P}}$ and $T_1,\ldots,T_n \in \mathbf{M}(\Omega,\mathbf{A})$ such that (EX) is fulfilled with

$$\tilde{f}(P,\cdot) = \sum_{k=1}^{n} \zeta_k(P)T_k \quad \text{for all} \quad P \in \mathcal{P}.$$

In case summands of the form $\zeta_k \otimes 1_\Omega$ or $1_{\mathcal{P}} \otimes T_k$ occur in (EX) we shall absorb those in functions $C \in \mathbb{R}^{\mathcal{P}}$ and $h \in \mathbf{M}(\Omega,\mathbf{A})$ respectively, and we obtain the representation

$$P = C(P)\exp\left(\sum_{k=1}^{m} \zeta_k(P)T_k\right)h\cdot\mu = C(P)\exp\langle\zeta(P),T\rangle h\cdot\mu \tag{EX'}$$

where $\zeta := (\zeta_1,\ldots,\zeta_m)$ and $T := (T_1,\ldots,T_m)$.

Remark 8.11. One notes that m is minimal if and only if the system $\{1_{\mathcal{P}},\zeta_1,\ldots,\zeta_m\}$ is linearly independent and the system $\{1_\Omega,T_1,\ldots,T_m\}$ is μ-a.e. linearly independent. In this case the dimension m of the exponential experiment is uniquely determined; the exponential experiment admitting the representation (EX') is called *m-dimensional*.

Remark 8.12. The statistic $T: (\Omega,\mathbf{A}) \to (\mathbb{R}^m,\mathbf{B}^m)$ appearing in the representation (EX') is sufficient for the exponential family $\mathcal{P}$.

This follows directly from Theorem 8.7(ii).

Example 8.13. The experiment $(\Omega,\mathbf{A},\mathcal{P})$ with $\Omega := \mathbb{R}$, $\mathbf{A} := \mathbf{B}$ and $\mathcal{P} := \{\chi_n^2 : n \in \mathbb{N}\}$ where for every $n \in \mathbb{N}$ the measure $\chi_n^2 \in \mathcal{M}^1(\mathbb{R},\mathbf{B})$ defined by $\chi_n^2 = g_n\cdot\lambda$ with

$$g_n(x) := \frac{1}{2^{\frac{n}{2}}\,\Gamma\left(\frac{n}{2}\right)}\, x^{\frac{n}{2}-1}\, e^{-\frac{x}{2}}\, 1_{\mathbb{R}_+}(x)$$

for all $x \in \mathbb{R}$, is an exponential experiment.

One just introduces the functions C, ζ_1 in $\mathbb{R}^{\mathcal{P}}$ by

$$C(\chi_n^2) := \frac{1}{2^{\frac{n}{2}}\,\Gamma\left(\frac{n}{2}\right)}$$

and

$$\zeta_1(\chi_n^2) := \frac{n}{2} - 1$$

respectively for all $n \geq 1$, and the functions $h, T_1 \in \mathfrak{M}(\Omega,\mathbf{A})$ by

$$h(x) := e^{-\frac{x}{2}}\,1_{\mathbb{R}_+}(x)$$

and

$$T_1(x) := (\log x)\cdot 1_{\mathbb{R}_+^*}(x)$$

respectively for all $x \in \mathbb{R}$.

Example 8.14. Let $(\Omega,\mathbf{A},\mathscr{P})$ be defined by $\Omega := \mathbb{R}$, $\mathbf{A} := \mathfrak{B}$ and $\mathscr{P} := \{\tau_n : n \in \mathbb{N}\}$ where for every $n \in \mathbb{N}$ the measure $\tau_n := t_n\cdot\lambda \in \mathscr{M}^1(\Omega,\mathbf{A})$ has λ-density t_n defined by

$$t_n(x) := \frac{1}{\sqrt{n\pi}}\,\frac{\Gamma\left(\frac{n+1}{2}\right)}{\Gamma\left(\frac{n}{2}\right)}\left(1 + \frac{x^2}{n}\right)^{-\frac{n+1}{2}}$$

for all $x \in \mathbb{R}$. Then $(\Omega,\mathbf{A},\mathscr{P})$ is *not* an exponential experiment.

Indeed, if it were, then there would exist a $k \in \mathbb{N}$ and functions $\zeta_i : \mathbb{N} \to \mathbb{R}$, $T_i : \mathbb{R} \to \mathbb{R}$ $(1 \leq i \leq k)$ satisfying

$$-\frac{n+1}{2}\log\left(1 + \frac{x^2}{n}\right) = \sum_{i=1}^{k} \zeta_i(n)T_i(x)$$

for all $x \in \mathbb{R}$, $n \in \mathbb{N}$. Consequently, the vector space spanned by the sequence $(g_n)_{n\geq 1}$ of functions g_n defined by

$$g_n(x) := \log\left(1 + \frac{x^2}{n}\right) \quad \text{for all } x \in \mathbb{R}$$

is finite-dimensional. Thus the vector space spanned by the sequence $(g_n')_{n\geq 1}$ of derivatives g_n' of g_n defined by

$$g_n'(x) = \frac{2}{n}\,x\,\frac{1}{1 + \frac{x^2}{n}} \quad \text{for all } x \in \mathbb{R}$$

is finite-dimensional, and so is the vector space spanned by the sequence $(h_n)_{n\geq 1}$ of functions h_n defined by

$$h_n(x) := \frac{1}{1 + \frac{x^2}{n}} \quad \text{for all} \quad x \in \mathbb{R}.$$

This, however, is obviously false.

In the remaining part of the section we shall deal with a very useful generalization of the concept of sufficiency.

Definition 8.15. Let $(\Omega, \mathfrak{A}, \mathcal{P})$ be an arbitrary experiment and let $\mathfrak{S}$ be a sub-σ-algebra of $\mathfrak{A}$.

$\mathfrak{S}$ is called *pairwise sufficient* for $\mathcal{P}$ if $\mathfrak{S}$ is sufficient for any two-element subset $\mathcal{P}'$ of $\mathcal{P}$.

Remark 8.16. By its very definition the notion of pairwise sufficiency will apply especially to those statistical methods which are based on the comparison of experiments involving only two probability measures. Moreover, pairwise sufficiency will be easier to handle than sufficiency, as the following result shows.

Theorem 8.17. (J. Pfanzagl). Let $(\Omega, \mathfrak{A}, \mathcal{P})$ be an experiment and let $\mathfrak{S}$ be a sub-σ-algebra of $\mathfrak{A}$. Moreover let the following condition be satisfied

(P) For every $A \in \mathfrak{A}$ there exists a $Q_A \in \mathfrak{M}_{(1)}(\Omega, \mathfrak{S})$ such that

$$\int Q_A \, dP = \int 1_A \, dP = P(A) \quad \text{for all} \quad P \in \mathcal{P}.$$

Then $\mathfrak{S}$ is pairwise sufficient for $\mathcal{P}$.

Proof: Let $\mathcal{P}' := \{P_1, P_2\} \subset \mathcal{P}$. We put $Q := P_1 + P_2$. Then for each $i = 1,2$ we have $P_i \ll Q$, and there exists an $h_i \in \mathfrak{M}_{(1)}(\Omega, \mathfrak{A})$ such that $P_i = h_i \cdot Q$. Moreover, the functions h_1, h_2 can be chosen so that $h_1 + h_2 = 1_\Omega$.

For every $\gamma \in \mathbb{R}_+$ we introduce the set $A_\gamma := [h_1 < \gamma h_2]$. By the hypothesis there exists a $Q_{A_\gamma} \in \mathfrak{M}_{(1)}(\Omega, \mathfrak{S})$ satisfying

$$\int Q_{A_\gamma} \, dP = \int 1_{A_\gamma} \, dP = P(A_\gamma) \quad \text{for all} \quad P \in \mathcal{P}.$$

1. First we show that $Q([Q_{A_\gamma} \neq 1_{A_\gamma}]) = 0$. We introduce functions $g := 1_{A_\gamma} - Q_{A_\gamma}$ and $f := g(\gamma h_2 - h_1)$. Since $0 \leq Q_{A_\gamma} \leq 1$, we get the implications

$$g(\omega) > 0 \Rightarrow \omega \in A_\gamma \Rightarrow \gamma h_2(\omega) - h_1(\omega) > 0$$

and

$$g(\omega) < 0 \Rightarrow \omega \in CA_\gamma \Rightarrow \gamma h_2(\omega) - h_1(\omega) \leq 0.$$

From this follows $f \geq 0$.

But we have

$$\int f\,dQ = \int (1_{A_\gamma} - Q_{A_\gamma})(\gamma h_2 - h_1)dQ$$
$$= \gamma \int (1_{A_\gamma} - Q_{A_\gamma})dP_2 - \int (1_{A_\gamma} - Q_{A_\gamma})dP_1 = 0,$$

whence $f = 0[Q]$. This implies $g \leq 0[Q]$, since $[g > 0] \subset [f > 0]$. Therefore, it follows from

$$\int g\,dQ = \int (1_{A_\gamma} - Q_{A_\gamma})dP_1 + \int (1_{A_\gamma} - Q_{A_\gamma})dP_2 = 0$$

that also $g = 0[Q]$. Thus we arrived at $Q_{A_\gamma} = 1_{A_\gamma}[Q]$ which was to be shown.

2. From 1. we deduce that the set

$$S := [h_2 > 0] = \bigcup_{n \geq 1} [h_1 < nh_2]$$

belongs to the Q-completion $\tilde{\mathfrak{S}}^Q := \mathfrak{S} \vee \mathfrak{N}_Q$ of $\mathfrak{S}$. Thus the functions $\frac{h_1}{h_2} 1_S$, $\tilde{f}_1 := \frac{h_1}{h_2} 1_S + 1_{CS}$ and $\tilde{f}_2 := 1_S$ are $\tilde{\mathfrak{S}}^Q$-measurable. It follows that for $i = 1,2$ there exists an $\mathfrak{S}$-measurable f_i on Ω satisfying $Q(\tilde{f}_i \neq f_i) = 0$. Putting $h := h_2 + h_1 1_{CS}$ we obtain for $i = 1,2$ that $P_i = f_i h \cdot Q$, and f_i is $\mathfrak{S}$-measurable. By the Halmos-Savage theorem 8.4, $\mathfrak{S}$ is sufficient for $\mathscr{P}' = \{P_1,P_2\}$. □

For dominated experiments the notions of pairwise sufficiency and sufficiency coincide.

Theorem 8.18. Let $(\Omega,\mathbf{A},\mathscr{P})$ be an experiment and let $\mathfrak{S}$ be a sub-σ-algebra of $\mathbf{A}$. The following statements are equivalent:

(i) $\mathfrak{S}$ is pairwise sufficient for $\mathscr{P}$.

(ii) $\mathfrak{S}$ is sufficient for every dominated subfamily $\mathscr{P}'$ of $\mathscr{P}$.

Proof: Since the implication (ii) ⇒ (i) is obvious, we are left with the proof of the implication (i) ⇒ (ii). Let $\mathfrak{S}$ be pairwise sufficient and let $\mathscr{P}' \subset \mathscr{P}$ such that $\mathscr{P}' \ll \mu \in \mathscr{M}_+^\sigma(\Omega,\mathbf{A})$. By Theorem 8.3 there exists a measure $P_0 := \Sigma_{n\geq 1} c_n P_n \in \mathrm{conv}_\sigma \mathscr{P}'$ such that $\mathscr{P}' \sim P_0$.

Now let $P \in \mathscr{P}'$. Since $\mathfrak{S}$ is pairwise sufficient for $\mathscr{P}$, there exists for every $A \in \mathbf{A}$ a function $Q_A^{P,n} \in \mathfrak{M}_{(1)}(\Omega,\mathfrak{S})$ satisfying

$$Q_A^{P,n} = E_P^{\mathfrak{S}}(1_A)[P] \quad \text{and} \quad Q_A^{P,n} = E_{P_n}^{\mathfrak{S}}(1_A)[P_n] \quad \text{for every } n \geq 1.$$

We put

$$Q_A^P := \sum_{n\geq 1} c_n Q_A^{P,n} \frac{d(P_n)_{\mathbf{S}}}{d(P_0)_{\mathbf{S}}} .$$

Then we obtain $Q_A^P = E_P^{\mathbf{S}}(1_A)[P]$ and at the same time

$$\sum_{n\geq 1} c_n E_{P_n}^{\mathbf{S}}(1_A) \frac{d(P_n)_{\mathbf{S}}}{d(P_0)_{\mathbf{S}}} = E_{P_0}^{\mathbf{S}}(1_A)[P_0].$$

Since $E_{P_0}^{\mathbf{S}}(1_A)$ is independent of P, $\mathbf{S}$ has been shown to be sufficient for $\mathcal{P}'$. □

Corollary 8.19. For dominated experiments $(\Omega,\mathbf{A},\mathcal{P})$ and sub-σ-algebras $\mathbf{S}$ of $\mathbf{A}$ the following statements are equivalent:

(i) $\mathbf{S}$ is pairwise sufficient for $\mathcal{P}$.

(ii) $\mathbf{S}$ satisfies Property (P) of Theorem 8.17.

(iii) $\mathbf{S}$ is sufficient for $\mathcal{P}$.

Proof: Clear. □

Remark 8.20. The hypothesis of domination posed in Theorem 8.18 and Corollary 8.19 cannot be dropped without replacement, as will be shown in the following section.

§9. EXAMPLES AND COUNTEREXAMPLES

In this section we will specify by examples the domain of validity of certain statements on sufficiency which have been proved in the previous sections. At the same time we are going to present various extremal examples of experiments that are typical in the general theory without additional assumptions. It can be useful for the reader to continue some of the examples or to construct experiments in which the structure of these or similar examples occurs as a sub-or quotient structure.

We shall fix the following notations: For a given set Ω we will use the power-σ-algebra $\mathfrak{P}(\Omega)$, the trivial σ-algebra $\mathbf{A}_0 := \{\emptyset,\Omega\}$ and the σ-algebra $\mathbf{A}_1(\Omega)$ generated by the one-point-subsets of Ω, i.e., the σ-algebra $\{A \subset \Omega$: Either A of CA is countable$\}$.

Example 9.1. In general the sufficiency of σ-algebras $\mathbf{S}_1$ and $\mathbf{S}_2$ does not imply the sufficiency of $\mathbf{S}_1 \cap \mathbf{S}_2$.

9.1.1. Let $\Omega := \mathbb{R}^2$, $\mathbf{A} := \mathbf{B}^2$, $\mathcal{P}$ an at least two-point subset of the set $\{P \in \mathcal{M}^1(\Omega,\mathbf{A}): P(D) = 1\}$ where D denotes the diagonal of $\mathbb{R}^2$, $\mathbf{S}_1 := \mathbf{B} \otimes \{\emptyset, \mathbb{R}\}$ and $\mathbf{S}_2 := \{\emptyset, \mathbb{R}\} \otimes \mathbf{B}$, i.e., the σ-algebras generated by the projections $X_1, X_2 \colon \Omega = \mathbb{R}^2 \to \mathbb{R}$ respectively. Then $\mathfrak{S}_1$ and $\mathfrak{S}_2$ are sufficient for $\mathscr{P}$.

Indeed, for every $A \in \mathfrak{A}$ and $i = 1,2$ we have $D \cap X_i^{-1}(X_i(A \cap D)) = A \cap D$. Since every $P \in \mathscr{P}$ is concentrated on D, we conclude that

$$\begin{aligned}\int_{X_i^{-1}(B)} 1_{X_i(A\cap D)} \circ X_i \, dP &= P(X_i(B) \cap X_i^{-1}(X_i(A \cap D))) \\ &= P(X_i^{-1}(B) \cap X_i^{-1}(X_i(A \cap D)) \cap D) \\ &= P(X_i^{-1}(B) \cap A \cap D) \\ &= P(X_i^{-1}(B) \cap A) = \int_{X_i^{-1}(B)} 1_A \, dP,\end{aligned}$$

whenever $B \in \mathfrak{A}$. Hence the function $1_{X_i(A\cap D)} \circ X_i$ is a version of $E_P^{\mathfrak{S}_i}(1_A)$ which shows the sufficiency of $\mathfrak{S}_i$ for $\mathscr{P}$ ($i = 1,2$). On the other hand, $\mathfrak{S}_1 \cap \mathfrak{S}_2 = \{\emptyset, \Omega\}$ is *not* sufficient for $\mathscr{P}$, since $\mathscr{P}$ admits more than one element.

9.1.2. Let $\Omega := \{1,2,3,4\}$, $\mathfrak{A} := \mathfrak{P}(\Omega)$, $\mathscr{P} := \{\varepsilon_1, \varepsilon_4\}$, $\mathfrak{S}_1 := \{\emptyset, \Omega, \{1,2\}, \{3,4\}\}$ and $\mathfrak{S}_2 := \{\emptyset, \Omega, \{1,3\}, \{2,4\}\}$. Then $\mathfrak{S}_1$ and $\mathfrak{S}_2$ are sufficient for $\mathscr{P}$, but $\mathfrak{S}_1 \cap \mathfrak{S}_2$ is not.

This example can be made a sub-example of 9.1.1 by applying the mapping $\Omega \to \mathbb{R}^2$ defined by the assignments $1 \to (0,0)$, $2 \to (0,1)$, $3 \to (1,0)$ and $4 \to (1,1)$.

Example 9.2. In general the sufficiency of σ-algebras $\mathfrak{S}_1$ and $\mathfrak{S}_2$ does not imply the sufficiency of $\mathfrak{S}_1 \vee \mathfrak{S}_2$.

Let $\Omega := \{\omega = (\omega_1, \omega_2) \in \mathbb{R}^2 \colon |\omega_1| = |\omega_2| \neq 0\}$. We introduce the mappings $\sigma_1, \sigma_2 \colon \Omega \to \Omega$ by $\sigma_1(\omega_1, \omega_2) := (-\omega_1, \omega_2)$ and $\sigma_2(\omega_1, \omega_2) = (\omega_1, -\omega_2)$ for all $(\omega_1, \omega_2) \in \Omega$ respectively.

Now let $\mathfrak{S}_1 := \{S \in \mathfrak{A}_1(\Omega) \colon \sigma_1(S) = S\}$ and $\mathfrak{S}_2 := \{S \in \mathfrak{A}_1(\Omega) \colon \sigma_2(S) = S\}$, and put $A_0 := \{\omega = (\omega_1, \omega_2) \in \Omega \colon \omega_1 = \omega_2\}$. Then we define $\mathfrak{A} := \mathfrak{A}(\mathfrak{S}_1, \mathfrak{S}_2, A_0)$ and $\mathscr{P} := \{P_\omega \colon \omega \in \Omega\}$, where for each $\omega \in \Omega$,

$$P_\omega := \frac{1}{4}(\varepsilon_\omega + \varepsilon_{\sigma_1(\omega)} + \varepsilon_{\sigma_2(\omega)} + \varepsilon_{\sigma_1\sigma_2(\omega)}).$$

For $i = 1,2$ the σ-algebra $\mathfrak{S}_i$ is sufficient for $\mathscr{P}$ with $\mathscr{P}^{\mathfrak{S}_i}(A) = \frac{1}{2}(1_A + 1_A \circ \sigma_i)$ for each $A \in \mathfrak{A}$. Moreover. $\mathfrak{N}_{\mathscr{P}} \subset \mathfrak{S}_1 \cap \mathfrak{S}_2$ since $\mathfrak{N}_{\mathscr{P}} = \{\emptyset, \Omega\}$. But $\mathfrak{S}_1 \vee \mathfrak{S}_2$ is *not* sufficient for $\mathscr{P}$.

Indeed, if $Q_{A_0} : \mathscr{P}^{\mathfrak{S}_1 \vee \mathfrak{S}_2}(A_0)$ exists, then for each $\omega \in \Omega$ we have

$Q_{A_0}(\omega) = 1_{A_0}(\omega)$ since P_ω is concentrated on the set $\Omega' := \{\omega, \sigma_1(\omega), \sigma_2(\omega), \sigma_1\sigma_2(\omega)\}$ and $\mathbf{A} \cap \Omega' = (\mathbf{S}_1 \vee \mathbf{S}_2) \cap \Omega'$. But $Q_{A_0} = 1_{A_0}$ cannot be true since $Q_{A_0} \in \mathfrak{M}^b(\Omega, \mathbf{S}_1 \vee \mathbf{S}_2)$ and $1_{A_0} \notin \mathfrak{M}^b(\Omega, \mathbf{S}_1 \vee \mathbf{S}_2)$.

Example 9.3. Let $\mathbf{S}_1$ be a σ-algebra sufficient for the experiment $(\Omega, \mathbf{A}, \mathscr{P})$ and $\mathbf{A}$ separable. Then, in general, a σ-algebra $\mathbf{S}_2$ satisfying $\mathbf{S}_1 \subset \mathbf{S}_2 \subset \mathbf{A}$ is not sufficient for $\mathscr{P}$.

Let $\Omega := \mathbb{R}$, $\mathbf{A} := \mathbf{B}$, $\mathscr{P} := \{P \in \mathscr{M}^1(\Omega, \mathbf{A}) : P(A) = P(-A)$ for all $A \in \mathbf{A}\}$, $\mathbf{S}_1 := \{A \in \mathbf{A} : A = -A\}$ and $\mathbf{S}_2 := \{S \cup A : S \in \mathbf{S}_1, A \in \mathbf{A}$ with $A \subset M\}$, where M denotes a non-$\mathbf{A}$-measurable subset of Ω satisfying $0 \in M$ and $-M = M$. Clearly $\mathbf{S}_1$ is a sufficient σ-algebra with $\mathscr{P}^{\mathbf{S}_1}(A) = \frac{1}{2}(1_A + 1_{-A})$ for all $A \in \mathbf{A}$.

It is easy to see that $\mathbf{S}_2$ is in fact a σ-algebra containing $\mathbf{S}_1$. Indeed, $\mathbf{S}_2$ is closed under the formation of countable unions. Moreover, let $S \cup A \in \mathbf{S}_2$. Clearly $C(S \cup A) = CS \cap CA = CS \cap (C(A \cup -A) \cup (-A \setminus A))$ where $C(A \cup -A) \in \mathbf{S}_1$ and $-A \setminus A \subset M$ since $-M = M$. Therefore $C(S \cup A) = (CS \cap C(A \cup -A)) \cup (S \cap (-A \setminus A)) \in \mathbf{S}_2$, and $\mathbf{S}_2$ is shown to be also closed under the formation of complements.

We still have to prove that $\mathbf{S}_2$ is not sufficient for $\mathscr{P}$. Assuming that $\mathbf{S}_2$ is sufficient for $\mathscr{P}$ we obtain for $\mathbb{R}_+ \in \mathbf{A}$ a function $Q_{\mathbb{R}_+} \in \mathfrak{M}^b(\Omega, \mathbf{S}_2)$ satisfying

$$P(S_2 \cap \mathbb{R}_+) = \int_{S_2} Q_{\mathbb{R}_+} \, dP \quad \text{for all } S_2 \in \mathbf{S}_2 \text{ and } P \in \mathscr{P}.$$

For $\omega \in M$ we choose $S_2 := \{\omega\}$ and $P := \frac{1}{2}(\varepsilon_\omega + \varepsilon_{-\omega})$ and obtain $Q_{\mathbb{R}_+}(\omega) = 1_{\mathbb{R}_+}(\omega)$; for $\omega \in CM$ we choose $S_2 := \{-\omega, \omega\}$ and $P := \frac{1}{2}(\varepsilon_\omega + \varepsilon_{-\omega})$ and since $Q_{\mathbb{R}_+} \in \mathfrak{M}(\Omega, \mathbf{S}_2)$, we get $Q_{\mathbb{R}_+}(\omega) = \frac{1}{2}$. Consequently

$$M = [Q_{\mathbb{R}_+} \in \{0,1\}] \in \mathbf{S}_2$$

which contradicts the assumption that M is not $\mathbf{A}$-measurable.

Example 9.4. Pairwise sufficiency of a σ-algebra $\mathbf{A}$ for an experiment $(\Omega, \mathbf{A}, \mathscr{P})$ does not necessarily imply sufficiency.

Let $\Omega := [0,1] \times \{0,1\}$, $\mathbf{A} := \mathbf{B}([0,1]) \otimes \mathbf{P}(\{0,1\})$, $\mathscr{P} := \{P_\alpha : \alpha \in [0,1]\}$ where for each $\alpha \in [0,1]$, $P_\alpha((A_0 \times \{0\}) \cup (A_1 \times \{1\})) := \frac{1}{2}(\lambda_{[0,1]}(A_0) + \varepsilon_\alpha(A_1))$ whenever $A_0, A_1 \in \mathbf{B}([0,1])$. Then the σ-algebra $\mathbf{S} := \mathbf{B}([0,1]) \otimes \{\emptyset, \{0,1\}\}$ is pairwise sufficient for $\mathscr{P}$.

Indeed, for $\alpha_1, \alpha_2 \in [0,1]$ and $A := (A_0 \times \{0\}) \cup (A_1 \times \{1\}) \in \mathbf{A}$ the set

$$A' := ((A_0 \setminus \{\alpha_1,\alpha_2\}) \cup (A_1 \cap \{\alpha_1,\alpha_2\})) \times \{0,1\}$$

belongs to $\mathfrak{S}$, and one has

$$1_{A'} = E_{P_{\alpha_i}}^{\mathfrak{S}}(1_A)[P_{\alpha_i}] \quad \text{for} \quad i = 1,2.$$

However, $\mathfrak{S}$ is not sufficient for $\mathscr{P}$.

Suppose there exists a function $Q \in \mathfrak{M}^b(\Omega,\mathfrak{S})$ (of the form $Q = q \otimes 1_{\{0,1\}}$ with $q \in \mathfrak{M}^b([0,1],\mathfrak{B}([0,1]))$ such that $Q = P^{\mathfrak{S}}([0,1] \times \{0\})$ for all $P \in \mathscr{P}$. Then for all $\alpha \in [0,1]$ we have

$$\int_{\{\alpha\}\times\{0,1\}} Q\, dP_\alpha = \int_{\{\alpha\}\times\{0,1\}} 1_{[0,1]\times\{0\}}\, dP_\alpha,$$

thus

$$\frac{1}{2}\left(\int_{\{\alpha\}} q\, d\lambda + \int_{\{\alpha\}} q\, d\varepsilon_\alpha\right) = \frac{1}{2}\left(\int_{\{\alpha\}} 1_{[0,1]}\, d\lambda\right),$$

whence $\frac{1}{2} q(\alpha) = 0$. But this implies $Q \equiv 0$ contradicting the equality $\int Q\, dP_\alpha = \frac{1}{2}$ for all $\alpha \in [0,1]$.

Example 9.5. There exists an experiment $(\Omega,\mathfrak{A},\mathscr{P})$ such that $\mathfrak{A}$ admits no minimal sufficient sub-σ-algebra. In order to see this, we put $\Omega := [0,1]$, $\mathfrak{A} := \mathfrak{B}([0,1])$ and

$$\mathscr{P} := \{\tfrac{1}{2}(\varepsilon_\alpha + \varepsilon_{\alpha+\frac{1}{2}}) : \alpha \in [0,\tfrac{1}{2}[\} \cup \{P_0\}$$

where P_0 denotes an atomless measure in $\mathscr{M}^1(\Omega,\mathfrak{A})$ such that $P_0([0,\frac{1}{2}[) > \frac{1}{2}$. Moreover, for every $\alpha \in [0,\frac{1}{2}[$ we introduce the sub-σ-algebra

$$\mathfrak{S}_\alpha := \{A \in \mathfrak{A} : A \cap \{\alpha,\alpha + \tfrac{1}{2}\} \in \{\emptyset,\{\alpha,\alpha + \tfrac{1}{2}\}\}\}$$

of $\mathfrak{A}$. Clearly $\mathfrak{S}_\alpha$ is sufficient for $\mathscr{P}$, since for every $A \in \mathfrak{A}$ we obtain

$$P^{\mathfrak{S}_\alpha}(A) = 1_{A\setminus\{\alpha,\alpha+\frac{1}{2}\}} + \tfrac{1}{2}(1_A(\alpha) + 1_A(\alpha + \tfrac{1}{2}))1_{\{\alpha,\alpha+\frac{1}{2}\}}$$

whenever $\alpha \in [0,\frac{1}{2}[$.

It remains to be shown that there is no minimal sufficient sub-σ-algebra of $\mathfrak{A}$.

We assume that $(\Omega,\mathfrak{A},\mathscr{P})$ does admit such a sub-σ-algebra $\mathfrak{S}$ and produce a contradiction.

Since $\mathfrak{N}_{\mathscr{P}} = \{\emptyset,\Omega\}$, we have $\mathfrak{S} \subset \bigcap_{\alpha\in[0,\frac{1}{2}]} \mathfrak{S}_\alpha$. Thus for all $S \in \mathfrak{S}$ we get $\frac{1}{2} + (S \cap [0,\frac{1}{2}[) = S \cap [\frac{1}{2},1[$, and every $f \in \mathfrak{M}(\Omega,\mathfrak{S})$ satisfies $f(\alpha + \frac{1}{2}) = f(\alpha)$ for all $\alpha \in [0,\frac{1}{2}[$.

Now let $Q \in \mathfrak{M}(\Omega,\mathfrak{S})$ such that $Q = \mathscr{P}^{\mathfrak{S}}([0,\frac{1}{2}[)$. Then for each $\alpha \in [0,\frac{1}{2}[$ we obtain

$$\int Q \, d(\tfrac{1}{2}(\varepsilon_\alpha + \varepsilon_{\alpha+\frac{1}{2}})) = \tfrac{1}{2}(\varepsilon_\alpha + \varepsilon_{\alpha+\frac{1}{2}})([0,\tfrac{1}{2}[),$$

whence $\frac{1}{2}(Q(\alpha) + Q(\alpha + \frac{1}{2})) = \frac{1}{2}$ and so $Q \equiv \frac{1}{2} 1_\Omega$. But this contradicts

$$\int Q \, dP_0 = P_0([0,\tfrac{1}{2}[) > \tfrac{1}{2}.$$

The construction of this counterexample shows in fact that there exists no sufficient sub-σ-algebra $\mathfrak{S} \subset \bigcap_{\alpha\in[0,\frac{1}{2}]} \mathfrak{S}_\alpha$.

Example 9.6. The intersection of an arbitrary descending family of sufficient σ-algebras is not necessarily sufficient itself.

An example supporting this fact can be obtained from Example 9.5 by taking the family

$$\Sigma := \left\{ \bigcap_{i=1}^{n} \mathfrak{S}_{\alpha_i} : n \in \mathbb{N},\ \alpha_i \in [0,\tfrac{1}{2}[\ \text{ for } \ i = 1,\ldots,n \right\}$$

of all finite intersections of the σ-algebra $\mathfrak{S}_\alpha$ $(\alpha \in [0,\frac{1}{2}[)$. The same example also shows that the intersection of a totally ordered family of sufficient σ-algebras in general fails to be sufficient.

One chooses a well ordering $<$ for $[0,\frac{1}{2}[$ and looks at the family $\{\mathfrak{S}'_k : k \in [0,\frac{1}{2}[\}$ of σ-algebras $\mathfrak{S}'_k := \bigcap_{i<k} \mathfrak{S}_i$. This family is clearly totally ordered and descending. If the σ-algebra $\mathfrak{S}'_k$ is sufficient for all $k \in [0,\frac{1}{2}[$, then the σ-algebra

$$\bigcap_{k\in[0,\frac{1}{2}[} \mathfrak{S}'_k = \bigcap_{\alpha\in[0,\frac{1}{2}[} \mathfrak{S}_\alpha$$

is not sufficient. On the other hand, if there is a non-sufficient $\mathfrak{S}'_k$, then from the well ordering of $[0,\frac{1}{2}[$ we deduce that there exists a minimal element $k_0 \in [0,\frac{1}{2}[$ with non-sufficient corresponding σ-algebra $\mathfrak{S}'_{k_0}$. But

$$\mathfrak{S}'_{k_0} = \bigcap_{i<k_0} \mathfrak{S}_i = \bigcap_{i<k_0} \mathfrak{S}'_i$$

is the intersection of a totally ordered family of sufficient σ-algebras.

Example 9.7. There exist an experiment $(\Omega,\mathbb{A},\mathscr{P})$ and a non-minimal sufficient statistic on $(\Omega,\mathbb{A})$ such that the σ-algebra $\mathbb{A}(T)$ is minimal sufficient (for $\mathscr{P}$).

Let Ω be a set with at least two elements, $\mathbb{A}:=\{\emptyset,\Omega\}$, $\mathscr{P}:=\mathscr{M}^1(\Omega,\mathbb{A})$ (a one-point set) and $T:=id_\Omega$. Then $\mathbb{A}(T)=\{\emptyset,\Omega\}$ is minimal sufficient for $\mathscr{P}$. But T is not a minimal sufficient statistic. Indeed, let Ω' be a one-element set and $\mathbb{A}':=\{\emptyset,\Omega'\}$, then the constant mapping $T': \Omega \to \Omega'$ is measurable and sufficient for $\mathscr{P}$. But there does not exist any mapping $S: \Omega' \to \Omega$ such that $T = S \circ T'[\mathscr{P}]$.

Example 9.8. There exist an experiment $(\Omega,\mathbb{A},\mathscr{P})$ and a minimal sufficient statistic T on $(\Omega,\mathbb{A})$ such that $\mathbb{A}(T)$ is not a minimal sufficient σ-algebra (for $\mathscr{P}$).

Let $\Omega:=]0,1[$, let $\mathfrak{S}$ be the σ-algebra of all Lebesgue (λ-) measurable subsets of $]0,1[$ and B a subset of $]0,1[$ with $\lambda^*(B)=1$, $\lambda_*(B)=0$. Then we introduce the σ-algebra $\mathbb{A}:=\mathbb{A}(\mathfrak{S},B)$ consisting of exactly the sets of the form $(S_1 \cap B) \cup (S_2 \cap CB)$ where $S_1,S_2 \in \mathfrak{S}$, the family $\mathscr{P}:=\{P_1, 2fP_2\}$ where $P_1 \in \mathscr{M}^1(\Omega,\mathbb{A})$ is defined by

$$P_1((S_1 \cap B) \cup (S_2 \cap CB)) := \frac{1}{2}(\lambda(S_1) + \lambda(S_2))$$

for all $S_1,S_2 \in \mathfrak{S}$, and $f:=id_\Omega$, and finally let $T:=id_\Omega: (\Omega,\mathbb{A}) \to (\Omega,\mathbb{A})$ (considered as a measurable mapping).

We first note that P_1 is well-defined. Suppose, we have $(S_1 \cap B) \cup (S_2 \cap CB) = (S_1' \cap B) \cup (S_2' \cap CB)$ for $S_1,S_2,S_1',S_2' \in \mathfrak{S}$. Then $(S_1 \,\Delta\, S_1') \cap B = \emptyset$ and $(S_2 \,\Delta\, S_2') \cap CB = \emptyset$, whence $S_1 \,\Delta\, S_1' \subset CB$ and $S_2 \,\Delta\, S_2' \subset B$. Since by assumption B and CB have inner Lebesgue measure 0, we obtain $\lambda(S_i \,\Delta\, S_i') = 0$ for $i = 1,2$, thus

$$\frac{1}{2}(\lambda(S_1) + \lambda(S_2)) = \frac{1}{2}(\lambda(S_1') + \lambda(S_2')),$$

i.e., $P_1((S_1 \cap B) \cup (S_2 \cap CB)) = P_1((S_1' \cap B) \cup (S_2' \cap CB))$.

Now we show that T is minimal sufficient for $\mathscr{P}$. Suppose that $T': (\Omega,\mathbb{A}) \to (\Omega',\mathbb{A}')$ is another sufficient statistic for $\mathscr{P}$. Then by the Neyman criterion 8.7 there exist $g_1',g_2' \in \mathfrak{M}(\Omega',\mathbb{A}')$ and $h \in \mathfrak{M}_+(\Omega,\mathbb{A})$ satisfying $P_1 = (g_1' \circ T')h \cdot P_1$ and $2fP_1 = (g_2' \circ T')h \cdot P_1$. Thus we get $T = \frac{1}{2}\frac{g_2'}{g_1'} \circ T'[\mathscr{P}]$ where $\frac{1}{2}\frac{g_2'}{g_1'}$ is the desired mapping $\Omega' \to \Omega$.

But $\mathbb{A}(T) = \mathbb{A}$ is not minimal sufficient for $\mathscr{P}$. Since the measures P_1 and $2fP_1$ admit $\mathfrak{S}$-measurable P_1-densities, $\mathfrak{S}$ is sufficient for by the Halmos-Savage theorem 8.4. On the other hand, we have for any

$S \in \mathfrak{S}$, by the assumptions on $B \subset]0,1[$, $P_1(B \Delta S) = P_1((CS \cap B) \cup (S \cap CB)) = \frac{1}{2}(\lambda(CS) + \lambda(S)) = \frac{1}{2} > 0$, thus the negation of $\mathfrak{S} = \mathfrak{A}[\mathscr{P}]$.

CHAPTER IV

Testing Experiments

§10. FUNDAMENTALS

The theory of testing statistical hypotheses is based on the notions of testing experiments and tests, which will be introduced purely measure theoretically. Once first results have been established, these notions will gain their concrete statistical meaning. Till now they have been only roughly described, in Examples 3.10 and 3.13.1 of the game theoretical set-up.

A *testing experiment* will be a quintuple $(\Omega, \mathbf{A}, \mathcal{P}, \mathcal{P}_0, \mathcal{P}_1)$, consisting of an experiment $(\Omega, \mathbf{A}, \mathcal{P})$ and a partition $\{\mathcal{P}_0, \mathcal{P}_1\}$ of $\mathcal{P}$ in the sense that $\mathcal{P}_0, \mathcal{P}_1 \neq \emptyset$, $\mathcal{P}_0 \cap \mathcal{P}_1 = \emptyset$ and $\mathcal{P}_0 \cup \mathcal{P}_1 = \mathcal{P}$.

For a given experiment $(\Omega, \mathbf{A}, \mathcal{P})$ we consider the set $\mathfrak{M}_{(1)}(\Omega, \mathbf{A}) := \{t \in \mathfrak{M}(\Omega, \mathbf{A}) : 0 \leq t \leq 1\}$ of all *tests*. The notion of a test originates from the interpretation of $t \in \mathfrak{M}_{(1)}(\Omega, \mathbf{A})$ as a decision function assigning to every sample $\omega \in \Omega$ the probability $t(\omega)$ that a certain hypothesis will be rejected. Thus for the moment the set $\mathfrak{M}_{(1)}(\Omega, \mathbf{A})$ of tests corresponding to $(\Omega, \mathbf{A}, \mathcal{P})$ will just signify an order interval in the algebra $\mathfrak{M}^b(\Omega, \mathbf{A})$ which obtains a statistical structure via testing experiments. Given a testing experiment $(\Omega, \mathbf{A}, \mathcal{P}, \mathcal{P}_0, \mathcal{P}_1)$ we introduce some conventions determined by the particular application envisaged. The set $\mathcal{P}_0$ will be called the *hypothesis*, $\mathcal{P}_1$ the *alternative* of the testing experiment. For $i = 0,1$ the set $\mathcal{P}_i$ is said to be *simple* or *composite* if it contains one or more than one element respectively.

In the literature the functions of $\mathfrak{M}_{(1)}(\Omega, \mathbf{A})$ are also called *randomized tests*, while the indicator functions in $\mathfrak{M}_{(1)}(\Omega, \mathbf{A})$ are often considered as *deterministic tests* corresponding to the given experiment.

Definition 10.1. Let $(\Omega, \mathbf{A}, \mathscr{P}, \mathscr{P}_0, \mathscr{P}_1)$ be a testing experiment and t a test in $\mathfrak{M}_{(1)}(\Omega, \mathbf{A})$.

(a) The function $\beta_t : \mathscr{P} \to [0,1]$ defined by $\beta_t(P) := E_P(t)$ for all $P \in \mathscr{P}$ is called the *power function* of t.

(b) For any $P \in \mathscr{P}_1$ the number $\beta_t(P)$ is referred to as the *power of* t *at* P.

(c) The number $\sup_{P \in \mathscr{P}_0} \beta_t(P)$ is called the *size* of the test t.

The first task in the development of the theory of testing statistical hypotheses is to minimize for a given testing experiment $(\Omega, \mathbf{A}, \mathscr{P}, \mathscr{P}_0, \mathscr{P}_1)$ within some class of tests $t \in \mathfrak{M}_{(1)}(\Omega, \mathbf{A})$, the size $\sup_{P \in \mathscr{P}_0} \beta_t(P)$ (error of the first kind) as well as $\sup_{P \in \mathscr{P}_1} (1 - \beta_t(P))$ (error of the second kind). The minimization can be achieved for various classes of tests which we are going to introduce now.

Definition 10.2. Let $(\Omega, \mathbf{A}, \mathscr{P}, \mathscr{P}_0, \mathscr{P}_1)$ be a testing experiment and $\alpha \in [0,1]$.

(a) A *test of level* α (level α test) is a test $t \in \mathfrak{M}_{(1)}(\Omega, \mathbf{A})$ whose size is smaller or equal than α. By $\mathfrak{T}_\alpha$ we abbreviate the set of all tests of level α.

(b) A test $t \in \mathfrak{T}_\alpha$ is called *most powerful* (of level α) if for all $t' \in \mathfrak{T}_\alpha$ and $P \in \mathscr{P}_1$ we have $\beta_t(P) \geq \beta_{t'}(P)$.

(c) $t \in \mathfrak{T}_\alpha$ is said to be a *maximin test* (of level α) if for all $t' \in \mathfrak{T}_\alpha$ one has

$$\inf_{P \in \mathscr{P}_1} \beta_t(P) \geq \inf_{P \in \mathscr{P}_1} \beta_{t'}(P)$$

or equivalently if

$$\inf_{P \in \mathscr{P}_1} \beta_t(P) = \sup_{t' \in \mathfrak{T}_\alpha} \inf_{P \in \mathscr{P}_1} \beta_{t'}(P)$$

holds.

We note that maximin tests of level α are those tests, for which the error of the second kind is minimized under the condition that the error of the first kind remains bounded. In order to be more precise we shall sometimes emphasize that level α tests, most powerful tests and maximin tests are in fact defined for the testing problem $\mathscr{P}_0$ versus $\mathscr{P}_1$ or just for $\mathscr{P}_0$ versus $\mathscr{P}_1$. Plainly, if $\mathscr{P}_1$ is simple, maximin tests of level α and most powerful level α tests coincide.

Remark 10.3. In the terminology of Chapter I we note that $t \in \mathfrak{T}_\alpha$ is most powerful iff t dominates $\mathfrak{T}_\alpha$, and that $t \in \mathfrak{T}_\alpha$ is a maximin test iff t is a maximin strategy with respect to the game $\Gamma := (\mathfrak{T}_\alpha, \mathscr{P}_1, M)$, where $M(t,P) := \beta_t(P)$ for all $(t,P) \in \mathfrak{T}_\alpha \times \mathscr{P}_1$.

In the following we want to discuss the question under what general conditions on the hypothesis and on the alternative of the given testing experiment $(\Omega, \mathbf{A}, \mathscr{P}, \mathscr{P}_0, \mathscr{P}_1)$, most powerful and maximin tests of level α exist.

Theorem 10.4. Let $(\Omega, \mathbf{A}, \mathscr{P}, \mathscr{P}_0, \mathscr{P}_1)$ be a testing experiement such that $\mathscr{P}_1$ is μ-dominated by a measure $\mu \in \mathscr{M}_+^\sigma(\Omega, \mathbf{A})$. Then for every $\alpha \in [0,1]$ there exists a maximin test of level α for $\mathscr{P}_0$ versus $\mathscr{P}_1$.

Proof: By Lemma 8.2 we may assume without loss of generality that $\mu \in \mathscr{M}^1(\Omega, \mathbf{A})$. Let q_μ denote the canonical projection from $\mathscr{L}^1(\Omega, \mathbf{A}, \mu)$ onto $L^1(\Omega, \mathbf{A}, \mu)$. Moreover, let $\alpha \in [0,1]$.

1. $q_\mu(\mathfrak{T}_\alpha)$ is a norm-closed subset of $L^1(\Omega, \mathbf{A}, \mu)$. In fact, let $(t_n)_{n \geq 1}$ be a sequence in $\mathfrak{T}_\alpha$ with $\lim_{n\to\infty} q(t_n) = f$. Then $(t_n)_{n\geq 1}$ converges μ-stochastically, and there exists a subsequence $(t_{n_k})_{k\geq 1}$ of $(t_n)_{n\geq 1}$ which converges μ-a.s. Let $A \in \mathbf{A}$ with $\mu(A) = 1$ be a set on which $(t_{n_k})_{k \geq 1}$ converges in all points, and let $t := \lim_{k\to\infty} t_{n_k} \cdot 1_A$. Clearly $0 \leq t \leq 1$. The Lebesgue dominated convergence theorem yields for all $P \in \mathscr{P}_0$ that

$$\int t \, dP = \lim_{k\to\infty} \int t_{n_k} \cdot 1_A \, dP \leq \overline{\lim_{k\to\infty}} \int t_{n_k} \, dP \leq \alpha.$$

Therefore we obtain that $f := q_\mu(t) \in q_\mu(\mathfrak{T}_\alpha)$.

2. $q_\mu(\mathfrak{T}_\alpha)$ is $\sigma(L^1, L^\infty)$-compact. In fact, by the weak convergence lemma $q_\mu(\mathfrak{T}_1)$ is weakly compact. Since the set $q_\mu(\mathfrak{T}_\alpha)$ is convex, by 1. it is a weakly closed subset of $q_\mu(\mathfrak{T}_1)$.

3. For every $P \in \mathscr{P}_1$ the function $M_\mu(\cdot, P)$ on $L^1(\Omega, \mathbf{A}, \mu)$ defined by

$$M_\mu(f,P) := \int f \frac{dP}{d\mu} \, d\mu$$

is upper semicontinuous. In fact, from $dP/d\mu \in L^1(\Omega, \mathbf{A}, \mu)$ we conclude that there exists a sequence $(f_n)_{n\geq 1}$ of functions in $L^\infty(\Omega, \mathbf{A}, \mu)$ converging isotonically to $dP/d\mu$. For every $n \geq 1$ the mapping $f \to \int (1-f) f_n d\mu$ from $q_\mu(\mathfrak{T}_\alpha)$ into $\mathbb{R}$ is continuous. Since $1-f \geq 0$ we have

$$\int (1-f) \frac{dP}{d\mu} \, d\mu = \sup_{n \geq 1} \int (1-f) f_n \, d\mu,$$

which implies that the mapping $f \to \int (1-f)\frac{dP}{d\mu}\, d\mu$ from $q_\mu(\mathfrak{T}_\alpha)$ into $\mathbb{R}$ is lower semicontinuous, and the mapping $f \to \int f \frac{dP}{d\mu}\, d\mu = 1 - \int (1-f)\frac{dP}{d\mu}\, d\mu$ from $q_\mu(\mathfrak{T}_\alpha)$ into $\mathbb{R}$ is upper semicontinuous.

4. Since $q_\mu(\mathfrak{T}_\alpha)$ is $\sigma(L^1, L^\infty)$-compact by 2. and $M_\mu(\cdot,P)$ is upper semicontinuous on $L^1(\Omega,\mathbf{A},\mu)$ by 3, $\inf_{P\in\mathscr{P}_1} M_\mu(\cdot,P)$ attains its maximum on $q_\mu(\mathfrak{T}_\alpha)$, i.e., there exists a maximin test of level α for $\mathscr{P}_0$ versus $\mathscr{P}_1$. □

Corollary 10.5. (D. Landers, L. Rogge). Let $(\Omega,\mathbf{A}, \mathscr{P}, \mathscr{P}_0, \mathscr{P}_1)$ be a testing experiment with simple alternative $\mathscr{P}_1 = \{P_1\}$. Then for every $\alpha \in [0,1]$ there exists a most powerful level α test for $\mathscr{P}_0$ versus $\mathscr{P}_1$.

Proof: Since $\mathscr{P}_1$ is assumed to be simple, it is dominated. But for simple alternatives the notions of maximin tests and most powerful test (of level α) coincide. □

Definition 10.6. Let $(\Omega,\mathbf{A}, \mathscr{P}, \mathscr{P}_0, \mathscr{P}_1)$ be a testing problem and $\alpha \in [0,1]$. For every deterministic test $t \in \mathfrak{T}_\alpha$ of the form $t := 1_A$ for $A \in \mathbf{A}$, the set A is called the *rejection region* or *critical region* of t. Clearly a critical region $A \in \mathbf{A}$ satisfies $P(A) \leq \alpha$ for all $P \in \mathscr{P}_0$.

Example 10.7. Within the framework developed we will discuss a basic testing problem concerning the mean of a normal distribution under the hypothesis that the variance is known. We are given the testing experiment $(\Omega,\mathbf{A}, \mathscr{P}, \mathscr{P}_0, \mathscr{P}_1)$ with $\Omega := \mathbb{R}^n$, $\mathbf{A} := \mathfrak{B}^n$, $\mathscr{P} := \{\nu_{a,\sigma^2}^{\otimes n} : a \in \mathbb{R}\}$, $\mathscr{P}_0 := \{\nu_{a_0,\sigma^2}^{\otimes n}\}$ and $\mathscr{P}_1 := \{\nu_{a,\sigma^2}^{\otimes n} : a \in \mathbb{R} \setminus \{a_0\}\}$. This means that on the basis of a random sample $X = (X_1,\ldots,X_n)$ of size n with $P_{X_k} := \nu_{a,\sigma^2}$ for all $k = 1,\ldots,n$ ($a \in \mathbb{R}$, $\sigma^2 \in \mathbb{R}_+^*$) we wish to test the hypothesis H_0: $a = a_0$ by the following procedure: H_0 will be rejected if

$$Y := \left|\frac{\overline{X} - a_0}{\sigma}\right| \geq \frac{\kappa_\alpha}{\sqrt{n}}$$

and accepted otherwise, where the critical value κ_α is determined by the equation

$$\alpha = 1 - \int_{-\kappa_\alpha}^{\kappa_\alpha} n_{0,1}\, d\lambda.$$

The set $A := \{x \in \mathbb{R}^n : |Y(x)| \geq \kappa_\alpha/\sqrt{n}\}$ is the critical region of the test $t := 1_A$. By

$$\beta_t(\nu^{\otimes n}_{a_0,\sigma^2}) = \nu^{\otimes n}_{a_0,\sigma^2}(A) = \nu^{*n}_{a_0,\sigma^2}(C[na_0 - \sqrt{n}\,\sigma\kappa_\alpha, na_0 + \sqrt{n}\,\sigma\kappa_\alpha])$$
$$= \nu_{0,1}(C[-\kappa_\alpha,\kappa_\alpha]) = \alpha$$

t is a level α test for $\mathscr{P}_0$ versus $\mathscr{P}_1$.

For practical purposes we can describe this procedure as follows: Given a sample $(x_1,\ldots,x_n)$ corresponding to the random sample $X = (X_1,\ldots,X_n)$ of size n we have to decide whether $(x_1,\ldots,x_n)$ lies in the critical region A or not. In the case that $(x_1,\ldots,x_n) \in A$, the hypothesis H_0: $a = a_0$ will be rejected, otherwise it will be accepted.

Example 10.8. (Computation of a power function). Let $(\Omega,\mathfrak{A},\mathscr{P},\mathscr{P}_0,\mathscr{P}_1)$ be a testing experiment with $\Omega := \mathbb{R}^n$, $\mathfrak{A} := \mathfrak{B}^n$,

$$\mathscr{P} := \{P_a : P_{(a_1,\ldots,a_n)} := \nu_{a_1,1} \otimes \ldots \otimes \nu_{a_n,1} : a = (a_1,\ldots,a_n) \in \mathbb{R}^n\},$$

$\mathscr{P}_0 := \{P_{a_0}\}$ with $a_0 := (a_1^0,\ldots,a_n^0) \in \mathbb{R}^n$ and $\mathscr{P}_1 := \mathscr{P} \setminus \mathscr{P}_0$. For any $r \in \mathbb{R}_+^*$ and $a_0 \in \mathbb{R}^n$ the open ball with radius r and center a_0 will be denoted by $B_r(a_0)$. For $\alpha \in [0,1]$ we choose a test $t_\alpha \in \mathfrak{T}_\alpha$ of the form $t_\alpha := 1_{CB_{r_\alpha}(a_0)}$ such that $P_{a_0}(CB_{r_\alpha}(a_0)) = \alpha$ holds. Then for all $a := (a_1,\ldots,a_n) \in \mathbb{R}^n$ we obtain

$$\beta_{t_\alpha}(P_a) = P_a(CB_{r_\alpha}(a_0))$$
$$= P_0(CB_{r_\alpha}(a_0-a)) = P_{a_0-a}(CB_{r_\alpha}(0)).$$

In addition we see that by the rotation symmetry of P_0 the equality $||a_0-a|| = ||a_0-b||$ implies $\beta_{t_\alpha}(P_a) = \beta_{t_\alpha}(P_b)$ for all $b \in \mathbb{R}$.

In order to obtain more precise information on the power $\beta_{t_\alpha}(P_a)$ of t_α in P_a for $a \in \mathbb{R} \setminus \{0\}$, we have to compute the exact distributions. We define the *non-central* χ^2*-distribution with* n *degrees of freedom and non-centrality parameter* $\gamma \in \mathbb{R}^n \setminus \{0\}$ as the measure $\chi^2_{n,\gamma} := h_{n,\gamma}\cdot\lambda \in \mathscr{M}^1(\mathbb{R},\mathfrak{B})$ with

$$h_{n,\gamma}(\xi) = 2^{-\frac{n}{2}} e^{-\frac{1}{2}(\xi+||\gamma||^2)} \xi^{\frac{n}{2}-1} \sum_{j\geq 0} \frac{(||\gamma||^2\xi)^j}{4^j j!\,\Gamma(j+\frac{n}{2})} 1_{\mathbb{R}_+}(\xi)$$
$$= \sum_{j\geq 0} \frac{1}{j!} \left(\frac{||\gamma||^2}{2}\right)^j e^{-(||\gamma||^2/2)} h_{2j+n,0}(\xi) \quad \text{for all } \xi \in \mathbb{R},$$

and note that for $\gamma = 0$ this measure coincides with the χ^2-distribution

χ_n^2 with n degrees of freedom.

By means of the Fourier transform it is shown that $\chi_{n,\gamma}^2$ is the distribution of $\Sigma_{k=1}^n X_k^2$ where $X_1,\ldots,X_n$ are independent real random variables with $P_{X_k} := \nu_{\gamma_k,1}$.

Let X_k denote the k-th coordinate projection of $\mathbb{R}^n$. Then $(P_{a_0-a})_{X_k} = \nu_{a_k-a_k^0,1}$ and therefore

$$\beta_t(P_a) = P_{a_0-a}(CB_{r_\alpha}(0)) = P_{a_0-a}\left(\left[\sum_{k=1}^n X_k^2 \geq r_\alpha^2\right]\right)$$
$$= \chi_{n,a_0-a}^2([r_\alpha^2,\infty[) = \int_{r_\alpha^2}^\infty h_{n,a_0-a}(x)dx$$

for all $a \in \mathbb{R}^n$. Here r_α is determined by $\int_{r_\alpha^2} h_{n,0}(x)dx = \alpha$. The critical region corresponding to t_α is the set

$$CB_{r_\alpha}(a_0) = \{x \in \mathbb{R}^n : ||x-a_0|| \geq r_\alpha\}.$$

Our next aim will be a discussion of the *testing problem within the framework of optimization theory*.

Let $(\Omega,\mathfrak{A},\mathscr{P},\mathscr{P}_0,\mathscr{P}_1)$ be a testing experiment with a simple alternative $\mathscr{P}_1 := \{P_1\}$. We assume $\mathscr{P} \ll \mu$ for a measure $\mu \in \mathscr{M}_+^\sigma(\Omega,\mathfrak{A})$ such that to every $P \in \mathscr{P}$ there exists a function $f_P \in \mathfrak{M}_+(\Omega,\mathfrak{A})$ with $P = f_P\cdot\mu$. Furthermore we suppose that on $\mathscr{P}_0$ there exists a σ-algebra Σ_0 such that the mapping $(\omega,P) \to f_P(\omega)$ from $\Omega \times \mathscr{P}_0$ into $\mathbb{R}$ is $\mathfrak{A} \otimes \Sigma_0$-measurable. For given $\alpha \in [0,1]$ we keep the notation

$$\mathfrak{T}_\alpha := \{t \in \mathfrak{M}_{(1)}(\Omega,\mathfrak{A}) : \int t f_P d\mu \leq \alpha \text{ for all } P \in \mathscr{P}_0\}$$

and add

$$\mathscr{S} := \{(\pi,v) \in \mathscr{M}_+^\sigma(\mathscr{P}_0,\Sigma_0) \times \mathfrak{M}(\Omega,\mathfrak{A}) : \pi \geq 0,\ v \geq 0[\mu],$$
$$\int_{\mathscr{P}_0} f_P\, \pi(dP) + v \geq f_{P_1}[\mu]\}.$$

Thus, to the *primary program*

$$\int t\, f_{P_1} d\mu = \sup_{t'\in\mathfrak{T}_\alpha} \int t'\, f_{P_1} d\mu \qquad \text{(PP)}$$

there corresponds its *dual program*

$$\alpha\pi(\mathscr{P}_0) + \int v\, d\mu = \inf_{(\pi',v')\in\mathscr{S}} \left(\alpha\pi'(\mathscr{P}_0) + \int v' d\mu\right). \qquad \text{(DP)}$$

Special Case: Let $\mathscr{P}_0 := \{P_1,\ldots,P_m\}$ and $\mathscr{P}_1 := \{P_{m+1}\}$ be two sets of discrete probability measures $P_i \in \mathscr{M}^1(\Omega,\mathfrak{A})$ with finite support $\{\omega_1,\ldots,\omega_n\}$ for $i = 1,\ldots,m+1$. Then, with the abbreviations $\mu_j := \mu(\{\omega_j\})$, $\mu_{ij} := f_{P_i}(\omega_j)$, $t_j := t(\omega_j)$, $\pi_i := \pi(\{i\})$ and $v_j := v(\omega_j)$ for $j = 1,\ldots,n$; $i = 1,\ldots,m+1$, the programs attain the following form: The primary program (PP) translates into the task

of maximizing $\sum_{j=1}^{n} \mu_{m+1,j}\mu_j t_j$ under the conditions (PP')

$$\sum_{j=1}^{n} \mu_{ij}\mu_j t_j \leq \alpha \quad \text{for all} \quad i = 1,\ldots,m, \tag{P1'}$$

$$t_j \leq 1 \quad \text{for} \quad j = 1,\ldots,n, \tag{P2'}$$

$$t_j \geq 0 \quad \text{for} \quad j = 1,\ldots,n, \tag{P3'}$$

the dual program (DP) consists of

minimizing $\alpha \sum_{i=1}^{m} \pi_i + \sum_{j=1}^{n} v_j\mu_j$ under the conditions (DP')

$$\sum_{i=1}^{m} \mu_{ij}\pi_i + v_j \geq \mu_{m+1,j} \quad \text{for all} \quad j = 1,\ldots,n, \tag{D1'}$$

$$\pi_i \geq 0 \quad \text{for} \quad i = 1,\ldots,m \tag{D2'}$$

and

$$v_j \geq 0 \quad \text{for} \quad j = 1,\ldots,n. \tag{D3'}$$

Putting, in addition, $x_j := t_j$ for $j = 1,\ldots,n$, $y_k := \pi_k$ for $k = 1,\ldots,m$ and $y_k := v_{k-m}\mu_{k-m}$ for $k = m+1,\ldots,m+n$, we see that the duality of the linear programs (PP') and (DP') (under the given linear conditions) coincides with that commonly treated in optimization theory. The programs (PP) and (DP) admit an interrelationship which is made precise in the following

Theorem 10.9. Let the data of the programs (PP) and (DP) be given as before.

(i) For $t \in \mathfrak{T}_\alpha$ and $(\pi,v) \in \mathscr{S}$ we have

$$\int t\, f_{P_1}\, d\mu \leq \alpha\pi(\mathscr{P}_0) + \int v\, d\mu$$

with equality iff the conditions

(a) $\int_{\mathscr{P}_0} f_P \pi(dP) + v = f_{P_1}\,[\mu]$ on $[t > 0]$,

(b) $\int t f_P \, d\mu = \alpha\,[\pi]$ and

(c) $t = 1[\mu]$ on $[v > 0]$ are satisfied.

(ii) All tests $t' \in \mathfrak{T}_\alpha$ and pairs $(\pi', v') \in \mathscr{S}$ with

$$\int t' f_{P_1} d\mu = \alpha\pi'(\mathscr{P}_0) + \int v' \, d\mu$$

are solutions of the programs (PP) and (DP) respectively, and v' is of the form

$$v' = \max\left(0, f_{P_1} - \int_{\mathscr{P}_0} f_P \pi(dP)\right)[\pi'].$$

<u>Proof</u>: (i) is deduced from the inequalities

$$\int t\ f_{P_1} d\mu \leq \int_{\mathscr{P}_0}\int_\Omega t\ f_P d\mu d\pi + \int_\Omega tv\ d\mu \leq \alpha\pi(\mathscr{P}_0) + \int v\ d\ .$$

The rest of the proof is evident.

(ii) The first assertion is clear. Let $(\pi', v') \in \mathscr{S}_\alpha$ be a solution of (DP). Since (DP) will be improved whenever v' becomes smaller and since $\int_{\mathscr{P}_0} f_P d\pi + v \geq f_{P_1}\,[\mu]$ remains invariant with respect to forming the infimum, the remaining assertions follow. □

Now let $\pi \in \mathscr{M}_+(\mathscr{P}_0, \Sigma_0)$ be fixed. Then the function v corresponding to the solution $(\pi, v) \in \mathscr{S}$ of (DP) is of the form

$$v = v_\pi := \left(f_{P_1} - \int_{\mathscr{P}_0} f_P d\pi\right)^+ [\mu].$$

Defining for $\alpha \in \mathbb{R}_+^*$ and all $\pi \in \mathscr{M}_+^b(\mathscr{P}_0, \Sigma_0)$ the quantity

$$f_\alpha(\pi) := \alpha\pi(\mathscr{P}_0) - \int\left(f_{P_1} - \int_{\mathscr{P}_0} f_P d\pi\right)^+ d\mu$$

we try to find measures $\pi' \in \mathscr{M}_+^b(\mathscr{P}_0, \Sigma_0)$ with the property

$$f_\alpha(\pi') = \inf_{\pi \in \mathscr{M}_+^b(\mathscr{P}_0, \Sigma_0)} f_\alpha(\pi).$$

A sufficient condition for the optimality of $t' \in \mathfrak{T}_\alpha$ and $(\pi', v_\pi) \in \mathscr{S}$ is any of the two following statements:

(a) $E_{P_1}(t') = f_\alpha(\pi')$.

(b) $t' = \begin{cases} 1\ [\mu] & \text{on } \left[f_{P_1} > \int_{\mathscr{P}_0} f_P\ d\pi\right] \\ 0\ [\mu] & \text{on } \left[f_{P_1} < \int_{\mathscr{P}_0} f_P\ d\pi\right] \end{cases}$

and $E_P(t') = \alpha[\pi']$.

We collect the obtained results in

Theorem 10.10. Let $(\Omega, \mathbb{A}, \mathscr{P}, \mathscr{P}_0, \mathscr{P}_1)$ be a testing experiment with simple alternative $\mathscr{P}_1 := \{P_1\}$, which is μ-dominated by a measure $\mu \in \mathscr{M}_+^\sigma(\Omega, \mathbb{A})$. Let every $P \in \mathscr{P}$ be of the form $P = f_P \mu$ with $f_P \in \mathfrak{M}_+(\Omega, \mathbb{A})$. Moreover we assume that on $\mathscr{P}_0$ there exists a σ-algebra Σ_0 such that the mappings $(\omega, P) \to f_P(\omega)$ from $\Omega \times \mathscr{P}_0$ into $\mathbb{R}$ are $\mathbb{A} \otimes \Sigma_0$-measurable. For each $\alpha \in [0,1]$ let $\mathfrak{T}_\alpha$ and $\mathscr{S}$ be defined as above. Under the hypothesis that there exist $t' \in \mathfrak{T}_\alpha$ and $(\pi', v_{\pi'}) \in \mathscr{S}$ with

$$\beta_{t'}(P_1) = \alpha\pi'(\mathscr{P}_0) + \int v_{\pi'}\, d\mu$$

the following statements are equivalent:

(i) $t \in \mathfrak{T}_\alpha$ is a solution of (PP).

(ii) There exists a $\pi \in \mathscr{M}_+(\mathscr{P}_0, \Sigma_0)$ with

$$t = \begin{cases} 1\ [\mu] & \text{on } \left[f_{P_1} > \int_{\mathscr{P}_0} f_P\, d\pi\right] \\ 0\ [\mu] & \text{on } \left[f_{P_1} < \int_{\mathscr{P}_0} f_P\, d\pi\right] \end{cases}$$

and $\beta_t(P) = \alpha[\pi]$.

The idea of presenting this theorem at this point is to indicate that further studies within the theory of testing statistical hypotheses can be carried out in the directions of the following two sections, the first one devoted to the construction of optimal (most powerful level α) tests $t \in \mathfrak{M}_{(1)}(\Omega, \mathbb{A})$, the other one dealing with optimal (least favorable) apriori measures $\pi \in \mathscr{M}_+^\sigma(\mathscr{P}_0, \Sigma_0)$.

§11. CONSTRUCTION OF MOST POWERFUL TESTS

In constructing most powerful level α tests for a simple hypothesis versus a simple alternative one often applies the following purely measure theoretic result.

Theorem 11.1. (Fundamental Lemma of J. Neyman and E. S. Pearson). Let $(\Omega, \mathbb{A}, \mu)$ be a measure space and let f_0, f_1 be two μ-integrable numerical functions on Ω with $f_0 \geq 0$. For every $k \in \overline{\mathbb{R}}$ we define

$$M_k := [f_1 > kf_0]$$

and

$$M_k^+ := [f_1 \geq kf_0].$$

Using the notation $\nu := f_0 \cdot \mu$ we obtain the subsequent statements:

(i) For every $K \in [0,\nu(\Omega)]$ there exists a $k \in \overline{\mathbb{R}}$ such that

$$\nu(M_k) \leq K \leq \nu(M_k^+) \tag{1}$$

holds.

(ii) Define for every $K \in [0,\nu(\Omega)]$ the set

$$\mathfrak{T}_K := \left\{t \in \mathfrak{M}_{(1)}(\Omega,\mathfrak{A}) : \int t \, d\nu \leq K\right\}.$$

Let $t \in \mathfrak{T}_K$ satisfy the following two conditions:

$$\int t \, d\nu = K. \tag{2a}$$

There exists a $k \in \overline{\mathbb{R}}_+$ such that (2b)

$$t \cdot 1_{M_k} = 1_{M_k} [\mu] \quad \text{and} \quad t \cdot 1_{CM_k^+} = 0[\mu].$$

Then

$$\int t \, f_1 \, d\mu = \sup_{t' \in \mathfrak{T}_K} \int t' f_1 \, d\mu. \tag{3}$$

(iii) Let $K \in [0,\nu(M_0^+)] \neq \emptyset$ and let $k \geq 0$ be chosen for K according to (i). If γ is a number in $[0,1]$ such that

$$\gamma\nu(M_k^+ \setminus M_k) = K - \nu(M_k) \tag{4}$$

holds, then the function

$$t := 1_{M_k} + \gamma 1_{M_k^+ \setminus M_k} \tag{5}$$

satisfies the equalities (2a), (2b) and therefore (3).

(iv) Let under the assumptions of (iii) be $t' \in \mathfrak{T}_K$ another function having property (3). Then we have

$$k\left(K - \int t' f_0 \, d\mu\right) = 0, \quad \text{i.e., (2a) if } k \neq 0,$$

and $t = t'[\mu]$ on $M_k \cup CM_k^+$.

<u>Proof</u>: (i) The definition of M_k yields the following properties:

(a) $M_{k'} \subset M_k$ for $k,k' \in \overline{\mathbb{R}}$, $k \leq k'$.

(b) $M_k \cap [f_0 < \infty] = \bigcup_{n \geq 1} M_{k_n} \cap [f_0 < \infty]$ for $k,k_n \in \overline{\mathbb{R}}$ with $k_n \downarrow k$.

(c) $M_k^+ \cap [0 < f_0 < \infty] = \bigcap_{n \geq 1} M_{k'_n} \cap [0 < f_0 < \infty]$ for $k,k'_n \in \overline{\mathbb{R}}$

with $k'_n \neq k$, $k'_n \uparrow k$.

For $K \in [0,\nu(\Omega)]$ we set $S := \{r \in \overline{\mathbb{R}}: \nu(M_r) > K\}$ and $k := \sup S$ (with $k := -\infty$ if $S = \emptyset$). Since f_0 is μ-integrable and $\nu \ll \mu$ we obtain

$$\nu([f_0 = \infty]) = \nu([f_0 = 0]) = 0$$

and therefore

$$\nu(M_\infty) = \nu([f_1 > \infty\cdot f_0] \cap [f_0 > 0]) = \nu([f_1 > +\infty]) = \nu(\emptyset) = 0.$$

By this equality, CS is nonempty, whence there exist $k_n \in CS$ $(n \geq 1)$ satisfying $k_n \downarrow k$, and by (b) we get

$$\nu(M_k) = \nu(M_k \cap [f_0 < \infty]) = \nu\left(\bigcup_{n\geq 1} M_{k_n} \cap [f_0 < \infty]\right)$$

$$= \lim_{n\to\infty} \nu(M_{k_n}) \leq K.$$

If $k = -\infty$, we have

$$\nu(M_k^+) = \nu([f_1 \geq -\infty\cdot f_0] \cap [f_0 > 0]) = \nu([f_1 \geq -\infty])$$

$$= \nu(\Omega) \geq K \geq \nu(M_k).$$

If, on the other hand, $k > -\infty$, then there exist $k'_n \in S$ $(n \geq 1)$ satisfying $k'_n \neq k$, $k'_n \uparrow k$, and by (c) we get

$$\nu(M_k^+) = \nu(M_k^+ \cap [0 < f_0 < \infty]) = \nu\left(\bigcap_{n\geq 1} M_{k'_n} \cap [0 < f_0 < \infty]\right)$$

$$= \lim_{n\to\infty} \nu(M_{k'_n}) \geq K \geq \nu(M_k).$$

(ii) Let $t \in \mathfrak{T}_K$ be chosen such that condition (2) is satisfied, and let $t' \in \mathfrak{T}_K$ be arbitrary. Then

$$\int t(1-t')f_1\, d\mu = \int t(1_{M_k^+} + 1_{CM_k^+})(1-t')f_1\, d\mu$$

$$= \int t\, 1_{M_k^+}(1-t')f_1\, d\mu$$

$$\geq k \int t\, 1_{M_k^+}(1-t')f_0\, d\mu$$

$$= k \int t(1-t')f_0\, d\mu$$

and

$$-\int t'(1-t)f_1\, d\mu = -\int t'(1-t\, 1_{M_k} - t\, 1_{CM_k})f_1\, d\mu$$

$$= -\int t'(1-t)1_{CM_k} f_1\, d\mu$$

$$\geq -k \int t'(1-t)\, 1_{CM_k} f_0 \, d\mu$$
$$= -k \int t'(1-t) f_0 \, d\mu,$$

thus

$$\int t\, f_1 \, d\mu - \int t' f_1 \, d\mu = \int t(1-t') f_1 \, d\mu - \int t'(1-t) f_1 \, d\mu$$
$$\geq k\left[\int t(1-t')d\nu - \int t'(1-t)d\nu\right]$$
$$= k\left[\int t\, d\nu - \int t' d\nu\right] = k\left(K - \int t'\, d\nu\right) \geq 0,$$

which is the desired statement.

(iii) By assumptions (4) and (5) we get $\int t\, d\nu = \int (1_{M_K} + \gamma 1_{M_k^+ \setminus M_k}) d\nu = \nu(M_k) + \gamma\nu(M_k^+ \setminus M_k) = K$, and with the aid of (ii) this yields assertion (3).

(iv) Let t' be as in (3) and let t be the function defined in (5). Then we can supplement the inequality in the proof of (ii) as follows:

$$0 = \int t\, f_1 \, d\mu - \int t'\, f_1 \, d\mu \geq k\left(K - \int t'\, d\nu\right) \geq 0,$$

thus we get $k \int (t-t') f_0 \, d\mu = 0$.

Since, moreover, $\int (t-t') f_1 \, d\mu = 0$, we obtain

$$0 = \int (t-t')(f_1 - kf_0) d\mu = \int_{[f_1 \neq kf_0]} (t-t')(f_1 - kf_0) d\mu$$
$$= \int_{[f_1 > kf_0]} (t-t')(f_1 - kf_0) d\mu + \int_{[f_1 < kf_0]} (t-t')(f_1 - kf_0) d\mu$$
$$= \int_{[f_1 > kf_0]} (1-t')(f_1 - kf_0) d\mu + \int_{[f_1 < kf_0]} (0-t')(f_1 - kf_0) d\mu.$$

The integrands being ≥ 0 and the factors $f_1 - kf_0 \neq 0$ we conclude μ - a.e., $t' 1_{M_k} = 1_{M_k}$ and $t' 1_{CM_k^+} = 0$, thus $t' 1_{M_k \cup CM_k^+} = t\, 1_{M_k \cup CM_k^+}$, which is the desired result. □

<u>Theorem 11.2</u>. Let $(\Omega, \mathbb{A}, \mathscr{P}, \mathscr{P}_0, \mathscr{P}_1)$ be a testing experiment with simple hypothesis $\mathscr{P}_0 := \{P_0\}$ and simple alternative $\mathscr{P}_1 := \{P_1\}$. Then for every $\alpha \in [0,1]$ there exists a most powerful level α test t_α for $\mathscr{P}_0$ versus $\mathscr{P}_1$ satisfying $\beta_{t_\alpha}(P_0) = \alpha$.

<u>Proof</u>: We put $\mu := P_0 + P_1$,

$$f_0 := \frac{dP_0}{d\mu}, \quad f_1 := \frac{dP_1}{d\mu}$$

and $\nu := P_0$. With this notation adjusted to the Fundamental Lemma we obtain by (i) of Theorem 11.1 for each $\alpha \in [0,1]$ an extended real number $k_\alpha \in \overline{\mathbb{R}}_+$ satisfying

$$P_0[f_1 > k_\alpha f_0] \leq \alpha \leq P_0[f_1 \geq k_\alpha f_0]$$

and by (iii)(4) of the same theorem a number $\gamma_\alpha \in [0,1]$ such that

$$\gamma_\alpha P_0[f_1 = k_\alpha f_0] = \alpha - P_0[f_1 > k_\alpha f_0]$$

holds. Defining

$$t_\alpha := 1_{[f_1 > k_\alpha f_0]} + \gamma_\alpha 1_{[f_1 = k_\alpha f_0]}$$

we obtain by (ii)(2a) and (ii)(3) of Theorem 11.1 the desired equalities:

$$\beta_{t_\alpha}(P_0) = \int t_\alpha \, dP_0 = \int t_\alpha f_0 \, d\mu = \alpha$$

as well as

$$\beta_{t_\alpha}(P_1) = \int t_\alpha \, dP_1 = \int t_\alpha f_1 \, d\mu = \sup_{t' \in \mathfrak{T}_\alpha} \beta_{t'}(P_1). \qquad \square$$

Remark 11.3. Theorem 11.1 (iv) tells us that the test t_α constructed in Theorem 11.2 is (P_0+P_1)- a.s. uniquely determined on $[f_1 \neq k_\alpha f_0]$.

If $\beta_{t_\alpha}(P_1) < 1$, then $k_\alpha \neq 0$, and so by Theorem 11.1(iv) we have $\beta_{t'}(P_0) = \alpha$ for every most powerful level α test t' for $\mathscr{P}_0$ versus $\mathscr{P}_1$. Indeed, if $k_\alpha = 0$, then clearly

$$\beta_{t_\alpha}(P_1) = P_1[f_1 > 0 f_0] + \gamma_\alpha P_1[f_1 = 0 f_0] = P_1[f_1 > 0] = 1,$$

which contradicts the hypothesis.

Remark 11.4. For every sequence $(\alpha_n)_{n \geq 1}$ in $[0,1]$ such that $\lim_{n\to\infty} \alpha_n = \alpha$ we have $\lim_{n\to\infty} t_{\alpha_n} = t_\alpha [P_0+P_1]$.

In fact, without loss of generality we may assume that $\alpha_n < \alpha_{n+1}$ for all $n \geq 1$. Since $\beta < \gamma$ implies $t_\beta \leq t_\gamma$, we get $\lim_{n\to\infty} t_{\alpha_n} = \sup_{n \geq 1} t_{\alpha_n} \leq t_\alpha$ and

$$\lim_{n\to\infty} \int (t_\alpha - t_{\alpha_n}) dP_0 = \lim_{n\to\infty} (\alpha - \alpha_n) = 0,$$

whence $\lim_{n\to\infty} t_{\alpha_n} = t_\alpha [P_0]$. On the other hand, we deduce from $0 \leq t_\alpha - t_{\alpha_n} \leq t_\alpha - t_{\alpha_1}$ that

$$[t_\alpha - t_{\alpha_n} > 0] \subset [t_{\alpha_1} < 1] \subset [f_1 \le k_{\alpha_1} f_0].$$

Therefore

$$\lim_{n\to\infty} \int (t_\alpha - t_{\alpha_n}) dP_1 = \lim_{n\to\infty} \int (t_\alpha - t_{\alpha_n}) f_1 \, d(P_0+P_1)$$

$$\le \lim_{n\to\infty} k_{\alpha_1} \int (t_\alpha - t_{\alpha_n}) f_0 d(P_0+P_1) = 0,$$

whence $\lim_{n\to\infty} t_{\alpha_n} = t_\alpha[P_1]$.

<u>Example 11.5</u>. Let $\Omega: = \mathbb{R}^n$, $\mathfrak{A}: = \mathfrak{B}^n$, $P_0: = \nu_{0,1}^{\otimes n}$ and $P_1: = \nu_{\xi,1}^{\otimes n}$ where $\xi \in \mathbb{R}_+^*$. Clearly $P_0 = f_0 \cdot \lambda^n$ and $P_1 = f_1 \cdot \lambda^n$ with

$$f_0(x): = \frac{1}{(2\pi)^{n/2}} e^{-\frac{1}{2}\Sigma_{i=1}^n x_i^2}$$

and

$$f_1(x): = \frac{1}{(2\pi)^{n/2}} e^{-\frac{1}{2}\Sigma_{i=1}^n (x_i-\xi)^2}$$

defined for all $x = (x_1,\dots,x_n) \in \mathbb{R}^n$, respectively. By the Fundamental Lemma for each $\alpha \in [0,1]$ there exist $k \in \overline{\mathbb{R}}$ and $\gamma \in [0,1]$ such that the test

$$t_\alpha: = 1_{\left[\frac{f_1}{f_0} > k\right]} + \gamma 1_{\left[\frac{f_1}{f_0} = k\right]}$$

is most powerful of level α for P_0 versus P_1. As usual let $\overline{X} = \overline{X}_n: = \frac{1}{n}\Sigma_{k=1}^n X_k$, where $(X_1,\dots,X_n)$ denotes the sample of size n associated with our model. Then for every $k \in \mathbb{R}_+^*$ we get

$$\left[\frac{f_1}{f_0} > k\right] = \left[-\frac{n\xi}{2}(\xi - 2\overline{X}) > \ln k\right] = \left[\overline{X} > \frac{\ln k}{n\xi} + \frac{\xi}{2}\right],$$

whence there exists a number k_α such that

$$t_\alpha = 1_{[\overline{X}>k_\alpha]} + \gamma 1_{[\overline{X}=k_\alpha]},$$

where k_α is determined by $\beta_{t_\alpha}(P_0) = \alpha$ or equivalently by the inequalities

$$P_0[\overline{X} > k_\alpha] \le \alpha \le P_0[\overline{X} \ge k_\alpha].$$

Since $\overline{X}(P_0) = \nu_{0,\frac{1}{n}}$, the number k_α can be calculated (or looked up in tables) from the equalities

$$\alpha = \int_{k_\alpha}^{\infty} \sqrt{(n/2\pi)}e^{-\frac{n}{2}x^2}\,dx = \int_{k_\alpha\sqrt{n}}^{\infty} \frac{1}{\sqrt{2\pi}}e^{-\frac{x^2}{2}}\,dx.$$

It should be noted that the test t_α for $\mathscr{P}_0$ versus $\mathscr{P}_1$ constructed in this example appears to be independent of ξ, i.e., t_α is a most powerful level α test also for the hypothesis $\mathscr{P}_0 = \{\nu_{0,1}^{\otimes n}\}$ versus the *composite* alternative $\mathscr{P}_1' := \{\nu_{\xi,1}^{\otimes n} : \xi \in \mathbb{R}_+^*\}$.

After having established the existence of most powerful level α tests t_α for any $\alpha \in [0,1]$ we shall show how the power $\beta_{t_\alpha}(P_1)$ of such tests depends on the level α. The subsequent result contains a few general properties concerning this dependence.

Theorem 11.6. Let $(\Omega, \mathbf{A}, \mathscr{P}, \mathscr{P}_0, \mathscr{P}_1)$ be a testing experiment with simple hypothesis $\mathscr{P}_0 := \{P_0\}$ and simple alternative $\mathscr{P}_1 := \{P_1\}$, and let the mapping $\beta: [0,1] \to [0,1]$ be defined by $\beta(\alpha) := \beta_{t_\alpha}(P_1)$ for all $\alpha \in [0,1]$ and some most powerful level α test t_α for $\mathscr{P}_0$ versus $\mathscr{P}_1$. Then

(i) β is isotone, concave on $[0,1]$, and continuous on $]0,1]$.

(ii) The function $\alpha \to \frac{\beta(\alpha)}{\alpha}$ is antitone on $]0,1]$ and satisfies on $]0,1]$ the inequality

$$1 \leq \frac{\beta(\alpha)}{\alpha} \leq \frac{1}{\alpha}.$$

Moreover, we have

$$\lim_{\alpha \to 1} \frac{\beta(\alpha)}{\alpha} = 1.$$

(iii) If $P_0 \perp P_1$, then $\beta \equiv 1$.

Proof: First of all we note that the mapping β is well-defined. Indeed, if t_α' is another most powerful level α test for $\mathscr{P}_0$ versus $\mathscr{P}_1$, then by the very definition of the power we get $\beta_{t_\alpha}(P_1) = \beta_{t_\alpha'}(P_1)$.

(i) Now let $\alpha, \alpha' \in [0,1]$, $\alpha \leq \alpha'$. Then $\mathfrak{T}_\alpha \subset \mathfrak{T}_{\alpha'}$ and hence $\beta_{t_\alpha} \leq \beta_{t_{\alpha'}}$, which shows the isotonicity of β. Given $\alpha_1, \alpha_2, u \in [0,1]$ we have

$$\int (ut_{\alpha_1} + (1-u)t_{\alpha_2})\,dP_0 = u\alpha_1 + (1-u)\alpha_2,$$

thus

$$u\beta(\alpha_1) + (1-u)\beta(\alpha_2) = \int (ut_{\alpha_1} + (1-u)t_{\alpha_2})\,dP_1$$

$$\leq \int t_{u\alpha_1+(1-u)\alpha_2}\,dP_1 = \beta(u\alpha_1 + (1-u)\alpha_2),$$

which implies the concavity of β on $[0,1]$. This implies that β is continuous on $]0,1[$. Since β has been shown to be isotone on $[0,1]$, we may conclude that β is continuous even on $]0,1]$.

(ii) By the concavity of β on $[0,1]$ we obtain for all $\alpha_1,\alpha_2 \in]0,1[$ with $\alpha_1 \leq \alpha_2$ the inequality

$$\frac{\alpha_1}{\alpha_2}\beta(\alpha_2) \leq \frac{\alpha_2-\alpha_1}{\alpha_2}\beta(0) + \frac{\alpha_1}{\alpha_2}\beta(\alpha_2) \leq \beta\left(\frac{\alpha_2-\alpha_1}{\alpha_2}\cdot 0 + \frac{\alpha_1}{\alpha_2}\alpha_2\right) = \beta(\alpha_1).$$

i.e., the function $\alpha \to \frac{\beta(\alpha)}{\alpha}$ is antitone on $]0,1[$. The inequality asserted in the statement of the theorem follows from

$$\alpha = \alpha\beta(1) \leq \alpha\beta(1) + (1-\alpha)\beta(0) \leq \beta(\alpha) \leq 1$$

valid for all $\alpha \in [0,1]$. Consequently,

$$\lim_{\alpha\to 1} \frac{\beta(\alpha)}{\alpha} = 1.$$

(iii) Let $M \in \mathfrak{A}$ be such that $P_0(M) = 0$ and $P_1(M) = 1$. Then 1_M is a most powerful level α test for each $\alpha \in [0,1]$ (for $\mathscr{P}_0$ versus $\mathscr{P}_1$), thus $\beta \equiv 1$. □

We shall terminate the section by a generalization of the Neyman-Pearson Fundamental Lemma whose proof can be carried out in analogy to that of Theorem 11.1.

<u>Theorem 11.7</u>. Let $(\Omega,\mathfrak{A},\mu)$ be a measure space, $n \geq 1$ and let $f_0,\ldots,f_n$ be μ-integrable functions on Ω. For every n-tuple $(\alpha_0,\ldots,\alpha_{n-1}) \in \mathbb{R}^n$ we define the sets

$$\mathfrak{T}_{\alpha_0,\ldots,\alpha_{n-1}} := \{t \in \mathfrak{M}_{(1)}(\Omega,\mathfrak{A}): \int t\, f_i\, d\mu = \alpha_i \text{ for all } 0 \leq i \leq n-1\}$$

and

$$M_{k_0,\ldots,k_{n-1}} := \left[f_n \geq \sum_{i=0}^{n-1} k_i f_i\right].$$

Then

(i) If for a given n-tuple $(\alpha_0,\ldots,\alpha_{n-1}) \in \mathbb{R}^n$ there exists an n-tuple $(k_0,\ldots,k_{n-1}) \in \mathbb{R}^n$ satisfying

$$\int 1_{M_{k_0,\ldots,k_{n-1}}} f_i\, d\mu = \alpha_i \quad \text{for all } 0 \leq i \leq n-1,$$

then we have

$$\int 1_{M_{k_0,\ldots,k_{n-1}}} f_n\, d\mu \geq \int t\, f_n\, d\mu$$

for all $t \in \mathfrak{T}_{\alpha_0,\ldots,\alpha_{n-1}}$.

(ii) If $t' \in \mathfrak{T}_{\alpha_0,\ldots,\alpha_{n-1}}$ is another function satisfying

$$\int t' f_n \, d\mu = \sup_{t \in \mathfrak{T}_{\alpha_0,\ldots,\alpha_{n-1}}} \int t f_n \, d\mu,$$

then we obtain

$$t' = 1_{M_{k_0,\ldots,k_{n-1}}} \quad \mu\text{-a.s. on} \quad \left[f_n \neq \sum_{i=0}^{n-1} k_i f_i\right].$$

§12. LEAST FAVORABLE DISTRIBUTIONS AND BAYES TESTS

In this section we shall present two approaches in order to handle composite testing problems: least favorable apriori distributions (mixtures) and Bayes (π-) tests for the cases of composite hypotheses and composite alternatives respectively.

First let $(\Omega,\mathbb{A},\mathscr{P})$ be an arbitrary experiment. We introduce some more notation.

1. For every $A \in \mathbb{A}$ we define the mapping $w_A: \mathscr{P} \to \mathbb{R}$ by $w_A := P(A)$ for all $P \in \mathscr{P}$.
2. Let $\Sigma_{\mathbb{A}}$ be the σ-algebra in $\mathscr{P}$ generated by the family $\{w_A: A \in \mathbb{A}\}$.
3. Moreover let the mapping $N_{\mathscr{P}}: \mathscr{P} \times \mathbb{A} \to \mathbb{R}$ be defined by $N_{\mathscr{P}}(P,A) := P(A)$ for all $(P,A) \in \mathscr{P} \times \mathbb{A}$.

Since the mappings $w_A := N_{\mathscr{P}}(\cdot,A)$ $(A \in \mathbb{A})$ are $\Sigma_{\mathbb{A}}$-measurable by definition, $N_{\mathscr{P}} \in \mathrm{Stoch}((\mathscr{P},\Sigma),(\Omega,\mathbb{A}))$ for every σ-algebra Σ containing $\Sigma_{\mathbb{A}}$.

12.1. <u>Properties of</u> $N_{\mathscr{P}}$.

<u>12.1.1</u>. If $t \in \mathfrak{M}_{(1)}(\Omega,\mathbb{A})$ and if $\beta_t: \mathscr{P} \to \mathbb{R}$ denotes the power of t defined by $\beta_t(P) := \int t \, dP$ for all $P \in \mathscr{P}$, then $\beta_t = N_{\mathscr{P}} t$.

This follows from the equalities

$$(N_{\mathscr{P}} t)(P) = \int t(\omega) N_{\mathscr{P}}(P,d\omega) = \int t(\omega) P(d\omega)$$

valid for all $P \in \mathscr{P}$.

<u>12.1.2</u>. Let $\mathscr{P}' \subset \mathscr{P}$. Then for every $t \in \mathfrak{M}_{(1)}(\Omega,\mathbb{A})$ we have $\mathrm{Res}_{\mathscr{P}'}(N_{\mathscr{P}} t) = N_{\mathscr{P}'} t$.

<u>12.1.3</u>. Let $(\Omega,\mathbb{A},\mathscr{P})$ be μ-dominated by a measure $\mu \in \mathscr{M}^1(\Omega,\mathbb{A})$ and let there exist a function $\tilde{f} \in \mathfrak{M}_+(\Omega \times \mathscr{P}, \mathbb{A} \otimes \Sigma_{\mathbb{A}})$ such that

$P = \tilde{f}(\cdot,P)\cdot\mu$ holds for all $P \in \mathscr{P}$. Let moreover Σ be any σ-algebra containing $\Sigma_{\mathfrak{A}}$. Then for each (apriori measure) $\pi \in \mathscr{M}^1(\mathscr{P},\Sigma)$ the measure $\pi N_{\mathscr{P}}$ is the Ω-projection of $\tilde{f}\cdot(\mu \otimes \pi)$.

This assertion follows from the subsequent simple computation: For all $A \in \mathfrak{A}$ we have

$$\begin{aligned}(\pi N_{\mathscr{P}})(A) &= \int \pi(dP)N_{\mathscr{P}}(P,A) = \int \pi(dP)P(A)\\ &= \int \pi(dP)\left[\int 1_A(\omega)\tilde{f}(\omega,P)\mu(d\omega)\right]\\ &= \int 1_A(\omega)\tilde{f}(\omega,P)\pi \otimes \mu(d(P,\omega))\\ &= [\tilde{f}\cdot(\mu \otimes \pi)](A \times \mathscr{P}).\end{aligned}$$

Now let $(\Omega,\mathfrak{A},\mathscr{P},\mathscr{P}_0,\mathscr{P}_1)$ be a testing experiment, Σ_0 any σ-algebra in $\mathscr{P}_0$ such that the elements of the family $\{P \to P(A): A \in \mathfrak{A}\}$ are Σ_0-measurable, and let

$$\mathfrak{T}_\alpha := \mathfrak{T}_\alpha(\mathscr{P}_0) := \{t \in \mathfrak{M}_{(1)}(\Omega,\mathfrak{A}): N_{\mathscr{P}_0}t \le \alpha\}.$$

Definition 12.2. (i) The elements of $\mathscr{M}^1(\mathscr{P}_0,\Sigma_0)$ are called *mixtures on* $\mathscr{P}_0$.

(ii) Given a mixture π on $\mathscr{P}_0$ and a number $\alpha \in [0,1]$ we introduce the set

$$\mathfrak{T}_{\alpha,\pi} := \left\{t \in \mathfrak{M}_{(1)}(\Omega,\mathfrak{A}): \int N_{\mathscr{P}_0}t\, d\pi \le \alpha\right\}$$

of all *π-tests of level* α.

(iii) A mixture π on $\mathscr{P}_0$ is called *least favorable* for (testing) $\mathscr{P}_0$ versus $\mathscr{P}_1$ if for all mixtures π' on $\mathscr{P}$ we have

$$\sup_{t\in\mathfrak{T}_{\alpha,\pi}} N_{\mathscr{P}_1}t \le \sup_{t\in\mathfrak{T}_{\alpha,\pi'}} N_{\mathscr{P}_1}t.$$

(iv) $\pi \in \mathscr{M}^1(\mathscr{P}_0,\Sigma_0)$ is said to be *admissible of level* α if there exists a most powerful level α test t_π for $\{\pi N_{\mathscr{P}_0}\}$ versus $\mathscr{P}_1$ satisfying

$$N_{\mathscr{P}_0}t_\pi \le \alpha 1_{\mathscr{P}_0}.$$

Theorem 12.3. Let $(\Omega,\mathfrak{A},\mathscr{P},\mathscr{P}_0,\mathscr{P}_1)$ be a testing experiment, $\alpha \in [0,1]$ and let $\pi \in \mathscr{M}^1(\mathscr{P}_0,\Sigma_0)$ be a mixture on $\mathscr{P}_0$ which is admissible of level α. Then we have

(i) π is least favorable.

(ii) There exists a most powerful level α test for $\mathscr{P}_0$ versus $\mathscr{P}_1$.

Proof: Let t_π be the test corresponding to the given mixture $\pi \in \mathcal{M}^1(\mathcal{P}_0,\Sigma_0)$ by the very definition of admissibility. Then $N_{\mathcal{P}_0} t_\pi \leq \alpha$, and $N_{\mathcal{P}_1} t_\pi \geq N_{\mathcal{P}_1} t$ for all $t \in \mathfrak{M}_{(1)}(\Omega,\mathfrak{A})$ satisfying $\int N_{\mathcal{P}_0} t \, d\pi \leq \alpha$.

(i) For all $\pi' \in \mathcal{M}^1(\mathcal{P}_0,\Sigma_0)$ we obtain

$$\sup_{t\in\mathfrak{T}_{\alpha,\pi}} N_{\mathcal{P}_1} t = N_{\mathcal{P}_1} t_\pi \leq \sup_{t\in\mathfrak{T}_{\alpha,\pi'}} N_{\mathcal{P}_1} t,$$

since by $N_{\mathcal{P}_0} t_\pi \leq \alpha$ we have $\int N_{\mathcal{P}_0} t_\pi d\pi' \leq \alpha$ which means that $t_\pi \in \mathfrak{T}_{\alpha,\pi'}$.

(ii) t_π is most powerful for $\mathcal{P}_0$ versus $\mathcal{P}_1$.

Indeed, by assumption we have $N_{\mathcal{P}_0} t_\pi \leq \alpha$. If $t \in \mathfrak{T}_\alpha$, then the inequality $\int N_{\mathcal{P}_0} t \, d\pi \leq \alpha$ and the fact that t_π is most powerful of level α for $\{\pi N_{\mathcal{P}_0}\}$ versus $\mathcal{P}_1$ imply $N_{\mathcal{P}_1} t_\pi \geq N_{\mathcal{P}_1} t$ which proves the assertion. □

Example 12.4. Let $\Omega = \mathbb{R}^n$, $\mathfrak{A} := \mathfrak{B}^n$, $0 < \sigma_0^2 < \sigma_1^2$, $\mathcal{P}_0 := \{\nu^{\otimes n}_{\xi,\sigma^2}: \xi \in \mathbb{R}, 0 < \sigma^2 \leq \sigma_0^2\}$, $\mathcal{P}_1 := \{\nu^{\otimes n}_{\xi,\sigma_1^2}: \xi \in \mathbb{R}\}$ and $\mathcal{P} := \mathcal{P}_0 \cup \mathcal{P}_1$. Thus we are given a testing experiment $(\Omega,\mathfrak{A},\mathcal{P},\mathcal{P}_0,\mathcal{P}_1)$. Let $I_0 := \mathbb{R} \times]0,\sigma_0^2]$, $I_1 := \mathbb{R} \times \{\sigma_1^2\}$ and $I := I_0 \cup I_1$. Then we define a parametrization $\chi: \mathcal{P} \to I$ by $\chi(\nu^{\otimes n}_{\xi,\sigma^2}) := (\xi,\sigma^2)$ for all $\xi \in \mathbb{R}$, $\sigma^2 \in]0,\sigma_0^2] \cup \{\sigma_1^2\}$. This parametrization χ is obviously bijective. We therefore put $P_{\xi,\sigma^2} := \chi^{-1}(\xi,\sigma^2)$ for all $\xi \in \mathbb{R}$, $\sigma^2 \in]0,\sigma_0^2] \cup \{\sigma_1^2\}$. Clearly the measures P_{ξ,σ^2} are λ^n-absolutely continuous, and we have $P_{\xi,\sigma^2} := p_{\xi,\sigma^2}\cdot\lambda^n$ with

$$p_{\xi,\sigma^2}(x) := \frac{1}{(2\pi\sigma^2)^{n/2}} \exp\left(-\frac{1}{2\sigma^2}\sum_{i=1}^n (X_i(x) - \xi)^2\right)$$

$$= \frac{1}{(2\pi\sigma^2)^{n/2}} \exp\left(-\frac{n}{2\sigma^2}(\overline{X}(x) - \xi)^2 - \frac{1}{2\sigma^2}U(x)\right)$$

for all $x \in \mathbb{R}^n$, where as usual $\overline{X} := \frac{1}{n}\Sigma_{i=1}^n X_i$ and $U := \Sigma_{i=1}^n (X_i - \overline{X})^2$.

For each $A \in \mathfrak{A} = \mathfrak{B}^n$ the mapping

$$(\xi,\sigma^2) \to \int_A \frac{1}{(2\pi\sigma^2)^{n/2}} e^{-\frac{1}{2\sigma^2}\Sigma_{i=1}^n (x_i-\xi)^2} \lambda^n(d(x_1,\ldots,x_n))$$

from $\mathbb{R} \times \mathbb{R}_+^*$ into $\mathbb{R}$ is (Borel) measurable. Therefore $\Sigma_{\mathfrak{A}}$ is contained in the σ-algebra mapped by χ from I_0 into $\mathcal{P}_0$. Consequently, we may identify I_0 with $\mathscr{P}_0$ and probability measures on $I_0 = \mathbb{R} \times]0,\sigma_0^2]$ with mixtures on $\mathscr{P}_0$.

Now we fix $\xi_1 \in \mathbb{R}$. Our next aim will be to approximate the image measure $\overline{X}(P_{\xi_1,\sigma_1^2}) = \nu_{\xi_1,(\sigma_1^2/n)}$ of P_{ξ_1,σ_1^2} under $\overline{X}$ as good as possible by image measures of elements of $\mathscr{P}_0$ under $\overline{X}$. For every $\xi \in \mathbb{R}$ we put

$$\pi_\xi := \nu_{\xi,(\sigma_1^2-\sigma_0^2)/n} \otimes \varepsilon_{\sigma_0}.$$

We choose the mixture $\pi := \pi_{\xi_1}$. Then the λ^n-density p_π of $P_\pi := \pi N_{\mathscr{P}_0}$ is given by

$$p_\pi(x) = \int p_{\xi,\sigma^2}(x)\pi(d(\xi,\sigma)) = \int p_{\xi,\sigma_0^2}(x)\nu_{\xi_1,(\sigma_1^2-\sigma_0^2)/n}(d\xi)$$

$$= \frac{1}{(2\pi\sigma_0^2)^{n/2}} \sqrt{\frac{n}{2\pi(\sigma_1^2-\sigma_0^2)}}\, e^{-\frac{1}{2\sigma_0^2}U(x)} \cdot$$

$$\cdot \int_{\mathbb{R}} e^{-\frac{n}{2\sigma_0^2}(\overline{X}(x)-\xi)^2}\, e^{-\frac{n}{2(\sigma_1^2-\sigma_0^2)}(\xi-\xi_1)^2}\, d\xi$$

$$= \text{const} \cdot e^{-\frac{n}{2\sigma_1^2}(\overline{X}(x)-\xi_1)^2}\, e^{-\frac{1}{2\sigma_0^2}U(x)}$$

for all $x \in \mathbb{R}^n$.

Considering, in addition, the λ^n-density p_{ξ_1,σ_1^2} defined by

$$p_{\xi_1,\sigma_1^2}(x) := \text{const}\; e^{-\frac{n}{2\sigma_1^2}(\overline{X}(x)-\xi_1)^2}\, e^{-\frac{1}{2\sigma_1^2}U(x)}$$

for all $x \in \mathbb{R}^n$ we obtain from the Neyman-Pearson Fundamental Lemma that there exists a most powerful test t_π for testing P_π versus P_{ξ_1,σ_1^2} of the form

$$t_\pi(x) := \begin{cases} 1 & \text{if } p_{\xi_1,\sigma_1^2}(x) \geq k p_\pi(x) \\ 0 & \text{if } p_{\xi_1,\sigma_1^2}(x) < k p_\pi(x) \end{cases}$$

for all $x \in \mathbb{R}^n$. Since

$$t_\pi(x) = \begin{cases} 1 & \text{if } U(x) \geq k_\alpha \\ 0 & \text{if } U(x) < k_\alpha \end{cases}$$

for all $x \in \mathbb{R}^n$, t_π is most powerful also for testing $P_{\pi_{\xi_1}}$ versus P_{ξ_1,σ_1^2}. Here, k_α is determined by the equality

$$\alpha = \beta_{t_\pi}(P_\pi) = P_\pi[U \geq k_\alpha] = U(P_\pi)([k_\alpha,\infty[)$$

for a given level $\alpha \in [0,1]$.

In the following it will be shown, that t_π is in fact independent of ξ_1 and σ_1.

For every $\sigma^2 \in \mathbb{R}_+^*$ we introduce a real function A_{σ^2} on $\mathbb{R}$ by $A_{\sigma^2}(\eta) := \sigma^2\eta$ for all $\eta \in \mathbb{R}$. With this notation we get

$$U(P_{\xi,\sigma^2}) = A_{\sigma^2}(\chi^2_{n-1}),$$

$$U(P_\pi) = A_{\sigma_0^2}(\chi^2_{n-1})$$

as well as

$$\alpha = A_{\sigma_0^2}(\chi^2_{n-1})([k_\alpha,\infty[) = \chi^2_{n-1}\left(\left[\frac{k_\alpha}{\sigma_0^2},\infty\right[\right).$$

From this we conclude the admissibility of π by showing that $E_P(t_\pi) \leq \alpha$ holds for all $P \in \mathscr{P}_0$. Indeed, for every $(\xi,\sigma^2) \in \mathbb{R} \times]0,\sigma_0^2]$ we have

$$\begin{aligned} E_{P_{\xi,\sigma^2}}(t_\pi) &= P_{\xi,\sigma^2}[U \geq k_\alpha] = U(P_{\xi,\sigma^2})([k_\alpha,\infty[) \\ &= A_{\sigma^2}(\chi^2_{n-1})([k_\alpha,\infty[) = \chi^2_{n-1}\left(\left[\frac{k_\alpha}{\sigma^2},\infty\right[\right) \\ &\leq \chi^2_{n-1}\left(\left[\frac{k_\alpha}{\sigma_0^2},\infty\right[\right) = \alpha. \end{aligned}$$

The hypotheses of Theorem 12.3 being established we now obtain from the very theorem that π is least favorable and t_π is most powerful for $\mathscr{P}_0$ versus P_{ξ_1,σ_1^2} and hence for $\mathscr{P}_0$ versus the entire alternative $\mathscr{P}_1$.

Remark 12.5. Since t_π is independent of σ_1^2, it is a most powerful test for $\mathscr{P}_0$ versus $\{P_{\xi,\sigma_1^2}: \xi \in \mathbb{R}\}$ for every $\sigma_1^2 > \sigma_0^2$ and hence a most powerful test for $\mathscr{P}_0$ versus $\{P_{\xi,\sigma^2}: \xi \in \mathbb{R}, \sigma > \sigma_0\}$.

In the remaining part of this section we shall consider Bayes (π-) tests in order to handle composite alternatives. Let $(\Omega,\mathfrak{A},\mathscr{P},\mathscr{P}_0,\mathscr{P}_1)$ be a testing experiment and let Σ_1 be a σ-algebra in $\mathscr{P}_1$ containing the σ-algebra $\Sigma_{\mathfrak{A}} := \Sigma_{\mathfrak{A}}(\mathscr{P}_1)$.

Definition 12.6. Let $\pi \in \mathscr{M}^1(\mathscr{P}_1,\Sigma_1)$. A test $t_\pi \in \mathfrak{T}_\alpha(\mathscr{P}_0)$ is said to be a *Bayes π-test* if

$$\int N_{\mathscr{P}_1} t_\pi \, d\pi \geq \int N_{\mathscr{P}_1} t \, d\pi$$

for all $t \in \mathfrak{T}_\alpha$.

We note that a test $t_\pi \in \mathfrak{M}_{(1)}(\Omega,\mathfrak{A})$ is a Bayes π-test iff t is a most powerful level α test for testing $\mathscr{P}_0$ versus $\{\pi N_{\mathscr{P}_1}\}$. In particular we obtain the following

Theorem 12.7. Let there exist a measure $\mu \in \mathscr{M}^1(\Omega,\mathfrak{A})$ and a function $\tilde{f} \in \mathfrak{M}_+(\Omega \times \mathscr{P}_1, \mathfrak{A} \otimes \Sigma_1)$ such that $P = \tilde{f}(\cdot,P)\cdot\mu$ holds for all $P \in \mathscr{P}_1$, and let P_π denote the Ω-projection of $\tilde{f}\cdot(\mu \otimes \pi)$. Then every Bayes π-test is a most powerful level α test for testing $\mathscr{P}_0$ versus $\{P_\pi\}$.

The proof is obvious, since by Property 12.1.3 we have $P_\pi = \pi N_{\mathscr{P}_1}$. □

Theorem 12.8. Every Bayes π-test t_π satisfying

$$\int N_{\mathscr{P}_1} t_\pi \, d\pi = \inf_{\pi' \in \mathscr{M}^1(\mathscr{P}_1,\Sigma_1)} \int N_{\mathscr{P}_1} t_\pi \, d\pi' \tag{*}$$

is a maximin test.

Proof: We have to show that

$$\inf_{P \in \mathscr{P}_1} (N_{\mathscr{P}_1} t_\pi)(P) \geq \inf_{P \in \mathscr{P}_1} (N_{\mathscr{P}_1} t)(P)$$

holds for all $t \in \mathfrak{T}_\alpha$. This, however, follows from the subsequent chain of inequalities valid for any $t \in \mathfrak{T}_\alpha$:

$$\begin{aligned}
\inf_{P \in \mathscr{P}_1} (N_{\mathscr{P}_1} t_\pi)(P) &= \inf_{P \in \mathscr{P}_1} \int N_{\mathscr{P}_1} t_\pi \, d\varepsilon_P \\
&\geq \inf_{\pi' \in \mathscr{M}^1(\mathscr{P}_1,\Sigma_1)} \int N_{\mathscr{P}_1} t_\pi \, d\pi' \\
&= \int N_{\mathscr{P}_1} t_\pi \, d\pi \qquad \text{(by (*))} \\
&\geq \int N_{\mathscr{P}_1} t \, d\pi \qquad \text{(since } t_\pi \text{ is a Bayes } \pi\text{-test)} \\
&\geq \inf_{P \in \mathscr{P}_1} (N_{\mathscr{P}_1} t)(P). \qquad □
\end{aligned}$$

CHAPTER V

Testing Experiments Admitting an Isotone Likelihood Quotient

§13. ISOTONE LIKELIHOOD QUOTIENT

Testing experiments with an isotone likelihood quotient arise whenever one considers a special class of *parametrized experiments* $(\Omega, \mathfrak{A}, \mathscr{P}, \chi: \mathscr{P} \to \mathbb{R})$ and investigates those testing experiments $(\Omega, \mathfrak{A}, \mathscr{P}, \mathscr{P}_0, \mathscr{P}_1)$ which are consistent with the given parametrization. Here, parametrizations are understood to be injective mappings $\chi: \mathscr{P} \to \mathbb{R}$. We shall put $\Theta := \chi(\mathscr{P})$, and for every $\theta \in \Theta$ we write $P_\theta := \chi^{-1}(\theta)$. In this context, the mapping $\beta_t: \Theta \to [0,1]$ defined by

$$\beta_t(\theta) := \int t \, dP_\theta = E_\theta(t)$$

for all $\theta \in \Theta$ appears to be the power function of the test $t \in \mathfrak{M}_{(1)}(\Omega, \mathfrak{A})$.

Let $(\Omega, \mathfrak{A}, \mathscr{P}, \chi: \mathscr{P} \to \mathbb{R})$ be a parametrized experiment and let μ be a measure in $\mathscr{M}_+^\sigma(\Omega, \mathfrak{A})$ with $\mathscr{P} \ll \mu$. The mapping $\theta \to dP_\theta/d\mu$ from Θ into $L^1(\Omega, \mathfrak{A}, \mu)$ is known to be the *likelihood function with respect to* μ.

Definition 13.1. $(\Omega, \mathfrak{A}, \mathscr{P}, \chi: \mathscr{P} \to \mathbb{R})$ is said to *admit a (strictly) isotone likelihood quotient*, in symbols (S)ILQ, if there exists a statistic $T: (\Omega, \mathfrak{A}) \to (\mathbb{R}, \mathfrak{B})$ and if for every pair $\theta_1, \theta_2 \in \Theta$ with $\theta_1 < \theta_2$ there exists a (strictly) isotone function $H_{\theta_1,\theta_2}: \mathbb{R} \to \overline{\mathbb{R}}$ such that

$$H_{\theta_1,\theta_2} \circ T = \frac{p_{\theta_2}}{p_{\theta_1}} \, [P_{\theta_1} + P_{\theta_2}],$$

where p_{θ_1} and p_{θ_2} denote versions of the densities $dP_{\theta_1}/d(P_{\theta_1}+P_{\theta_2})$ and $dP_{\theta_2}/d(P_{\theta_1}+P_{\theta_2})$ respectively with the conventions

$$\frac{p_{\theta_2}}{p_{\theta_1}} := \infty \quad \text{on} \quad [p_{\theta_1} = 0] \cap [p_{\theta_2} > 0]$$

and

$$\frac{p_{\theta_2}}{p_{\theta_1}} := 0 \quad \text{on} \quad [p_{\theta_1} = 0] \cap [p_{\theta_2} = 0].$$

Remark 13.2. In the case of ILQ the statistic T can be assumed without loss of generality to be integrable with respect to the $P_\theta(\theta \in \Theta)$. Otherwise, one just replaces T by $\arctan \circ T$ and H_{θ_1,θ_2} by $H_{\theta_1,\theta_2} \circ \tan$.

Example 13.3. Let $\Omega := \mathbb{R}^n$, $\mathbf{A} := \mathbb{B}^n$, $\Theta := \mathbb{R}$, $\sigma^2 \in]0,\infty[$ a fixed number, and for each $\theta \in \Theta$ let $P_\theta := \nu_{\theta,\sigma^2}^{\otimes n}$. Then for all $\theta \in \Theta$ and $x = (x_1,\dots,x_n) \in \mathbb{R}^n$ we have

$$p_\theta(x) = \frac{dP_\theta}{d\lambda^n}(x) = \frac{1}{(2\pi\sigma^2)^{n/2}} e^{-\frac{1}{2\sigma^2}\sum_{k=1}^{n}(x_k-\theta)^2},$$

whence for all $\theta',\theta'' \in \Theta$ with $\theta' < \theta''$ and all $x \in \mathbb{R}^n$

$$\frac{p_{\theta''}(x)}{p_{\theta'}(x)} = e^{-\frac{n}{2\sigma^2}(\theta''^2-\theta'^2)+\frac{n}{\sigma^2}(\theta''-\theta')\overline{X}(x)},$$

where $\overline{X}(x) = \frac{1}{n}\Sigma_{k=1}^{n} X_k(x) = \frac{1}{n}\Sigma_{k=1}^{n} x_k = \overline{x}$ for all $x = (x_1,\dots,x_n) \in \mathbb{R}^n$. Now we put $T := \overline{X}$ and define $H_{\theta',\theta''}: \mathbb{R} \to \overline{\mathbb{R}}$ by

$$H_{\theta',\theta''}(\xi) := e^{-\frac{n}{2\sigma^2}(\theta''^2-\theta'^2) + \frac{n}{\sigma^2}(\theta''-\theta')\xi}$$

for all $\xi \in \mathbb{R}$. Thus $H_{\theta',\theta''}$ is a strictly isotone function, and we have

$$\frac{p_{\theta''}}{p_{\theta'}} = H_{\theta',\theta''} \circ T.$$

Theorem 13.4. (S. Karlin, E. L. Lehmann, H. Rubin). Let $(\Omega,\mathbf{A},\mathcal{P},\chi: \mathcal{P} \to \mathbb{R})$ be an experiment admitting ILQ. For every $\theta_0 \in \Theta$ we consider the sets $\mathcal{P}_0 := \{P_\theta: \theta \leq \theta_0\}$ and $\mathcal{P}_1 := \{P_\theta: \theta > \theta_0\}$.

(i) For all $\alpha \in [0,1]$ and $\theta_0 \in \Theta$ there exists a most powerful level α test $t_{\alpha,\theta_0} \in \mathfrak{M}_{(1)}(\Omega,\mathbf{A})$ for testing $\mathcal{P}_0$ versus $\mathcal{P}_1$.

(ii) Let $\alpha \in [0,1]$, $\theta_0 \in \Theta$ and $t := t_{\alpha,\theta_0}$. Then the power function $\beta_t: \Theta \to [0,1]$ is isotone. If, moreover, $\theta_1 < \theta_2$, then $\beta_t(\theta_1) = \beta_t(\theta_2)$ holds iff $\beta_t(\theta_1) = 1$ or $\beta_t(\theta_2) = 0$.

The proof of the Theorem will be preceded by a

Lemma 13.5. Let $(\Omega,\mathfrak{A},\{P_\theta: \theta \in \Theta\})$ be an experiment admitting ILQ. Then for every $\alpha \in [0,1]$ and $\theta \in \Theta$ there exists a test $t_{\alpha,\theta}$ satisfying $\beta_{t_{\alpha,\theta}}(\theta) = \alpha$ and the following properties:

(a) If $\theta_1 \in \Theta$ and $\beta_{t_{\alpha,\theta}}(\theta_1) > 0$, then for all $\theta_2 \in \Theta$ with $\theta_2 > \theta_1$ the test $t_{\alpha,\theta}$ is most powerful for $\{P_{\theta_1}\}$ versus $\{P_{\theta_2}\}$.

(b) If $\theta_1 \in \Theta$ and $\beta_{t_{\alpha,\theta}}(\theta_1) < 1$, then for all $\theta_2 \in \Theta$ with $\theta_2 < \theta_1$ the test $1 - t_{\alpha,\theta}$ is most powerful for $\{P_{\theta_1}\}$ versus $\{P_{\theta_2}\}$.

(c) If $\theta_1,\theta_2 \in \Theta$ and $\theta_2 > \theta_1$, then t_{0,θ_1} is most powerful level 0 for $\{P_{\theta_1}\}$ versus $\{P_{\theta_2}\}$.

(d) If $\theta_1,\theta_2 \in \Theta$ and $\theta_2 < \theta_1$, then $1 - t_{1,\theta}$ is most powerful level 0 for $\{P_{\theta_1}\}$ versus $\{P_{\theta_2}\}$.

Proof: By 13.2 we may assume without loss of generality that T is integrable and nonnegative. To $\alpha \in [0,1]$ and $\theta \in \Theta$ there exist by the Fundamental Lemma 11.1 (i) and (iii) numbers $k \in \overline{\mathbb{R}}$ and $\gamma \in [0,1]$ such that the test

$$t_{\alpha,\theta} := 1_{[T>k]} + \gamma 1_{[T=k]}$$

satisfies $\beta_{t_{\alpha,\theta}}(\theta) = \alpha$. Here we choose in the case $\alpha = 0$, k minimal and γ maximal, in the case $\alpha = 1$, k maximal and γ minimal.

Now, let $\theta_1,\theta_2 \in \Theta$, p_1 be a $(P_{\theta_1} + P_{\theta_2})$-density of P_{θ_1} and p_2 a $(P_{\theta_1} + P_{\theta_2})$-density of P_{θ_2}. With the conventions $H_{\theta_1,\theta_2}(\infty) := \infty$ and $H_{\theta_1,\theta_2}(-\infty) := -\infty$ we obtain from the isotonicity of H_{θ_1,θ_2} the inclusions

$$[T \geq k] \subset [H_{\theta_1,\theta_2} \circ T \geq H_{\theta_1,\theta_2}(k)]$$

and

$$[T \leq k] \subset [H_{\theta_1,\theta_2} \circ T \leq H_{\theta_1,\theta_2}(k)].$$

We restrict ourselves to proving statements (a) and (c), the assertions (b) and (d) are established analogously.

(a) Let $\theta_1 < \theta_2$, $\alpha \in [0,1]$, $\theta \in \Theta$ and let $t := t_{\alpha,\theta} = 1_{[T>k]} + \gamma 1_{[T=k]}$ satisfy $\beta_t(\theta_1) > 0$. For $k' := H_{\theta_1,\theta_2}(k)$ we get $k' < \infty$, since

$k' = \infty$ implies

$$0 < \beta_t(\theta_1) \leq P_{\theta_1}([T \geq k]) \leq P_{\theta_1}([H_{\theta_1,\theta_2} \circ T \geq \infty])$$

$$= P_{\theta_1}([p_1 = 0 < p_2]) = 0.$$

Therefore we have within $(P_{\theta_1} + P_{\theta_2})$-null sets,

$$[p_2 > k'p_1] = [p_1 = 0 < p_2] \cup [p_1 > 0 \text{ and } \frac{p_2}{p_1} > k']$$

$$\subset [H_{\theta_1,\theta_2} \circ T > k'] \subset [T > k] \subset [t = 1]$$

and

$$[p_2 < k'p_1] = [p_1 > 0 \text{ and } \frac{p_2}{p_1} < k'] \subset [H_{\theta_1,\theta_2} \circ T < k']$$

$$\subset [T < k] \subset [t = 0].$$

From the Fundamental Lemma 11.1 (ii) we get that $t = t_{\alpha,\theta}$ is most powerful for $\{P_{\theta_1}\}$ versus $\{P_{\theta_2}\}$.

(b) Since H_{θ_1,θ_2} is isotone, the set $J: = [H_{\theta_1,\theta_2} = \infty]$ is an upper unbounded interval such that

$$P_{\theta_1}([T \in J]) = P_{\theta_1}([H_{\theta_1,\theta_2} \circ T = \infty]) = P_{\theta_1}([p_1 = 0 < p_2]) = 0.$$

Since by construction $J': = \{T(\omega): t_{0,\theta_1}(\omega) = 1\}$ is the largest upper unbounded interval such that $P_{\theta_1}([T \in J']) = 0$ holds, we get $J \subset J'$. From $P_{\theta_1}([T \in J']) = 0$ and $P_{\theta_1} = p_1 \cdot (P_{\theta_1} + P_{\theta_2})$ we conclude

$$P_{\theta_2}([T \in J' \text{ and } p_1 > 0]) \leq (P_{\theta_1} + P_{\theta_2})([T \in J' \text{ and } p_1 > 0]) = 0,$$

whence

$$P_{\theta_2}([T \in J' \setminus J]) = P_{\theta_2}([T \in J' \text{ and } H_{\theta_1,\theta_2} \circ T < \infty])$$

$$\leq P_{\theta_2}([T \in J' \text{ and } p_1 > 0]) + P_{\theta_2}([T \in J' \text{ and } p_2 = 0]) = 0.$$

Thus we have proved that

$$[T \in J] = [T \in J'] [P_{\theta_1} + P_{\theta_2}].$$

For the test $t: = t_{0,\theta_1} = 1_{[T \in J']}$ we have $\beta_t(\theta_1) = 0$ as well as the equalities

$$[p_2 > \infty p_1] = [p_1 = 0 < p_2] = [H_{\theta_1,\theta_2} \circ T = \infty]$$

$$= [T \in J] = [T \in J'] = [t = 1][P_{\theta_1} + P_{\theta_2}]$$

and

$$[p_2 < \infty p_1] = [p_1 > 0 \text{ and } p_2 < \infty] = [H_{\theta_1,\theta_2} \circ T < \infty]$$

$$= [T \notin J] = [T \notin J'] = [t = 0][P_{\theta_1} + P_{\theta_2}].$$

Theorem 11.1(ii) now implies that t_{0,θ_1} is a most powerful level 0 test for $\{P_{\theta_1}\}$ versus $\{P_{\theta_2}\}$.

Proof of Theorem 13.4: We are given the situation of the Lemma with T, k, γ and the test

$$t_{\alpha,\theta} := 1_{[T>k]} + \gamma 1_{[T=k]}$$

satisfying properties (a) to (d): First we show assertion (ii). Let $\theta_1 < \theta_2$. If $\eta := \beta_t(\theta_1) = 0$, nothing remains to be proved. Thus, let $\beta_t(\theta_1) > 0$. Then by property (a) of the Lemma, t is most powerful level η for $\{P_{\theta_1}\}$ versus $\{P_{\theta_2}\}$. From $\beta_{\eta 1_\Omega}(\theta_1) = \eta$ we get

$$\beta_t(\theta_2) \geq \beta_{\eta 1_\Omega}(\theta_2) = \eta = \beta_t(\theta_1),$$

i.e., the isotonicity of β_t. The second statement of (ii) will be proved by establishing a contradiction to $0 < \beta_t(\theta_1) = \beta_t(\theta_2) < 1$. In fact, by $P_{\theta_1} \neq P_{\theta_2}$ there exists an $\varepsilon > 0$ such that

$$(P_{\theta_1} + P_{\theta_2})([p_1 \geq p_2 + \varepsilon]) > 0$$

and

$$(P_{\theta_1} + P_{\theta_2})([p_2 \geq p_1 + \varepsilon]) > 0.$$

Choosing $\gamma, \delta \in \mathbb{R}$ such that

$$\gamma P_{\theta_1}([p_1 \geq p_2 + \varepsilon]) = \delta P_{\theta_1}([p_2 \geq p_1 + \varepsilon]),$$

$$0 < \delta \leq 1-\eta, \text{ and } 0 \leq \gamma \leq \eta,$$

we obtain for $t' := \eta \cdot 1_\Omega + \delta 1_{[p_2 \geq p_1+\varepsilon]} - \gamma 1_{[p_1 \geq p_2+\varepsilon]}$ the equality $\beta_{t'}(\theta_1) = \eta$ and thus

$$\begin{aligned}\beta_t(\theta_2) &\geq \beta_{t'}(\theta_2)\\ &\geq \eta + \varepsilon\Big[\delta(P_{\theta_1} + P_{\theta_2})([p_2 \geq p_1 + \varepsilon]) + \gamma(P_{\theta_1} + P_{\theta_2})([p_1 \geq p_2 + \varepsilon\\ &> \eta = \beta_t(\theta_1)\end{aligned}$$

which is the desired contradiction.

(i) From (a) and (c) of the Lemma, for every $\alpha \in [0,1]$ and $\theta_0 \in \Theta$ the test t_{α,θ_0} is most powerful level α for $\{P_{\theta_0}\}$ versus $\{P_\theta\}$ whenever $\theta > \theta_0$. Therefore t_{α,θ_0} is most powerful level α for $\{P_{\theta_0}\}$ versus the full alternative $\mathscr{P}_1 = \{P_\theta: \theta > \theta_0\}$. Now we infer from (ii) that for all $\theta \leq \theta_0$ we have

$$\beta_{t_{\alpha,\theta_0}}(\theta) \leq \beta_{t_{\alpha,\theta_0}}(\theta_0) = \alpha,$$

whence t_{α,θ_0} is a most powerful level α test for $\mathscr{P}_0 = \{P_\theta: \theta \leq \theta_0\}$ versus $\mathscr{P}_1$. □

In the following we are going to discuss the converse of statement (i) of Theorem 13.4. We want to show that under quite general assumptions the existence of most powerful tests implies ILQ.

For later applications of the result we slightly extend the framework and modify the definition of an experiment admitting ILQ.

We shall be concerned with *ordered experiments* $(\Omega,\mathbf{A},\mathscr{P})$ in the sense that we are given an order relation $<$ in the set $\mathscr{P}$. Clearly all parametrized experiments are ordered with respect to the natural ordering in $\mathscr{P}$ induced by $\mathbb{R}$ via the parametrization $\chi: \mathscr{P} \to \mathbb{R}$. Another useful example is the following: Let $P_0 \in \mathscr{P}$. We put

$$P_1 < P_2 :\iff P_1 = P_0 \quad \text{and} \quad P_2 \neq P_0.$$

Defining $\mathscr{P}_0 := \mathscr{P}$ in the first example or $\mathscr{P}_0 := \{P_0\}$ in the second one we see that the following condition is satisfied: For every $P_0' \in \mathscr{P}_0$ and $P \in \mathscr{P}$ one has $P_0' < P$ or $P < P_0'$ or $P = P_0'$.

<u>Definition 13.6</u>. Let $\mathscr{P}_0$ be a subset of $\mathscr{P}$. The ordered experiment $(\Omega,\mathbf{A},\mathscr{P})$ is said to *admit a (strictly) isotone likelihood quotient with respect to* $\mathscr{P}_0$ if there exists a statistic $T: (\Omega,\mathbf{A}) \to (\mathbb{R},\mathbf{B})$ and if for every pair $(P_0,P) \in \mathscr{P}_0 \times \mathscr{P}$ with $P_0 < P$ there exists a (strictly) isotone function $H_{P_0,P}: \mathbb{R} \to \overline{\mathbb{R}}$ satisfying

$$H_{P_0,P} \circ T = \frac{p}{p_0}\,[P_0 + P],$$

where p_0 and p denote versions of the densities $dP_0/d(P_0+P)$ and $dP/d(P_0+P)$ respectively with the conventions

$$\frac{p}{p_0} := \infty \quad \text{on} \quad [p_0 = 0] \cap [p > 0]$$

and

$$\frac{p}{p_0} := 0 \quad \text{on} \quad [p_0 = 0] \cap [p = 0].$$

In the special case $\mathscr{P}_0 := \mathscr{P}$ we just talk about a (strictly) isotone likelihood quotient as in Definition 13.1.

Theorem 13.7. (J. Pfanzagl). Let $(\Omega,\mathbf{A},\mathscr{P})$ be an ordered experiment which is μ-dominated by a measure $\mu \in \mathscr{M}_+^\sigma(\Omega,\mathbf{A})$, and let $\mathscr{P}_0$ be a subset of $\mathscr{P}$ such that for any pair $(P,P_0) \in \mathscr{P} \times \mathscr{P}_0$ we have either $P \leq P_0$ or $P > P_0$.

We assume given a set $\mathbb{K} \subset \mathbb{M}_{(1)}(\Omega,\mathbf{A})$ of tests satisfying the following conditions:

(i) For all $t \in \mathbb{K}$, $P_0 \in \mathscr{P}_0$ with $\beta_t(P_0) > 0$ and $P > P_0$ the test t is most powerful for testing $\{P_0\}$ against $\{P\}$.

(ii) For all $t \in \mathbb{K}$, $P_0 \in \mathscr{P}_0$ with $\beta_t(P_0) < 1$ and $P < P_0$ the test $1-t$ is most powerful for testing $\{P_0\}$ against $\{P\}$.

(iii) Let $P_0 \in \mathscr{P}_0$. Then there exists a test $t \in \mathbb{K}$ with $\beta_t(P_0) = 0$ which is most powerful for testing $\{P_0\}$ against $\{P\}$ whenever $P > P_0$.

(iv) Let $P_0 \in \mathscr{P}_0$. Then there exists a test $t \in \mathbb{K}$ with $\beta_t(P_0) = 1$ such that $1-t$ is most powerful for testing $\{P_0\}$ against $\{P\}$ whenever $P < P_0$.

(v) For all $\alpha \in]0,1[$ and $P_0 \in \mathscr{P}_0$ there exists a test $t \in \mathbb{K}$ satisfying $\beta_t(P_0) = \alpha$.

Then $(\Omega,\mathbf{A},\mathscr{P})$ admits ILQ with respect to $\mathscr{P}_0$.

The proof of the theorem will be prepared by two lemmas.

Lemma 13.8. Let $(\Omega,\mathbf{A},P)$ be a probability space and $\mathbf{D}$ a subsystem of $\mathbf{A}$ such that for every pair $A,B \in \mathbf{D}$ we have either $A \subset B[P]$ or $B \subset A[P]$. Then there exists a function $f \in \mathbb{M}_{(1)}(\Omega,\mathbf{A})$ satisfying for all $A \in \mathbf{D}$ the relation $A = [f \leq P(A)][P]$.

Proof: Without loss of generality we assume $\Omega \in \mathbf{D}$. In fact, if this assumption is not fulfilled, then we consider the subsystem $\mathbf{D} \cup \{\Omega\}$ of $\mathbf{A}$ so that the function f constructed for $\mathbf{D} \cup \{\Omega\}$ possesses the properties required for $\mathbf{D}$.

We now choose a countable subsystem $\mathfrak{D}_0$ of $\mathfrak{D}$ such that $\Omega \in \mathfrak{D}_0$ and $\{P(A): A \in \mathfrak{D}_0\}$ is a dense subset of $\{P(A): A \in \mathfrak{D}\}$. We define the function $f: \Omega \to \overline{\mathbb{R}}$ by

$$f(\omega) := \inf\{P(A): A \in \mathfrak{D}_0, A \ni \omega\} \quad \text{for all } \omega \in \Omega.$$

Since $\Omega \in \mathfrak{D}_0$, f is well-defined for every $\omega \in \Omega$, and we have $0 \leq f \leq 1$. An easy computation shows

$$[f < k] = \bigcup\{A' \in \mathfrak{D}_0: P(A') < k\} \quad \text{for all } k \in \mathbb{R}.$$

From this follows the measurability of f. For $A' \in \mathfrak{D}_0$ and $A \in \mathfrak{D}$ with $P(A') < P(A)$ we get by assumption that $A' \subset A[P]$ holds. This implies $[f < P(A)] \subset A[P]$ for all $A \in \mathfrak{D}$.

1. We first assume $A \in \mathfrak{D}_0$ and define

$$B_n := \bigcup\{A' \in \mathfrak{D}_0: P(A') < P(A) + \tfrac{1}{n}\} \quad \text{for all } n \geq 1.$$

Then $A \subset B_n$ for all $n \geq 1$, whence $A \subset \bigcap_{n\geq 1} B_n = [f \leq P(A)]$. Moreover, $B_1 \supset B_2 \supset \ldots \supset B_n \supset \ldots$, and so $P(\bigcap_{n\geq 1} B_n) = \lim_{n\to\infty} P(B_n)$. Since $\mathfrak{D}_0$ is countable, we can number the sets $A' \in \mathfrak{D}_0$ with $P(A') < P(A) + \frac{1}{n}$ such that we obtain a sequence $(A_i')_{i\geq 1}$ in $\mathfrak{D}_0$. By hypothesis on $\mathfrak{D}$ for every $k \geq 1$ there exists an $i_k \in \{1,\ldots,k\}$ satisfying $\bigcup_{i=1}^{k} A_i' \subset A_{i_k}'[P]$. Then from $A_{i_k}' \subset A_{i_{k+1}}'[P]$ and $A \subset B_n[P]$ follows

$$P(A) \leq P(B_n) = \lim_{k\to\infty} P(A_{i_k}') \leq P(A) + \frac{1}{n} \quad \text{for all } n \geq 1,$$

whence $P(\bigcap_{n\geq 1} B_n) = P(A)$. But then $A \subset \bigcap_{n\geq 1} B_n[P]$ implies $A = \bigcap_{n\geq 1} B_n = [f \leq P(A)][P]$.

2. Now let $A \in \mathfrak{D}$.

(a) Let $(A_i)_{i\geq 1}$ be a sequence in $\mathfrak{D}_0$ satisfying $P(A) = \inf_{i\geq 1} P(A_i)$. Then $A \subset A_i[P]$ for all $i \geq 1$, and for every $k \geq 1$ there exists an $i_k \in \{1,\ldots,k\}$ such that $A_{i_k} \subset \bigcap_{i=1}^{k} A_i[P]$ holds. This implies $A_{i_{k+1}} \subset A_{i_k}$ for all $k \geq 1$, thus $P(A) = \lim_{n\to\infty} P(A_{i_k}) = P(\bigcap_{k\geq 1} A_{i_k})$ and so $A = \bigcap_{k\geq 1} A_{i_k}[P]$, whence

$$[f \leq P(A)] = \left[f \leq \inf_{k\geq 1} P(A_{i_k})\right] = \bigcap_{k\geq 1}\left[f \leq P(A_{i_k})\right] = \bigcap_{k\geq 1} A_{i_k} = A[P].$$

(b) If there exists no sequence $(A_i)_{i\geq 1}$ in $\mathfrak{D}_0$ as in (a), then there are $A_i \in \mathfrak{D}_0$ with $P(A) = \sup_{i\geq 1} P(A_i)$, since $\{P(A'): A' \in \mathfrak{D}_0\}$ is dense in $\{P(A): A \in \mathfrak{D}\}$, and $P(A') \neq P(A)$ for all $A' \in \mathfrak{D}_0$, since otherwise (a) would hold. Consequently $A_i \subset A[P]$ and $P(A_i) < P(A)$ for all $i \geq 1$. As above we now establish the existence of an isotone sequence $(A_{i_k})_{k\geq 1}$ in $\mathfrak{D}_0$ satisfying $P(A) = \lim_{k\to\infty} P(A_{i_k})$, whence $P(A) = P(\bigcup_{k\geq 1} A_{i_k})$ and so $A = \bigcup_{k\geq 1} A_{i_k}\,[P]$.

From $P(A_{i_k}) < P(A)$ we first conclude $[f \leq P(A_{i_k})] \subset [f < P(A)]$ and hence $A \subset [f < P(A)][P]$. Thus, by the initial remarks we get $A = [f < P(A)][P]$.

But clearly $[f = P(A)] = \emptyset$. For, if $f(\omega_0) = P(A)$, then $P(A) = f(\omega_0) = \inf P(A'): A' \in \mathfrak{D}_0, A \ni \omega_0\}$ which is a contradiction of the hypothesis of (b). The proof is terminated. □

Lemma 13.9. Let $[a,b]$ and $[c,d]$ be closed intervals of $\overline{\mathbb{R}}$ and $G: [a,b] \to [c,d]$ right continuous and isotone with $G(b) = d$. Then there exists a left continuous and isotone function $u_G: [c,d] \to [a,b]$ with $u_G(c) = a$ such that

$$y \leq G(x) \Longleftrightarrow u_G(y) \leq x$$

for all $x \in [a,b]$, $y \in [c,d]$.

Proof: Since G is right continuous and $G(b) = d$ holds, the set $\{x \in [a,b]: G(x) \geq y\}$ is non-empty and compact for every $y \in [c,d]$. Thus there exists

$$u_G(y) := \min\{x \in [a,b]: G(x) \geq y\}.$$

From this definition follows immediately that

$$G(u_G(y)) \geq y \quad \text{for all} \quad y \in [c,d].$$

To show the isotonicity of u_G we consider $c \leq y_1 \leq y_2 \leq d$. Then

$$\{x \in [a,b]: G(x) \geq y_1\} \supset \{x \in [a,b]: G(x) \geq y_2\},$$

whence

$$\begin{aligned} u_G(y_1) &= \min\{x \in [a,b]: G(x) \geq y_1\} \\ &\leq \min\{x \in [a,b]: G(x) \geq y_2\} = u_G(y_2). \end{aligned}$$

For all $x' \in [a,b]$ we have $u_G(G(x')) \leq x'$, since $x' \in \{x \in [a,b]: G(x) \geq G(x')\}$. Therefore $y \leq G(x)$ implies $u_G(y) \leq u_G(G(x)) \leq x$, and conversely $u_G(y) \leq x$ implies $y \leq G(u_G(y)) \leq G(x)$. It remains to show the left continuity of u_G. Let $(y_n)_{n\geq 1}$ be an isotone sequence in $[c,d]$ with $\sup_{n\geq 1} y_n = y$ and let $x' := \sup_{n\geq 1} u_G(y_n)$. Since u_G is isotone, we get $x' \leq u_G(y)$, and from the isotonicity of G we conclude that for every $n \geq 1$, $y_n \leq G(u_G(y_n)) \leq G(x')$. But this implies $y \leq G(x')$. Then the above equivalence yields $u_G(y) \leq x'$, whence $\lim_{n\to\infty} u_G(y_n) = x' = u_G(y)$, and the left continuity of u_G has been proved. □

Proof of Theorem 13.7: By Theorem 8.3 there exist measures $P_n \in \mathscr{P}$ and numbers $c_n \in \mathbb{R}_+^*$ $(n \geq 1)$ with $\Sigma_{n\geq 1} c_n = 1$ such that $\nu := \Sigma_{n\geq 1} c_n P_n \sim \mathscr{P}$. For each $P \in \mathscr{P}$ let g_P denote a density of P with respect to ν. For all $P_0 \in \mathscr{P}_0$ we introduce the set

$$A(P_0) := [g_{P_0} = 0] \cap \bigcup_{P_n < P_0} [g_{P_n} > 0].$$

A test $t \in \mathfrak{M}_{(1)}(\Omega,\mathfrak{A})$ is called *two-sided most powerful in* P_0 if for every $P > P_0$, t is most powerful for testing $\{P_0\}$ against $\{P\}$, and for every $P < P_0$, $1-t$ is most powerful for testing $\{P_0\}$ against $\{P\}$.

1. Let $P_0 \in \mathscr{P}_0$ and $P \in \mathscr{P}$. Then the following statements are true:

1a. If $P > P_0$ and t is most powerful for testing $\{P_0\}$ against $\{P\}$, then

$$\nu([g_{P_0} = 0] \cap [g_P > 0] \cap [t < 1]) = 0.$$

1b. If $P < P_0$ and $1-t$ is most powerful for testing $\{P_0\}$ against $\{P\}$, then

$$\nu([g_{P_0} = 0] \cap [g_P > 0] \cap [t > 0]) = 0.$$

It suffices to show 1a, since 1b can be proved analogously.

If for $B := [g_{P_0} = 0] \cap [g_P > 0] \cap [t < 1]$ we suppose that $\nu(B) > 0$, then with $t' := t \cdot 1_{CB} + 1_B$ we obtain

$$\beta_t(P_0) = \beta_{t'}(P_0)$$

and

$$\beta_t(P) < \beta_{t'}(P)$$

which contradicts the hypothesis that t is most powerful for testing $\{P_0\}$ against $\{P\}$.

2. Let $P_0 \in \mathscr{P}$. Then

2a. $\nu(A(P_0) \cap [t > 0]) = 0$

whenever $1-t$ is most powerful for testing $\{P_0\}$ against $\{P\}$ for all $P < P_0$.

2b. $\nu\Big(CA(P_0) \cap [t = 0] \cap [g_{P_0} = 0]\Big) = 0$

whenever t is most powerful for testing $\{P_0\}$ against $\{P\}$ for all $P > P_0$.

The proof of 2a follows from

$$\begin{aligned}\nu(A(P_0) \cap [t > 0]) &= \nu\Big([g_{P_0} = 0] \cap \bigcup_{P_n < P_0} [g_{P_n} > 0] \cap [t > 0]\Big)\\ &\leq \sum_{P_n < P_0} \nu\Big([g_{P_0} = 0] \cap [g_{P_n} > 0] \cap [t > 0]\Big) = 0\end{aligned}$$

with the aid of 1b.

For the proof of 2b it suffices by $\nu \sim \{P_m : m \geq 1\}$ to show that

$$\nu\Big([g_{P_m} > 0] \cap CA(P_0) \cap [t = 0] \cap [g_{P_0} = 0]\Big) = 0$$

holds for all $m \geq 1$. For $P_m < P_0$ this is clear, since

$$[g_{P_m} > 0] \cap CA(P_0) \cap [g_{P_0} = 0] = \emptyset.$$

Under the assumption of the order structure of $\mathscr{P}$ it remains to consider the case that $P_m > P_0$. But then 1a yields

$$\begin{aligned}&\nu\Big([g_{P_m} > 0] \cap CA(P_0) \cap [t = 0] \cap [g_{P_0} = 0]\Big)\\ &\qquad \leq \nu\Big([g_{P_m} > 0] \cap [t < 1] \cap [g_{P_0} = 0]\Big) = 0.\end{aligned}$$

3. For all $P_0 \in \mathscr{P}_0$ and $t \in \mathfrak{K}$ we have the following implications:

3a. $P_0([t = 0]) = 0 \Rightarrow [t = 0] \subset A(P_0)\,[\nu]$.

3b. $P_0([t = 0]) = 1 \Rightarrow [g_{P_0} > 0] \cup A(P_0) \subset [t = 0]\,[\nu]$.

It suffices to show 3a, since 3b can be proved analogously. Let therefore $P_0([t = 0]) = 0$. Then we have

$$\nu([t = 0] \cap [g_{P_0} > 0]) = 0$$

and

$\beta_t(P_0) > 0.$

Thus by assumption (i) we can apply 2b, and the assertion follows from

$$\nu([t = 0] \cap CA(P_0))$$

$$\leq \nu([t = 0] \cap CA(P_0) \cap [g_{P_0} = 0]) + \nu([t = 0] \cap [g_{P_0} > 0]) = 0.$$

4. For all $P_0 \in \mathscr{P}_0$, $P \in \mathscr{P}$ with $P_0 < P$ and $c \in \overline{\mathbb{R}}_+$ let

$$A(P_0,P,c) := A(P_0) \cup ([g_P \leq cg_{P_0}] \cap [g_{P_0} > 0]).$$

Then there exists a test $t \in \mathfrak{M}_{(1)}(\Omega,\mathbf{A})$ which is two-sided most powerful in P_0 and satisfies the equivalence

$$[t = 0] = A(P_0,P,c)\,[\nu].$$

In order to see this we put $A := A(P_0,P,c)$ and $\alpha := 1 - P_0(A)$. Then there exists a test $t \in \mathbb{K}$ which is two-sided most powerful in P_0 and satisfies $\beta_t(P_0) = \alpha$. This follows for $\alpha = 0$ by assumptions (iii) and (ii), for $\alpha \in]0,1[$ by (v), (i) and (ii) and for $\alpha = 1$ by (iv) and (i). Now we infer from the Fundamental Lemma that

$$[t = 0] \subset [g_P \leq cg_{P_0}]\,[P_0],$$

and from

$$1-\alpha = \beta_{1-t}(P_0) \leq P_0([g_P \leq cg_{P_0}]) = P_0(A) = 1-\alpha$$

we deduce

$$[t = 0] = [g_P \leq cg_{P_0}]\,[P_0]. \qquad (*)$$

Suppose now that

$$[t = 0] \subset A[\nu]$$

is not fulfilled. Then

$$0 < \nu\Big([t = 0] \cap CA(P_0) \cap (C[g_P \leq cg_{P_0}] \cup [g_{P_0} = 0])\Big)$$

$$\leq \nu\Big([t = 0] \cap [g_{P_0} > 0] \cap C[g_P \leq cg_{P_0}]\Big)$$

contradicts (*).

Analogously one shows that

$$[t = 0] \supset A[\nu].$$

5. The system

$$\mathfrak{C} := \{A(P_0,P,c): P_0 \in \mathscr{P}_0, P \in \mathscr{P}, P > P_0, c \in \overline{\mathbb{R}}_+\}$$

can be totally ordered by ν-a.s. inclusion.

In fact, let $A := A(P_0,P,c)$ and $B \in \mathfrak{C}$. By 4 there exists a test $t \in \mathfrak{K}$ such that $B = [t = 0][\nu]$. If $P_0(B) = 0$ or 1, then the assertion follows from 3a or 3b. In all other cases we have $0 < \beta_t(P_0) < 1$, which means by assumptions (i) and (ii) that t is two-sided most powerful in P_0. But then the Fundamental Lemma yields a $c \in \overline{\mathbb{R}}_+$ such that

$$[g_P < c'g_{P_0}] \subset [t = 0] \subset [g_P \leq c'g_{P_0}][P_0].$$

In the case $c' > c$ we have $A \subset B[\nu]$ as follows by 2a from the subsequent chain of inequalities:

$$\begin{aligned}\nu(A \setminus B) &= \nu\Big((A(P_0) \cup ([g_{P_0} > 0] \cap [g_P \leq cg_{P_0}])) \cap [t > 0]\Big)\\ &\leq \nu(A(P_0) \cap [t > 0]) + \nu([t > 0] \cap [g_{P_0} > 0] \cap [g_P \leq cg_{P_0}])\\ &\leq \nu([g_P \geq c'g_{P_0}] \cap [g_P \leq cg_{P_0}] \cap [g_{P_0} > 0])\\ &= \nu([g_P \leq cg_{P_0} < c'g_{P_0} \leq g_P]) = 0.\end{aligned}$$

In the case $c' \leq c$ we get $B \subset A[\nu]$ as follows by 2b from

$$\begin{aligned}\nu(B \setminus A) &= \nu([t = 0] \cap CA(P_0) \cap ([g_{P_0} = 0] \cup [g_P > cg_{P_0}])\\ &\leq \nu([t = 0] \cap CA(P_0) \cap [g_{P_0} = 0])\\ &\quad + \nu([t = 0] \cap [g_{P_0} > 0] \cap [g_P > cg_{P_0}])\\ &\leq \nu([g_P \leq c'g_{P_0}] \cap [g_P > cg_{P_0}])\\ &= \nu([g_P \leq c'g_{P_0} \leq cg_{P_0} < g_P]) = 0.\end{aligned}$$

6. The statement in 5 enables us to apply Lemma 13.8 to the system $\mathfrak{C}$, and we obtain a statistic $T: (\Omega,\mathfrak{A}) \to ([0,1],\mathfrak{B}([0,1]))$ satisfying

$$A = [T \leq \nu(A)][\nu]$$

for all $A \in \mathfrak{C}$.

Since for all $P_0 \in \mathscr{P}_0$ and $P \in \mathscr{P}$ with $P > P_0$ the function

$c \to \nu(A(P_0,P,c))$ from $[0,\infty]$ into $[0,\nu(A(P_0,P,\infty))]$ is isotone and right continuous, an application of Lemma 13.9 provides us with an isotone function $H_{P_0,P}\colon \mathbb{R} \to [0,\infty]$ satisfying

$$H_{P_0,P}(y) \le c \iff y \le \nu(A(P_0,P,c))$$

for all $y \in \mathbb{R}$, $c \in \mathbb{R}_+$. It follows that

$$[H_{P_0,P} \circ T \le c] = [T \le \nu(A(P_0,P_1,c))] = A(P_0,P,c)\,[\nu]$$

for all $c \in \mathbb{R}_+$, whence

$$\begin{aligned}\left[H_{P_0,P} \circ T < \frac{g_P}{g_{P_0}}\right] &= \bigcup_{q \in \mathbb{Q}_+}\left([H_{P_0,P} \circ T \le q] \cap \left[\frac{g_P}{g_{P_0}} > q\right]\right)\\ &= \bigcup_{q \in \mathbb{Q}_+}\left(A(P_0) \cup \left([g_{P_0} > 0] \cap \left[\frac{g_P}{g_{P_0}} \le q\right]\right)\right) \cap \left[\frac{g_P}{g_{P_0}} > q\right]\\ &= \bigcup_{q \in \mathbb{Q}_+} A(P_0) \cap \left[\frac{g_P}{g_{P_0}} > q\right]\\ &= A(P_0) \cap [g_P > 0]\,[\nu],\end{aligned}$$

and similarly,

$$\left[H_{P_0,P} \circ T > \frac{g_P}{g_{P_0}}\right] = \bigcap_{P_n \le P_0} [g_{P_n} = 0] \cap [g_P = 0]\,[\nu].$$

Now, let $t \in \mathbb{K}$ be a test $t \in \mathfrak{M}_{(1)}(\Omega,\mathfrak{A})$ which is two-sided in P_0 and satisfies

$$[t = 0] = A(P_0,P,\infty) = \bigcup_{P_n \le P_0} [g_{P_n} > 0]\,[\nu].$$

Such tests exist by 4. Using 1a we get

$$\nu(A(P_0) \cap [g_P > 0]) = \nu(A(P_0) \cap [g_P > 0] \cap [t = 0]) = 0,$$

i.e.,

$$H_{P_0,P} \circ T \ge \frac{g_P}{g_{P_0}}\,[\nu].$$

On the other hand $\bigcap_{P_n \le P_0} [g_{P_n} = 0] \cap [g_P = 0]$ is a (P_0+P)-null set, which implies the inequality

$$H_{P_0,P} \circ T \le \frac{g_P}{g_{P_0}}\,[P_0 + P].$$

Altogether we obtain that

$$H \quad H_{P_0,P} \circ T = \frac{g_P}{g_{P_0}} [P_0 + P],$$

which completes the proof of the theorem. □

§14. ONE-DIMENSIONAL EXPONENTIAL EXPERIMENTS

In this section we continue the preceding discussion by studying in more detail one-dimensional exponential experiments which are easily shown to admit ILQ. The aim of our analysis will be a characterization of one-dimensional exponential experiments on the basis of their intrinsic properties.

At the end of the section we shall summarize the results of this and the preceding section. It turns out that either the existence of most powerful level α tests or the property of admitting an ILQ, which are essentially equivalent, reduce the initial experiments to one-dimensional exponential experiments. This result appears to be a fundamental motivation for the highly developed extension of the framework of parametric methods to that of non-parametric methods.

Preparations 14.1. Let $(\Omega, \mathfrak{A}, \mathscr{P}, \chi: \mathscr{P} \to \Theta)$ be an injectively parametrized experiment which is dominated by a measure $\mu \in \mathscr{M}_+^\sigma(\Omega, \mathfrak{A})$.

We recall that $(\Omega, \mathfrak{A}, \mathscr{P}, \chi: \mathscr{P} \to \Theta)$ is a one-dimensional exponential experiment (and $\mathscr{P}$ a one-dimensional exponential family) if there are two measurable mappings $T, h: (\Omega, \mathfrak{A}) \to (\mathbb{R}, \mathfrak{B})$ and two functions $C, \zeta: \Theta \to \mathbb{R}$ such that for all $\theta \in \Theta$ we have

$$P_\theta = C(\theta) e^{\zeta(\theta) T} h \cdot \mu.$$

$(\Omega, \mathfrak{A}, \mathscr{P}, \chi: \mathscr{P} \to \mathbb{R})$ admits an (S) ILQ with respect to μ if there exists a real statistic $T: (\Omega, \mathfrak{A}) \to (\mathbb{R}, \mathfrak{B})$ and if for every $\theta \in \Theta$ there is a (strictly) isotone function $H_\theta: \mathbb{R} \to \overline{\mathbb{R}}$ satisfying

$$P_\theta = (H_\theta \circ T) \cdot \mu.$$

We collect a few obvious *properties*.

1. The mapping $\zeta: \Theta \to \mathbb{R}$ of a one-dimensional exponential experiment is clearly injective. Therefore it becomes obvious that all one-dimensional exponential experiments are of the above form with

$$P_\theta = c(\theta) e^{\theta T} h \cdot \mu,$$

where $\theta \in Z \subset \mathbb{R}$ and $c(\theta) := (\int e^{\theta T} h \, d\mu)^{-1}$.

2. The measures of a one-dimensional exponential family are pairwise equivalent.

3. For the given one-dimensional exponential experiment and any $n \in \mathbb{N}$ we put

$$\mathscr{P}^{\otimes n} := \{P^{\otimes n} : P \in \mathscr{P}\}$$

and introduce the mapping $\chi^{\otimes n} : \mathscr{P}^{\otimes n} \to \Theta$ by

$$\chi^{\otimes n}(P^{\otimes n}) := \chi(P)$$

for all $P \in \mathscr{P}$.

If the mapping $\zeta : \Theta \to \mathbb{R}$ is isotone, then the experiment $(\Omega^n, \mathbf{A}^{\otimes n}, \mathscr{P}^{\otimes n}, \chi^{\otimes n} : \mathscr{P}^{\otimes n} \to \Theta)$ admits an SILQ in the sense of 13.

If $\zeta(\Theta)$ has a minimal (or maximal) element $z := \zeta(\theta)$, then the experiment $(\Omega^n, \mathbf{A}^{\otimes n}, \mathscr{P}^{\otimes n}, \chi^{\otimes n} : \mathscr{P}^{\otimes n} \to \Theta)$ admits an ILQ with respect to the measure $P_\theta^{\otimes n}$.

The following theorem concerns the converse of the properties 2 and 3.

<u>Theorem 14.2.</u> (R. Borges, J. Pfanzagl). Let $(\Omega, \mathbf{A}, \mathscr{P}, \chi : \mathscr{P} \to \Theta \cup \{*\})$ with $\Theta \subset \mathbb{R}$ be an injectively parametrized experiment which is dominated by a measure $\mu \in \mathscr{M}_+^\sigma(\Omega, \mathbf{A})$. Let $(\Omega, \mathbf{A}, \mathscr{P}_0, \mathscr{P}_1)$ be the associated testing experiment with $\mathscr{P}_0 := \{P_*\}$ and $\mathscr{P}_1 := \{P_\theta : \theta \in \Theta\}$. We suppose

(i) The measures $P_\theta (\theta \in \Theta)$ and P_* are pairwise equivalent.

(ii) $(\Omega^n, \mathbf{A}^{\otimes n}, \mathscr{P}^{\otimes n}, \chi^{\otimes n} : \mathscr{P}_1^{\otimes n} \to \Theta)$ admits an ILQ with respect to $P_*^{\otimes n}$ for all $n \in \mathbb{N}$.

Then $\mathscr{P}_1$ is a one-dimensional exponential family.

The <u>proof of the Theorem</u> will be preceded by two lemmas.

<u>Lemma 14.3.</u> Let $(\Omega, \mathbf{A}, \nu)$ be a probability space and let $\{f_\theta : \theta \in \Theta\}$ be a family of functions in $\mathfrak{M}_+(\Omega, \mathbf{A})$ such that the system $\{[f_\theta \leq \delta] : \theta \in \Theta, \delta \in \mathbb{R}_+\}$ is totally ordered with respect to inclusion. Then there exists a function $S \in \mathfrak{M}_+(\Omega, \mathbf{A})$ and for every $\theta \in \Theta$ there is a left-continuous, isotone function G_θ on $\mathbb{R}$ with $G_\theta(0) := 0$ such that

(i) for all $\omega \in \Omega$ with $S(\omega) > 0$ we have

$$S(\omega) = \nu\text{-}\operatorname{ess\,sup} S \cdot 1_{[S \leq S(\omega)]},$$

and

(ii) for all $\theta \in \Theta$ we have

$$G_\theta \circ S = f_\theta[\nu].$$

<u>Proof</u>: For every $\theta \in \Theta$ and $\delta \in \mathbb{R}_+$ we put

$$A_\theta^\delta := [f_\theta \leq \delta].$$

Moreover, let

$$\mathfrak{D} := \{A_\theta^\delta : \theta \in \Theta, \delta \in \mathbb{R}_+\}.$$

For each $\eta \in \{\nu(C): C \in \mathfrak{D}\}$ we choose a $D \in \mathfrak{D}$ satisfying $\nu(D) = \eta$. Let the totality of such sets D be denoted by $\mathfrak{U}$. Finally, for every $\omega \in \Omega$ we introduce the set

$$D_\omega := \bigcap\{D \in \mathfrak{U}: \omega \in D\}.$$

It follows that $D_\omega \in \mathfrak{A}$ for all $\omega \in \Omega$, since every intersection or union of an arbitrary subsystem of $\mathfrak{U}$ equals with the exception of ν-null sets, an intersection or union of a countable subsystem.

Now we define a mapping $S: \Omega \to \mathbb{R}_+$ by $S(\omega) := \nu(D_\omega)$. We obtain the following sequence of statements which in total completes the proof of the lemma.

1. S is $\mathfrak{A}$-measurable.

This follows from the identities

$$\begin{aligned}[S < \delta] &= \{\omega \in \Omega: \text{There exists a } D \in \mathfrak{U} \text{ such that } \nu(D) < \delta \text{ and } \omega \in D\} \\ &= \bigcap\{D \in \mathfrak{U}: \nu(D) < \delta\}.\end{aligned}$$

2. For every $B \in \mathfrak{D}$ or $B = D_\omega$ for some $\omega \in \Omega$ we have $B = [S \leq \nu(B)][\nu]$.

In order to prove this statement we give three cases:

2a. Let $B := D \in \mathfrak{U}$. Then $\omega \in D$ implies $D_\omega \subset D$, whence $S(\omega) \leq \nu(D)$ and therefore $D \subset [S \leq \nu(D)]$. The system $\{D_\omega: \omega \in \Omega\}$ is totally ordered. From the definition of S we conclude

$$\bigcup\{D_\omega: S(\omega) \leq \nu(D)\} \subset D[\nu].$$

Moreover, $\omega \in D_\omega$ implies

$$[S \leq \nu(D)] \subset \bigcup\{D_\omega: S(\omega) \leq \nu(D)\}.$$

2b. The case of a set $B \in \mathfrak{D}$ is now clear, since every element of $\mathfrak{D}$ equals ν-a.e. an element of $\mathfrak{U}$, and we can apply 2a.

2c. Let $B := D_\omega$ for $\omega \in \Omega$. Then the assertion follows from the representation of D_ω as a countable intersection of elements of $\mathfrak{U}$.

3. Let $\omega \in \Omega$. An application of 2 to the set $B := D_\omega$ yields

$$S(\omega) = \nu([S \leq S(\omega)]).$$

Therefore, for arbitrary $\varepsilon > 0$ we get

$$\nu([S(\omega) - \varepsilon < S \leq S(\omega)])$$
$$= S(\omega) - \sup\{S(\tau): S(\tau) \leq S(\omega) - \varepsilon\} \geq \varepsilon > 0,$$

whence

$$\nu\text{ - ess sup } S \cdot 1_{[S \leq S(\omega)]} \geq S(\omega) - \varepsilon,$$

which implies statement (i) of the lemma. □

4. Since the mappings $\delta \to \nu([f_\theta \leq \delta])$ from $[0,\infty]$ into $[0,1]$ are right continuous and isotone for all $\theta \in \Theta$, by Lemma 13.9 there exist left continuous and isotone functions $G_\theta: [0,1] \to [0,\infty]$ with $G_\theta(0) = 0$ and

$$G_\theta(s) \leq \delta \iff s \leq \nu([f_\theta \leq \delta])$$

for all $s \in [0,1]$, $\delta \in [0,\infty]$, $\theta \in \Theta$.

Let $\varepsilon > 0$. Then we have

$$[G_\theta \circ S < f_\theta - \varepsilon] = \bigcup_{q \in \mathbb{Q}} [G_\theta \circ S \leq q] \cap [q < f_\theta - \varepsilon]$$
$$= \bigcup_{q \in \mathbb{Q}} [S \leq \nu([f_\theta \leq q])] \cap [f_\theta > q + \varepsilon]$$
$$= \bigcup_{q \in \mathbb{Q}} [f_\theta \leq q] \cap [f_\theta > q + \varepsilon] = \emptyset[\nu],$$

whence $[G_\theta \circ S < f_\theta] = \emptyset[\nu]$. In the same way one proves $[G_\theta \circ S > f_\theta] = \emptyset[\nu]$. Thus we have (ii). □

<u>Corollary 14.4</u>. Assertion (ii) of the theorem holds for every $\omega \in \Omega$ if for any sequence $(C_k)_{k \geq 0}$ in $\mathfrak{D}$ we have the implication

$$\nu(C_0) \geq \nu(\bigcap_{k \geq 1} C_k) \Rightarrow C_0 \supset \bigcap_{k \geq 1} C_k.$$

<u>Proof</u>: Under the hypothesis of the corollary $\nu(D_\omega) \leq \nu(D)$ implies $D_\omega \subset D$ for all $\omega \in \Omega$. Hence by part 2 of the proof of the theorem we get $B = [S \leq \nu(B)]$ for all $B \in \mathfrak{D}$ which is the desired statement. □

Lemma 14.5. Let Ω be a set and f_1, f_2 two real-valued functions on Ω with the following properties

(i) For all $\omega_1,\ldots,\omega_r, \tau_1,\ldots,\tau_r \in \Omega$ the inequality

$$\sum_{i=1}^{r} f_1(\omega_i) < \sum_{i=1}^{r} f_1(\tau_i)$$

implies the inequality

$$\sum_{i=1}^{r} f_2(\omega_i) \le \sum_{i=1}^{r} f_2(\tau_i).$$

(ii) There are elements $\omega_1, \omega_2 \in \Omega$ satisfying

$$f_1(\omega_1) < f_1(\omega_2).$$

Then there exists a function $\rho\colon \Omega \to \mathbb{R}$ and for $k = 1,2$ there exist constants $a_k, b_k \in \mathbb{R}$, $a_k \ge 0$ such that

$$f_k(\omega) = a_k \rho(\omega) + b_k$$

holds for all $\omega \in \Omega$.

Proof: From (ii) we infer that for every $\omega \in \Omega$ and every $n \ge 1$ there exists an integer $m_n(\omega)$ satisfying

$$\frac{1}{n} m_n(\omega)[f_1(\omega_2) - f_1(\omega_1)]$$
$$< f_1(\omega) - f_1(\omega_1) < \frac{1}{n}[m_n(\omega) + 2][f_1(\omega_2) - f_1(\omega_1)].$$

This implies

$$\sum_{i=1}^{r} f_1(\omega_i) < \sum_{i=1}^{r} f_1(\tau_i)$$

for a suitable choice of ω_i, τ_i $(i = 1,2)$ from the set $\{\omega, \omega_1, \omega_2\}$.

Now (i) yields

$$\sum_{i=1}^{r} f_2(\omega_i) \le \sum_{i=1}^{r} f_2(\tau_i)$$

with a similar choice of ω_i, τ_i $(i = 1,2)$. We define for every $\omega \in \Omega$

$$\rho(\omega) := \lim_{n\to\infty} \frac{1}{n} m_n(\omega).$$

Then we obtain for $k = 1,2$ and $\omega \in \Omega$

$$f_k(\omega) = \rho(\omega)[f_k(\omega_2) - f_k(\omega_1)] + f_k(\omega_1).$$

But putting for $k = 1,2$

$$a_k := f_k(\omega_2) - f_k(\omega_1)$$

and

$$b_k := f_k(\omega_1),$$

and observing that (i) and (ii) imply $a_k \geq 0$ for $i = 1,2$ we conclude the desired assertion. □

Proof of Theorem 14.2: We start by choosing μ-densities $p_\theta := dP_\theta/d\mu$ and $p_* := dP_*/d\mu$ for $\theta \in \Theta$ and $*$.

1. We fix an $n \geq 1$ and define $P_*^{\otimes n}$ - a.s. the mapping

$$(\omega_1,\ldots,\omega_n) \to \prod_{i=1}^{n} \frac{p_\theta(\omega_i)}{p_*(\omega_i)}$$

from Ω^n into $\mathbb{R}$. By assumption (ii) of the theorem this mapping is an isotone function of an $\mathbf{A}^{\otimes n}$-measurable real-valued function T_n on Ω^n which is independent of θ. More precisely there exists a function $T_n \in \mathfrak{M}(\Omega^n, \mathbf{A}^{\otimes n})$ and for every $\theta \in \Theta$ there exists an isotone numerical function $H_\theta^{(n)}$ on $\mathbb{R}$ such that

$$(H_\theta^{(n)} \circ T_n)(\omega_1,\ldots,\omega_n) = \prod_{i=1}^{n} \frac{p_\theta(\omega_i)}{p_*(\omega_i)} \; [P_*^{\otimes n}].$$

Putting $H_\theta := H_\theta^{(1)}$ and $T := T_1$ we rewrite this expression as

$$\begin{aligned}(H_\theta^{(n)} \circ T_n)(\omega_1,\ldots,\omega_n) &= \prod_{i=1}^{n} (H_\theta^{(1)} \circ T_1)(\omega_i) \\ &= \prod_{i=1}^{n} (H_\theta \circ T)(\omega_i)\,[P_*^{\otimes n}].\end{aligned}$$

Our next aim in the proof of the theorem will be to replace this $P_*^{\otimes n}$-a.s. equality by a sure equality.

2. We apply Lemma 14.3 to $\nu := P_*$ and $f_\theta := H_\theta \circ T$ for all $\theta \in \Theta$. Let

$$A_\theta^\delta := [H_\theta \circ T \leq \delta] \quad \text{for each} \quad \theta \in \Theta,\ \delta \in \mathbb{R}_+,$$

and

$$\mathfrak{D} := \{A_\theta^\delta : \theta \in \Theta,\ \delta \in \mathbb{R}_+\}.$$

Since the experiment $(\Omega, \mathbf{A}, \mathscr{P}_1, X: \mathscr{P}_1 \to \Theta)$ admits an ILQ with respect to the measure P_*, every set A_θ^δ (for $\theta \in \Theta$) is of the form $[T < \delta']$ or $[T \leq \delta']$ with $\delta' \in \mathbb{R}$. Therefore the system $\mathfrak{D}$ is totally ordered.

Then Lemma 14.3 implies that there exists a function $S \in \mathbb{M}_+(\Omega,\mathbb{A})$ such that for all $\omega \in \Omega$ with $S(\omega) > 0$ we get

$$S(\omega) = P_* \text{ - ess sup } S \cdot 1_{[S \leq S(\omega)]}.$$

Moreover we obtain by this very lemma that for every $\theta \in \Theta$ there is a left continuous, isotone function G_θ on $\mathbb{R}$ satisfying

$$G_\theta \circ S = H_\theta \circ T[P_*]$$

or, using 1 of this proof,

$$(H_\theta^{(n)} \circ T_n)(\omega_1,\ldots,\omega_n) = \prod_{i=1}^n (G_\theta \circ S)(\omega_i)[P_*^{\otimes n}].$$

3. We shall now apply Lemma 14.3 to the measure space $(\Omega^n,\mathbb{A}^{\otimes n})$, the measure $\nu := P_*^{\otimes n}$ and the family $\{f_\theta: \theta \in \Theta\}$ of function f_θ on Ω^n defined by

$$f_\theta(\omega_1,\ldots,\omega_n) := \prod_{i=1}^n (G_\theta \circ S)(\omega_i) \quad \text{for all} \quad (\omega_1,\ldots,\omega_n) \in \Omega^n.$$

For every $\theta \in \Theta$ and $\delta \in \mathbb{R}_+$ let

$$C_\theta^\delta := \{(\omega_1,\ldots,\omega_n) \in \Omega^n: \prod_{i=1}^n (G_\theta \circ S)(\omega_i) \leq \delta\},$$

and let

$$\mathfrak{D}^{(n)} := \{C_\theta^\delta: \theta \in \Theta,\ \delta \in \mathbb{R}_+\}.$$

In analogy to 2 we obtain that the sets C_θ^δ are $P_*^{\otimes n}$-a.s. of the form $[T_n < \delta']$ or $[T_n \leq \delta']$ for $\delta' \in \mathbb{R}_+$. We shall show that the system $\mathfrak{D}^{(n)}$ is totally ordered with respect to inclusion. This will be done by verifying that $\mathfrak{D}^{(n)}$ satisfies the hypothesis of Corollary 14.4.

Indeed, let $D := \bigcap_{i \geq 1} C_{\theta_i}^{\delta_i}$ for $\theta_i \in \Theta$, $\delta_i \in \mathbb{R}_+$ $(i \geq 1)$ and $P_*^{\otimes n}(C_\theta^\delta) \geq P_*^{\otimes n}(D)$ for some $\theta \in \Theta$ and $\delta \in \mathbb{R}_+$. Then $C_\theta^\delta \supset D[P_*^{\otimes n}]$. Let $(\tau_1,\ldots,\tau_n) \in D$. For every $i \geq 1$ the isotonicity of G_{θ_i} implies that

$$M := \{(\omega_1,\ldots,\omega_n) \in \Omega^n: S(\omega_j) \leq S(\tau_j) \quad \text{for all} \quad j = 1,\ldots,n\} \subset C_{\theta_i}^{\delta_i},$$

whence $C_\theta^\delta \supset D \supset M[P_*^{\otimes n}]$.

But now we infer from

$$S(\omega) = P_* \text{ - ess sup } S \cdot 1_{[S \leq S(\omega)]}$$

and from the left continuity of G_θ that $C_\theta^\delta \supset D$ without any restriction.

4. By Corollary 14.4 there exists for every $n \geq 1$ a function $S_n \in \mathfrak{M}(\Omega^n, \mathbf{A}^{\otimes n})$ independent of $\theta \in \Theta$, and for every $\theta \in \Theta$ an isotone function $G_\theta^{(n)}$ on $\mathbb{R}$ such that

$$G_\theta^{(n)} \circ S_n(\omega_1,\ldots,\omega_n) = \prod_{i=1}^{n} (G_\theta \circ S)(\omega_i)$$

for all $(\omega_1,\ldots,\omega_n) \in \Omega^n$.

Let $\theta_0, \theta \in \Theta$. Since $G_{\theta_0}^{(n)}$ and $G_\theta^{(n)}$ are isotone, the inequality

$$\prod_{i=1}^{n} (G_{\theta_0} \circ S)(\omega_i) < \prod_{i=1}^{n} (G_{\theta_0} \circ S)(\tau_i)$$

implies

$$S_n(\omega_1,\ldots,\omega_n) < S_n(\tau_1,\ldots,\tau_n),$$

and hence

$$\prod_{i=1}^{n} (G_\theta \circ S)(\omega_i) \leq \prod_{i=1}^{n} (G_\theta \circ S)(\tau_i).$$

5. By hypothesis (i) of the theorem the measures P_θ $(\theta \in \Theta)$ and P_* are pairwise equivalent (and non identical). Therefore, for any fixed $\theta_0 \in \Theta$ there exist $\omega_1, \omega_2 \in \Omega$ such that

$$0 < (G_{\theta_0} \circ S)(\omega_1) < (G_{\theta_0} \circ S)(\omega_2) < \infty.$$

Applying Lemma 14.5 to the functions

$$f_1 := \log \circ G_{\theta_0} \circ S$$

and

$$f_2 := \log \circ G_\theta \circ S \quad (\text{for any } \theta \in \Theta)$$

we obtain

$$G_\theta \circ S = b(\theta) e^{a(\theta)\rho}$$

where $a(\theta) \in \mathbb{R}_+$, $b(\theta) \in \mathbb{R}$ and ρ is a real-valued function on Ω.

In 2 and 1 we have established that

$$G_\theta \circ S = H_\theta \circ T [P_*]$$

and

$$H_\theta \circ T = \frac{P_\theta}{P_*} \left[\tfrac{1}{2}(P_\theta + P_*)\right]$$

respectively. These relationships together with the above exponential representation imply

$$p_\theta = b(\theta)e^{a(\theta)\rho}p_*[\mu]$$

for all $\theta \in \Theta$.

Introducing the functions $\omega \to U(\omega) := \rho(\omega)$, $\omega \to h(\omega) := p_*(\omega)$ on Ω and $\theta \to \zeta(\theta) := a(\theta)$, $\theta \to C(\theta) := b(\theta)$ on Θ we end up with the representation

$$P_\theta = C(\theta)e^{\zeta(\theta)U}h \cdot \mu$$

valid for all $\theta \in \Theta$, and this proves the theorem. □

Combining the statements of Theorems 14.2 and 13.7 we immediately obtain

Theorem 14.6. Let $(\Omega, \mathbb{A}, \mathcal{P}, \chi: \mathcal{P} \to \mathbb{R} \cup \{*\})$ an injectively parametrized experiment which is dominated by a measure $\mu \in \mathcal{M}_+^\sigma(\Omega, \mathbb{A})$. Let $(\Omega, \mathbb{A}, \mathcal{P}, \mathcal{P}_0, \mathcal{P}_1)$ be the associated testing experiment with $\mathcal{P}_0 := \{P_*\}$ and $\mathcal{P}_1 := \{P_\theta: \theta \in \mathbb{R}\}$. We suppose

(i) The measures P_θ $(\theta \in \mathbb{R})$ and P_* are pairwise equivalent.

(ii) For all $n \geq 1$ and every $\alpha \in [0,1]$ there exists a most powerful level α test for the hypothesis $\mathcal{P}_0^{\otimes n}$ versus the alternative $\mathcal{P}_1^{\otimes n}$.

Under these conditions $\mathcal{P}_1$ is a one-dimensional exponential family.

Theorem 14.7. Let $(\Omega, \mathbb{A}, \mathcal{P}, \chi: \mathcal{P} \to \mathbb{R})$ be an injectively parametrized experiment such that the measures of $\mathcal{P}$ are pairwise equivalent. We further suppose that for all $n \geq 1$, every $\alpha \in [0,1]$ and for each $\theta_0 \in \mathbb{R}$ there exists a most powerful level α test for the hypothesis $\{P_{\theta_0}^{\otimes n}\}$ versus the alternative $\{P_\theta^{\otimes n}: \theta > \theta_0\}$.

Then $\mathcal{P}$ is a one-dimensional exponential family.

Proof: Let μ denote a measure in $\mathcal{M}^1(\Omega, \mathbb{A})$ such that $\mu \sim \mathcal{P}$. For every $\theta \in \mathbb{R}$ we introduce $p_\theta := dP_\theta/d\mu$. From Theorem 14.6 we infer that for every $\theta_0 \in \mathbb{R}$ there exists an $\mathbb{A}$-measurable function T_{θ_0} on Ω and there are functions $\theta \to C(\theta, \theta_0)$ and $\theta \to \zeta(\theta, \theta_0)$ on $]\theta_0, \infty[$ satisfying

$$p_\theta = p_{\theta_0} C(\theta, \theta_0) e^{\zeta(\theta, \theta_0)T_{\theta_0}}[\mu]$$

for all $\theta > \theta_0$.

Now we fix $\theta_0, \theta_1 \in \mathbb{R}$ with $\theta_0 < \theta_1$. For every $\theta < \theta_0$ we can apply the above representation to the pairs (θ, θ_0) equal to (θ_1, θ_0), (θ_1, θ) and (θ_0, θ). Then, outside a μ-null set, we get

$$[\zeta(\theta_1,\theta) - \zeta(\theta_0,\theta)]T_\theta - \zeta(\theta_1,\theta_0)T_{\theta_1} = \log \frac{C(\theta_1,\theta_0)C(\theta_0,\theta)}{C(\theta_1,\theta)}$$

Since $P_{\theta_1} \neq P_{\theta_2}$ implies $\zeta(\theta_1,\theta) \neq \zeta(\theta_0,\theta)$, we may choose the functions T_θ pairwise affinely dependent in the sense that there are $B(\theta), D(\theta) \in \mathbb{R}$ with

$$T_\theta = B_\theta T_{\theta_0} + D(\theta)$$

for every $\theta < \theta_0$.

Now we define

$$C(\theta) := \begin{cases} C(\theta,\theta_0) & \text{for } \theta > \theta_0 \\ 1 & \text{for } \theta = \theta_0 \\ \left[C(\theta_0,\theta)e^{\zeta(\theta_0,\theta)D_0(\theta)}\right]^{-1} & \text{for } \theta < \theta_0 \end{cases}$$

and

$$\zeta(\theta) := \begin{cases} \zeta(\theta,\theta_0) & \text{for } \theta > \theta_0 \\ 0 & \text{for } \theta = \theta_0 \\ -\zeta(\theta_0,\theta)B_0(\theta) & \text{for } \theta < \theta_0 \end{cases}$$

as well as

$$h := p_{\theta_0} \quad \text{and} \quad T := T_{\theta_0}.$$

With these definitions the assertion follows. □

We collect the most important results of Sections 13 and 14 in the following

<u>Theorem 14.8</u>. Let $(\Omega, \mathfrak{A}, \mathscr{P}, \chi: \mathscr{P} \to \mathbb{R})$ be an injectively parametrized experiment such that the measures of $\mathscr{P}$ are pairwise equivalent. The following statements are equivalent:

(i) For all $n \geq 1$, every $\alpha \in [0,1]$ and for each $\theta_0 \in \mathbb{R}$ there exists a most powerful level α test for $\{P_{\theta_0}^{\otimes n}\}$ versus $\{P_\theta^{\otimes n}: \theta > \theta_0\}$.

(ii) $\mathscr{P}$ is a one-dimensional exponential family with

$$P_\theta = C(\theta)e^{\zeta(\theta)T}h \cdot \mu \quad \text{for all} \quad \theta \in \Theta,$$

where ζ has the property

(*) There are no $\theta_1,\theta_2,\theta_3 \in \Theta$, $\theta_1 < \theta_2$, $\theta_1 < \theta_3$ satisfying $\zeta(\theta_2) < \zeta(\theta_1) < \zeta(\theta_3)$.

(iii) For all $n \geq 1$ and for every $\theta_0 \in \mathbb{R}$ the experiment $(\Omega^n, \mathbf{A}^{\otimes n}, \{P_\theta^{\otimes n}: \theta > \theta_0\})$ admits an ILQ with respect to $P_{\theta_0}^{\otimes n}$.

Proof: 1. (i) ⇒ (*). We suppose that (*) is false, i.e., that there exist $\theta_1,\theta_2,\theta_3 \in \Theta$, $\theta_1 < \theta_2$, $\theta_1 < \theta_3$ such that $\zeta(\theta_2) < \zeta(\theta_1) < \zeta(\theta_3)$. By (i) there is a most powerful level $\frac{1}{2}$ test t for $\{P_{\theta_1}\}$ versus $\{P_{\theta_2}, P_{\theta_3}\}$. We put

$$p_\theta := C(\theta)e^{\zeta(\theta)\cdot T}$$

for all $\theta \in \Theta$. By Remark 11.3 we obtain for suitable $k,k' \in \mathbb{R}_+^*$ that the inclusions

$$[p_{\theta_2} > kp_{\theta_1}] \subset [t = 1],$$

$$[p_{\theta_2} < kp_{\theta_1}] \subset [t = 0],$$

$$[p_{\theta_3} > k'p_{\theta_1}] \subset [t = 1] \quad \text{and}$$

$$[p_{\theta_3} < k'p_{\theta_1}] \subset [t = 0]$$

hold $[h\cdot\mu]$. Since $\zeta(\theta_1) > \zeta(\theta_2)$ and $\zeta(\theta_1) < \zeta(\theta_3)$, we get

$$[p_{\theta_2} > kp_{\theta_1}] = [T < k_2],$$

$$[p_{\theta_2} < kp_{\theta_1}] = [T > k_2],$$

$$[p_{\theta_3} > kp_{\theta_1}] = [T > k_3] \quad \text{and}$$

$$[p_{\theta_3} < kp_{\theta_1}] = [T < k_3]$$

with

$$k_2 := \frac{1}{\zeta(\theta_1)-\zeta(\theta_2)} \log \frac{C(\theta_2)}{k\cdot C(\theta_1)}$$

and

$$k_3 := \frac{1}{\zeta(\theta_3)-\zeta(\theta_1)} \log \frac{k'\cdot C(\theta_1)}{C(\theta_3)},$$

and therefore $[h\cdot\mu]$

$$[T < k_2] \cup [T > k_3] \subset [t = 1]$$

and

$$[T > k_2] \cup [T < k_3] \subset [t = 0] .$$

All three possible relations between k_2 and k_3 yield contradictions.

(a) For $k_2 < k_3$ we get $[T > k_2] \cup [T < k_3] = \Omega$, whence $t = 1[h\cdot\mu]$ and thus $\alpha = 1 \neq \frac{1}{2}$.

(b) The case $k_2 > k_3$ is treated analogous to (a).

(c) Let $k_2 = k_3$. Then $\emptyset = [t = 1] \cap [t = 0] \supset [T \neq k_2][h\cdot\mu]$, i.e., $T = k_2[h\cdot\mu]$ or $p_{\theta_2} = kp_{\theta_1}[h\cdot\mu]$ which implies $P_{\theta_1} = P_{\theta_2}$ contradicting the injectivity of the parametrization.

2. (ii) $\Rightarrow$ (iii). Let $\theta_0 \in \Theta$. If there exists a $\theta_1 > \theta_0$ with $\zeta(\theta_1) > \zeta(\theta_2)$, then by (*) of (ii) we have $\zeta(\theta) < \zeta(\theta_0)$ for all $\theta > \theta_0$. Otherwise, $\zeta(\theta) > \zeta(\theta_0)$ for all $\theta > \theta_0$. We retain the notation of the proof of Theorem 14.2. In the first above case we choose

$$H_\theta^{(n)}(x) := \left(\frac{C(\theta)}{C(\theta_0)}\right)^n e^{(\zeta(\theta_0)-\zeta(\theta))x}$$

and

$$T_n(\omega) := -\sum_{j=1}^{n} T(\omega_j),$$

in the second case,

$$H_\theta^{(n)}(x) := \left(\frac{C(\theta)}{C(\theta_0)}\right)^n e^{(\zeta(\theta)-\zeta(\theta_0))x}$$

and

$$T_n(\omega) := +\sum_{j=1}^{n} T(\omega_j). \qquad \square$$

§15. SIMILARITY, STRINGENCY AND UNBIASEDNESS

In the preceding sections we presented several aspects of the theory of testing whose formal contents consists of exhibiting for a given measurable space $(\Omega,\mathfrak{A})$, subsets of $\mathfrak{M}_{(1)}(\Omega,\mathfrak{A})$ and of analyzing them with respect to two subsets $\mathscr{P}_0$ and $\mathscr{P}_1$ of $\mathscr{M}^1(\Omega,\mathfrak{A})$. Here the selection of a subset of $\mathfrak{M}_{(1)}(\Omega,\mathfrak{A})$ means the introduction of a particular notion of an optimal test.

We are going to proceed in this spirit and to enhance the theory by defining further notions of optimality for tests, which are of special importance in applications.

Definition 15.1. Let $(\Omega, \mathbf{A}, \mathscr{P}, \mathscr{P}_0, \mathscr{P}_1)$ be a testing experiment. A test $t \in \mathfrak{M}_{(1)}(\Omega, \mathbf{A})$ is called *similar of level* $\alpha \in [0,1]$ if we have for all $P \in \mathscr{P}_0$,

$$E_P(t) = \alpha.$$

A set $A \in \mathbf{A}$ is called a *critical region similar of level* $\alpha \in [0,1]$ if the test 1_A is similar of level α.

Definition 15.2. Let $(\Omega, \mathbf{A}, \mathscr{P}, \mathscr{P}_0, \mathscr{P}_1)$ be a testing experiment. A test $t \in \mathfrak{M}_{(1)}(\Omega, \mathbf{A})$ is said to have *Neyman structure with respect to a statistic* $T: (\Omega, \mathbf{A}) \to (\Omega', \mathbf{A}')$ if there exists an $\alpha \in [0,1]$ such that

$$E_P^T(t) = \alpha \cdot 1_{\Omega'}[T(P)]$$

for all $P \in \mathscr{P}_0$.

Theorem 15.3. Let $(\Omega, \mathbf{A}, \mathscr{P}, \mathscr{P}_0, \mathscr{P}_1)$ be a testing experiment and let $\mathbf{S}$ be a sub-σ-algebra of $\mathbf{A}$ which is sufficient for $\mathscr{P}_0$. The following statements are equivalent:

(i) $\mathbf{S}$ is boundedly complete for $\mathscr{P}_0$.

(ii) If a test $t \in \mathfrak{M}_{(1)}(\Omega, \mathbf{A})$ is similar of level $\alpha \in [0,1]$, then

$$E_P^{\mathbf{S}}(t) = \alpha\, 1_\Omega[P]$$

for all $P \in \mathscr{P}_0$.

Proof: 1. (i) $\Rightarrow$ (ii). Let $t \in \mathfrak{M}_{(1)}(\Omega, \mathbf{A})$ be a test similar of level α, and let $\mathbf{S}$ be boundedly complete for $\mathscr{P}_0$. First of all there exists a function $Q_t \in \mathfrak{M}_{(1)}(\Omega, \mathbf{S})$ satisfying

$$Q_t = E_P^{\mathbf{S}}(t)[P] \quad \text{for all} \quad P \in \mathscr{P}_0.$$

Since $Q_t - \alpha 1_\Omega$ is a bounded $\mathbf{S}$-measurable function and since for each $P \in \mathscr{P}_0$ we have

$$\int (Q_t - \alpha 1_\Omega)\,dP = \int Q_t\,dP - \alpha = \int E_P^{\mathbf{S}}(t)\,dP - \alpha$$
$$= \int t\,dP - \alpha = \alpha - \alpha = 0,$$

the bounded completeness of $\mathbf{S}$ for $\mathscr{P}_0$ implies

$$Q_t = \alpha\, 1_\Omega[P],$$

whence

$$E_P^{\mathfrak{S}}(t) = \alpha\, 1_\Omega[P] \quad \text{for all} \quad P \in \mathscr{P}_0.$$

2. (ii) ⇒ (i). Let $\mathfrak{S}$ fail to be boundedly complete for $\mathscr{P}_0$. Then there exist a bounded $\mathfrak{S}$-measurable function f and a measure $P_0 \in \mathscr{P}_0$ such that the statements

$$\int f\, dP = 0 \quad \text{for all} \quad P \in \mathscr{P}_0$$

and

$$P_0[f \neq 0] > 0$$

are true. Since f is bounded, one can find real numbers $c \in \mathbb{R}_+^*$ and $\alpha \in]0,1[$ such that $t := cf + \alpha\, 1_\Omega \in \mathfrak{M}_{(1)}(\Omega,\mathfrak{S})$. Then t is similar of level α, and

$$E_{P_0}^{\mathfrak{S}}(t) = t \neq \alpha \cdot 1_\Omega[P_0].$$

This is a contradiction of (ii). □

Corollary 15.4. Let $(\Omega,\mathfrak{A},\mathscr{P},\mathscr{P}_0,\mathscr{P}_1)$ be a testing experiment and let $\mathfrak{A}$ be boundedly complete for $\mathscr{P}_0$. Then for every $\alpha \in [0,1]$ there exists a similar test of level α which is unique $[\mathscr{P}_0]$. Consequently, for every $\alpha \in]0,1[$ there is no critical region similar of level α.

Proof: The function $\alpha\, 1_\Omega$ is a test which is similar of level α. Since $\mathfrak{A}$ is boundedly complete and sufficient for $\mathscr{P}_0$, Theorem 15.3 yields the $[\mathscr{P}_0]$ uniqueness of this test. □

Corollary 15.5. Let $(\Omega,\mathfrak{A},\mathscr{P},\mathscr{P}_0,\mathscr{P}_1)$ be a testing experiment and let $T: (\Omega,\mathfrak{A}) \to (\Omega',\mathfrak{A}')$ be a statistic which is sufficient for $\mathscr{P}_0$. Let, moreover, $\mathfrak{A}'$ be boundedly complete for $T(\mathscr{P}_0)$. Then every test $t \in \mathfrak{M}_{(1)}(\Omega,\mathfrak{A})$ which is similar of level $\alpha \in [0,1]$ has Neyman structure with respect to T.

Proof: The σ-algebra $T^{-1}(\mathfrak{A}')$ is boundedly complete for $\mathscr{P}_0$. Indeed, every $f \in \mathfrak{M}^b(\Omega,T^{-1}(\mathfrak{A}'))$ is of the form $f = f' \circ T$ with $f' \in \mathfrak{M}^b(\Omega',\mathfrak{A}')$. If $\int f\, dP = 0$ for all $P \in \mathscr{P}_0$, then

$$\int f'\, dT(P) = \int f\, dP = 0 \quad \text{for all} \quad P \in \mathscr{P}_0.$$

By assumption we therefore obtain

$$0 = T(P)([f' \neq 0]) = P([f \neq 0]).$$

Since $T^{-1}(\mathfrak{A}')$ is also sufficient for $\mathscr{P}_0$, the theorem implies

$$E_P^{T^{-1}(\mathbf{A}')}(t) = \alpha \cdot 1_\Omega[P] \quad \text{for all} \quad P \in \mathscr{P}_0,$$

whence

$$E_P^T(t) = \alpha \cdot 1_\Omega[T(P)] \quad \text{for all} \quad P \in \mathscr{P}_0. \qquad \square$$

Definition 15.6. Let $(\Omega, \mathbf{A}, \mathscr{P}, \mathscr{P}_0, \mathscr{P}_1)$ be a testing experiment and $\alpha \in [0,1]$. For every $P \in \mathscr{P}_1$ we put

$$\beta(P) := \sup_{t \in \mathfrak{T}_\alpha} \beta_t(P).$$

A test $t \in \mathfrak{T}_\alpha$ is called *stringent* (at level α) if

$$\sup_{P \in \mathscr{P}_1} (\beta(P) - \beta_t(P)) \leq \sup_{P \in \mathscr{P}_1} (\beta(P) - \beta_{t'}(P))$$

for all $t' \in \mathfrak{T}_\alpha$.

Obviously, every most powerful level α test is stringent. Thus the tests discussed in Sections 13 and 14 are necessarily stringent, and under the assumptions of these sections there exist in fact stringent tests.

Theorem 15.7. Let $(\Omega, \mathbf{A}, \mathscr{P}, \mathscr{P}_0, \mathscr{P}_1)$ be a testing experiment, and let $\alpha \in [0,1]$. Furthermore let $\{\mathfrak{Q}_i : i \in I\}$ be a partition of $\mathscr{P}_1$ such that the function $\beta : \mathscr{P}_1 \to \mathbb{R}$ introduced in the above definition is constant on $\mathfrak{Q}_i$ for each $i \in I$. Then a test $t \in \mathfrak{T}_\alpha$ is stringent if for all $i \in I$ the equality

$$\inf_{P \in \mathfrak{Q}_i} \beta_t(P) = \sup_{t' \in \mathfrak{T}_\alpha} \inf_{P \in \mathfrak{Q}_i} \beta_{t'}(P) \tag{ST}$$

holds.

Proof: Let $t \in \mathfrak{T}_\alpha$ be a test satisfying condition (ST). Then for every test $t' \in \mathfrak{T}_\alpha$ we have the following chain of inequalities which yields the assertion:

$$\sup_{P \in \mathscr{P}_1} (\beta(P) - \beta_t(P))$$

$$= \sup_{i \in I} \sup_{P \in \mathfrak{Q}_i} (\beta(P) - \beta_t(P))$$

$$= \sup_{i \in I} (\beta(P_i) - \inf_{P \in \mathfrak{Q}_i} \beta_t(P)) \quad (\text{with } P_i \in \mathfrak{Q}_i \text{ chosen arbitrarily})$$

$$\leq \sup_{i \in I} (\beta(P_i) - \inf_{P \in \mathfrak{Q}_i} \beta_{t'}(P))$$

$$= \sup_{P \in \mathscr{P}_1} (\beta(P) - \beta_{t'}(P)). \qquad \square$$

Example 15.8. Let $(\Omega,\mathbf{A}) := (\mathbb{R}^n,\mathfrak{B}^n)$ and let $\Sigma \in \mathbb{M}(n, \mathbb{R})$ denote a symmetric positive-definite matrix. We consider $\mathscr{P} := \{\nu_{a,\Sigma}: a \in \mathbb{R}^n\}$, $\mathscr{P}_0 := \{\nu_{a,\Sigma} \in \mathscr{P}: a = a_0\}$ for some fixed $a_0 \in \mathbb{R}^n$, and $\mathscr{P}_1 := \mathscr{P} \setminus \mathscr{P}_0$. Here $\nu_{a,\Sigma} = n_{a,\Sigma}\cdot\lambda^n$ denotes the n-dimensional normal distribution with mean vector a and covariance matrix Σ.

Now, let $\alpha \in]0,1[$. In order to determine a stringent level α test for $\mathscr{P}_0$ versus $\mathscr{P}_1$ we may restrict ourselves to the class

$$\mathfrak{T}^*_\alpha := \left\{t \in \mathfrak{T}_\alpha: \int t \, d\nu_{a_0,\Sigma} = \alpha\right\}.$$

For a suitable number $k := k_\alpha$ we define a set

$$S_0 := \{a \in \mathbb{R}^n: <\Sigma^{-1}(a-a_0), a-a_0> \geq k\}$$

which can be transformed into

$$S := \{a \in \mathbb{R}^n: ||a-a_0|| \geq k\}$$

such that 1_S is a stringent level α test for $\mathscr{P}_0$ versus $\mathscr{P}_1$.

The proof of this statement requires the application of Theorem 15.7.

We define a parametrization $\chi: \mathscr{P} \to \mathbb{R}^n =: \Theta$ by $\chi(\nu_{a,\Sigma}) := a$ for all $a \in \mathbb{R}^n$ and positive-definite matrices $\Sigma \in \mathbb{M}(n, \mathbb{R})$. Moreover, we put $\Theta_0 := \{a_0\}$ and $\Theta_1 := \Theta \setminus \Theta_0$ and consider the partition $\Theta_1 = \bigcup_{t\in \mathbb{R}^*_+}$ of Θ_1 into shells of the form $E_r := \{a \in \mathbb{R}^n: ||a-a_0||^2 = r\}$ admitting a surface measure $\sigma_{\sqrt{r}}$.

For each $a \in \Theta_1$ let

$$\beta(a) := \sup_{t\in\mathfrak{T}^*_\alpha} \beta_t(a).$$

Then we have

(1) The function $\beta: \Theta_1 \to \mathbb{R}$ is constant on E_r for every $r \in \mathbb{R}^*_+$.

(2) For every $r \in \mathbb{R}^*_+$, 1_S satisfies the condition

$$\inf_{a\in E_r} \beta_{1_S}(a) = \sup_{t\in\mathfrak{T}^*_\alpha} \inf_{a\in E_r} \beta_t(a).$$

We shall show property (2). First one observes that by (1), β_{1_S} is constant on E_r for every $r \in \mathbb{R}^*_+$. It remains to be shown that for all $t \in \mathfrak{T}^*_\alpha$ and $r \in \mathbb{R}^*_+$ one has

$$\int_{E_r} \beta_{1_S} \, d\sigma_{\sqrt{r}} \geq \int_{E_r} \beta_t \, d\sigma_{\sqrt{r}}.$$

This inequality, however, follows from the Fundamental Lemma 11.1 after rewriting S as

$$\left\{a \in \mathbb{R}^n: \int_{E_r} e^{-\frac{1}{2}||a-x||^2} \sigma_{\sqrt{r}}(dx) \Big/ e^{-\frac{1}{2}||a-a_0||} \geq \gamma := \gamma_\alpha\right\}$$

$$= \left\{a \in \mathbb{R}^n: \text{const} \int_0^\pi e^{-||a-a_0||\sqrt{r}\cos \upsilon} \sin^{n-2}\upsilon \, d\upsilon \geq \delta := \delta_\alpha\right\}$$

(the γ and δ chosen appropriately) and noting that the function

$$a \to \int_0^\pi e^{-||a-a_0||\sqrt{r}\cos \upsilon} \sin^{n-2}\upsilon \, d\upsilon$$

admits strictly isotone "projections".

Definition 15.9. Let $(\Omega, \mathbf{A}, \mathscr{P}, \mathscr{P}_0, \mathscr{P}_1)$ be a testing experiment and let $\alpha \in [0,1]$.

(i) A test $t \in \mathfrak{T}_\alpha$ is called *unbiased* (of level α) if

$$\beta_t(P) \geq \alpha \quad \text{for all} \quad P \in \mathscr{P}_1.$$

(ii) $t \in \mathfrak{T}_\alpha$ is said to be *most powerful unbiased* (of level α) if for every unbiased test $t' \in \mathfrak{T}_\alpha$,

$$\beta_t(P) \geq \beta_{t'}(P) \quad \text{for all} \quad P \in \mathscr{P}_1.$$

Remark 15.10. Since for any $\alpha \in [0,1]$, $\alpha \cdot 1_\Omega \in \mathfrak{T}_\alpha$ is an unbiased level α test, every most powerful test is necessarily unbiased and hence most powerful unbiased.

The subsequent example, however, shows, that there exist testing problems which admit most powerful *unbiased* tests, but resist to admit most powerful tests.

Example 5.11. Let $(\Omega, \mathbf{A}, \mathscr{P}, \chi: \mathscr{P} \to \Theta)$ be a one-dimensional exponential experiment with parameter set $\Theta = \mathbb{R}$, dominating measure $\nu \in \mathscr{M}_+^\sigma(\Omega, \mathbf{A})$, and corresponding functions $C: \Theta \to \mathbb{R}$ and $T, h \in \mathfrak{M}(\Omega, \mathbf{A})$. With this notation we have for every $\theta \in \Theta$,

$$P_\theta := C(\theta) e^{\theta T} h \cdot \nu.$$

Let $\theta_1, \theta_2 \in \Theta$ with $\theta_1 < \theta_2$, and write $\mathscr{P} := \{P_\theta: \theta \in \Theta\}$, $\mathscr{P}_0 := \{P_\theta: \theta \in [\theta_1, \theta_2]\}$ and $\mathscr{P}_1 := \mathscr{P} \setminus \mathscr{P}_0$.

Then, in general, there exists no most powerful level α test for $\mathscr{P}_0$ versus $\mathscr{P}_1$. But there exists always a most powerful unbiased test for

$\mathscr{P}_0$ versus $\mathscr{P}_1$ of the form

$$t(\omega) := \begin{cases} 1 & \text{if } T(\omega) \in C[k_1,k_2] \quad \text{where } k_1 < k_2 \\ \gamma_i & \text{if } T(\omega) = k_i \quad (i = 1,2) \\ 0 & \text{if } T(\omega) \in]k_1,k_2[\quad (\text{for all } \omega \in \Omega) \end{cases}$$

satisfying $\beta_t(\theta_1) = \beta_t(\theta_2) = \alpha$.

CHAPTER VI

Estimation Experiments

§16. MINIMUM VARIANCE UNBIASED ESTIMATORS

In this section we shall deal with parametrized experiments (not necessarily injectively parametrized) whose parameter set is $\mathbb{R}^k$ for $k \geq 1$. As in Chapter IV we are going to study properties of these parametrized experiments with respect to a given class of functions. While the class of functions considered previously - we chose the set $\mathfrak{M}_{(1)}(\Omega,\mathbb{A})$ of all test functions on $(\Omega,\mathbb{A})$ - depends only on the underlying measurable space $(\Omega,\mathbb{A})$, we shall now admit more specific classes of functions which are more closely adapted to the given experiment.

Definition 16.1. Any parametrized experiment $(\Omega,\mathbb{A},\mathscr{P},g:\mathscr{P}\to\mathbb{R}^k)$ with $\mathbb{R}^k$ as its parameter set (for $k \geq 1$) will be called an *estimation experiment*. The mapping $g:\mathscr{P}\to\mathbb{R}^k$ is said to be the k-*dimensional parameter* corresponding to $(\Omega,\mathbb{A},\mathscr{P},g) := (\Omega,\mathbb{A},\mathscr{P},g:\mathscr{P}\to\mathbb{R}^k)$.

Definition 16.2. Let $(\Omega,\mathbb{A},\mathscr{P},g)$ be an estimation experiment (with parameter set $\mathbb{R}^k$). Any measurable mapping $s: \Omega \to \mathbb{R}^k$ is called an *estimator* for $(\Omega,\mathbb{A},\mathscr{P},g)$ (or for g).

The values $s(\omega)$ of s at $\omega \in \Omega$ are said to be the *estimates* (based on the sample ω) for g.

Remark 16.3. Considering estimators s for g whose probability distributions are λ^k-absolutely continuous one notes that every given value, in particular the value $g(P)$ of g at $P \in \mathscr{P}$, is attained with probability zero. That is to say that in this case one makes the wrong decision almost surely.

Therefore the choice of estimators has to be made in a more sophisticated fashion, i.e., by taking into account the size of the (expected) error.

We shall concentrate on two sizes of errors: the *distortion* $E_P(s) - g(P)$ and the *variance* $V_P(s) := E_P[(s - E_P(s))^2]$. The obvious aim of an optimal decision process will be the search for estimators s for g

(1) of vanishing distortion, i.e., such that

$$E_P(s) = g(P) \quad \text{for all} \quad P \in \mathscr{P}, \text{ and}$$

(2) of uniformly minimal variance, i.e., such that

$$V_P(s) = \inf V_P(s') \quad \text{for all} \quad P \in \mathscr{P},$$

where the infimum is taken over all estimators s' having property (1).

Definition 16.4. An estimator s for $(\Omega, \mathfrak{A}, \mathscr{P}, g)$ is called *unbiased* if

(i) s is (componentwise) P-integrable, and

(ii) $E_P(s) = g(P)$ for all $P \in \mathscr{P}$.

Let $\mathscr{S}_U := \mathscr{S}_U(g)$ denote the totality of unbiased estimators for g. Clearly, $\mathscr{S}_U$ is a convex subset of the space $\mathfrak{M}(\Omega, \mathfrak{A}, \mathbb{R}^k)$ of all $\mathfrak{A}$-$\mathfrak{B}^k$-measurable mappings from Ω into $\mathbb{R}^k$.

Example 16.5. Let $\mathscr{P}'$ be a set of measures $\mu \in \mathscr{M}^1(\mathbb{R}, \mathfrak{B})$ satisfying $\int |\xi| \mu(d\xi) < \infty$, let $\Omega := \mathbb{R}^n$, $\mathfrak{A} := \mathfrak{B}^n$, and $\mathscr{P} := \{\mu^{\otimes n} : \mu \in \mathscr{P}'\}$. We are given a parameter g defined by

$$g(P) := g(\mu^{\otimes n}) := \int \xi \mu(d\xi) =: \bar{\mu}$$

for all $P \in \mathscr{P}$. The mapping $s: \mathbb{R}^n \to \mathbb{R}$ defined by $s(x) := \bar{x} := \frac{1}{n} \Sigma_{k=1}^n x_k$ for all $x = (x_1, \ldots, x_n) \in \mathbb{R}^n$ is an unbiased estimator for g.

Indeed, for every $P \in \mathscr{P}$ we have

$$\int s \, dP = \frac{1}{n} \int \sum_{k=1}^{n} x_k \, \mu^{\otimes n}(dx) = \frac{1}{n}(\bar{\mu} + \ldots + \bar{\mu}) = \bar{\mu}.$$

Example 16.6. Let $\mathscr{P}'$ be a set of measures $\mu \in \mathscr{M}^1(\mathbb{R}, \mathfrak{B})$ satisfying $\int \xi^2 \mu(d\xi) < \infty$, let $\Omega := \mathbb{R}^n$, $\mathfrak{A} := \mathfrak{B}^n$ (for $n \geq 2$), and $\mathscr{P} := \{\mu^{\otimes n} : \mu \in \mathscr{P}'\}$. We introduce a parameter g by

$$g(P) := g(\mu^{\otimes n}) := \sigma^2(\mu) := \int \xi^2 \mu(d\xi) - \left(\int \xi \mu(d\xi)\right)^2$$

$$= \int (\xi - \bar{\mu})^2 \mu(d\xi)$$

for all $P \in \mathscr{P}$. Now we consider the estimator $s: \Omega \to \mathbb{R}$ defined by

$$s(x) := \frac{1}{n-1} \sum_{k=1}^{n} (x_k - \bar{x})^2$$

for all $x = (x_1,\ldots,x_n) \in \mathbb{R}^n$. We shall show that s is an unbiased estimator for g.

Indeed, for all $x = (x_1,\ldots,x_n) \in \mathbb{R}^n$ we obtain

$$s(x) = \frac{1}{n} \sum_{k=1}^{n} x_k^2 - \frac{2}{n(n-1)} \sum_{\substack{k<\ell \\ k,\ell=1}}^{n} x_k x_\ell,$$

whence for all $P \in \mathscr{P}$

$$\begin{aligned} \int s\, dP &= \int \cdot \ldots \cdot \int s(x_1,\ldots,x_n)\mu(dx_1)\cdot \ldots \cdot \mu(dx_n) \\ &= \int \xi^2 \mu(d\xi) - \bar{\mu}^2 = \sigma^2(\mu). \end{aligned}$$

The subsequent result indicates the relationship between unbiased estimators and tests.

<u>Theorem 16.7</u>. Let $(\Omega,\mathbf{A},\mathscr{P},\mathscr{P}_0,\mathscr{P}_1)$ be a dominated testing experiment with dominating measure $\mu \in \mathscr{M}_+^\sigma(\Omega,\mathbf{A})$. We shall consider this experiment as an estimation experiment $(\Omega,\mathbf{A},\mathscr{P},g: \mathscr{P} \to \mathbb{R})$ with parameter $g := 1_{\mathscr{P}_1}$. The following statements are equivalent:

(i) There exists a test $t \in \mathfrak{M}_{(1)}(\Omega,\mathbf{A})$ which viewed as an estimator, is unbiased for g.

(ii) $\mathscr{P}_0 \perp \mathscr{P}_1$.

<u>Proof</u>: 1. (i) ⇒ (ii). Let $t \in \mathfrak{M}_{(1)}(\Omega,\mathbf{A})$ be an unbiased estimator for g. Then the set $A := [t > 0]$ has the properties $P(A) = 0$ for all $P \in \mathscr{P}_0$ and $P(CA) = 0$ for all $P \in \mathscr{P}_1$. But this implies $\mathscr{P}_0 \perp \mathscr{P}_1$.

2. (ii) ⇒ (i). Let $\mathscr{P}_0 \perp \mathscr{P}_1$. By Theorem 8.3 there exist measures $\nu_0 := \sum_{k\geq 1} \alpha_k P_k \in \mathrm{conv}_\sigma(\mathscr{P}_0)$ and $\nu_1 := \sum_{j\geq 1} \beta_j Q_j \in \mathrm{conv}_\sigma(\mathscr{P}_1)$ satisfying $\nu_0 \sim \mathscr{P}_0$ and $\nu_1 \sim \mathscr{P}_1$ respectively. We have $\nu_0 \perp \nu_1$.

Indeed, let $\nu \in \mathscr{M}_+^b(\Omega,\mathbf{A})$ such that $\nu \ll \nu_i$ for $i = 0,1$. Then an iterated application of the Radon-Nikodym theorem yields

$$\sum_{k\geq 1} \alpha_k f_0 P_k = \nu = \sum_{j\geq 1} \beta_j f_1 Q_j$$

with ν_j-density f_j of ν $(j = 0,1)$. Thus, for all $k \geq 1$ we obtain

$$\alpha_k f_0 P_k \leq \sum_{j\geq 1} \beta_j f_1 Q_j,$$

whence

$$\alpha_k f_0 P_k = \sum_{j\geq 1} \beta_j f_1 h_k Q_j$$

with an $(\Sigma_{j\geq 1} \beta_j f_1 Q_j)$-density h_k. This implies $\beta_j f_1 h_k Q_j \leq \alpha_k f_0 P_k$ and so

$$f_1 h_k = 0[Q_j] \quad \text{for all} \quad k,j \geq 1,$$

whence

$$f_0 = 0[P_k] \quad \text{for all} \quad k \geq 1,$$

and therefore,

$$\nu = \sum_{k\geq 1} \alpha_k f_0 P_k = 0.$$

By the Hahn decomposition theorem we obtain a set $A \in \mathbf{A}$ such that $\nu_0(A) = 0$ and $\nu_1(CA) = 0$. The test $t := 1_A$ enjoys the desired property. □

Remark 16.8. The implication (ii) ⇒ (i) of the theorem becomes false as soon as we drop the hypothesis of domination for the testing experiment $(\Omega, \mathbf{A}, \mathscr{P}, \mathscr{P}_0, \mathscr{P}_1)$. To see this, it suffices to choose $\mathscr{P}_0$ as the set of all purely atomic measures and $\mathscr{P}_1$ as the set of all diffuse measures in $\mathscr{M}^1(\mathbb{R}, \mathfrak{B})$.

The parametrization $g: \mathscr{P} \to \mathbb{R}^k$ of the given estimation experiment $(\Omega, \mathbf{A}, \mathscr{P}, g: \mathscr{P} \to \mathbb{R}^k)$ will be used in order to determine the set $\mathscr{S}_U(g)$ of unbiased estimators for g.

Definition 16.9. A mapping $V: \mathbb{R}^k \times \mathscr{P} \to \mathbb{R}_+$ is called a *loss function* for $(\Omega, \mathbf{A}, \mathscr{P}, \mathbb{R}^k, g)$ if its section $x \to V(x,P)$ on $\mathbb{R}^k$ is measurable for all $P \in \mathscr{P}$.

Definition 16.10. Given an estimator s and a loss function V for $(\Omega, \mathbf{A}, \mathscr{P}, \mathbb{R}^k, g)$ the mapping $R_s^V: \mathscr{P} \to \overline{\mathbb{R}}_+$ given by

$$R_s^V(P) := \int V(s(\omega), P) P(d\omega)$$

for all $P \in \mathscr{P}$ is called the *risk function* corresponding to the experiment $(\Omega, \mathbf{A}, \mathscr{P}, \mathbb{R}^k, g)$, the estimator s and the loss function V. For s, V and P the symbol $R_s^V(P)$ denotes the *average loss* produced by s under the assumption that P is the "true distribution". Once the loss function has been fixed we shall abbreviate R_s^V and R_s.

In the sequel we shall deal exclusively with loss functions V such that the section $x \to V(x,P)$ is strictly convex for every $P \in \mathscr{P}$. This restriction is a grave one, but suitable for the discussion in the remaining part of the section.

Example 16.11. Let $(\Omega, \mathbf{A}, \mathscr{P}, \mathbb{R}^k, g)$ be an estimation experiment. The *mean square loss function* V is defined by

$$V(x,P) := ||x - g(P)||^2$$

for all $x \in \mathbb{R}^k$, $P \in \mathscr{P}$. Then for any $s \in \mathscr{S}_U(g)$ the corresponding risk function is given by

$$R_s(P) = \int ||s(\omega) - g(P)||^2 P(d\omega) = \int ||s - E_P(s)||^2 dP$$

for all $P \in \mathscr{P}$. In the special case $k = 1$ we obtain

$$R_s(P) = V_P(s)$$

for all $P \in \mathscr{P}$.

Let $(\Omega, \mathbf{A}, \mathscr{P}, \mathbb{R}^k, g)$ be an estimation experiment. By $\mathscr{V} := \mathscr{V}(\mathscr{P}, \mathbb{R}^k)$ we denote the set of all strictly convex loss functions for $(\Omega, \mathbf{A}, \mathscr{P}, \mathbb{R}^k, g)$. Given $V \in \mathscr{V}$ we introduce the set

$$\mathscr{S}_U^V := \{s \in \mathscr{S}_U : R_s^V(P) < \infty \text{ for all } P \in \mathscr{P}\}.$$

Definition 16.12. A function $s \in \mathscr{S}_U^V$ is called a *minimum variance unbiased estimator* (MVU estimator) for $(\Omega, \mathbf{A}, \mathscr{P}, \mathbb{R}^k, g)$ (for g) if for all $s' \in \mathscr{S}_U^V$ and all $P \in \mathscr{P}$ we have

$$R_s^V(P) \leq R_{s'}^V(P).$$

Let $\mathscr{S}_M^V := \mathscr{S}_M^V(g)$ denote the set of all MVU estimators for g. If $V \in \mathscr{V}$ is fixed, we shall write $\mathscr{S}_M$ in place of $\mathscr{S}_M^V$. Clearly $\mathscr{S}_M^V \subset \mathscr{S}_U$ for all $V \in \mathscr{V}$.

Theorem 16.13. Let $(\Omega, \mathbf{A}, \mathscr{P}, \mathbb{R}^k, g)$ be an estimation experiment, and let $V \in \mathscr{V}(\mathscr{P}, \mathbb{R}^k)$. Then for all estimators $s_1, s_2 \in \mathscr{S}_M^V(g)$ we have $s_1 = s_2 [\mathscr{P}]$.

Proof: We choose $\alpha \in]0,1[$ and put $s := \alpha s_1 + (1-\alpha)s_2$. Let P be an arbitrary element of $\mathscr{P}$. Since $V(\cdot,P)$ is convex, we have on Ω

$$f := \alpha V(s_1,P) + (1-\alpha)V(s_2,P) - V(s,P) \geq 0,$$

whence

$$\int f\, dP = \alpha R_{s_1}^V(P) + (1-\alpha)R_{s_2}^V(P) - R_s^V(P) \geq 0.$$

By assumption the estimators s_1 and s_2 are MVU, thus $\int f dP \leq 0$ which implies $f = 0[\mathscr{P}]$, i.e.,

$$V(\alpha s_1 + (1-\alpha)s_2, P) = \alpha V(s_1,P) + (1-\alpha)V(s_2,P)[\mathscr{P}].$$

But now we infer from the strict convexity of V that $s_1 = s_2[\mathscr{P}]$. □

Theorem 16.14. (D. Blackwell, C. R. Rao). Let $(\Omega, \mathfrak{A}, \mathscr{P}, \mathbb{R}^k, g)$ be an estimation experiment, $V \in \mathscr{V}(\mathscr{P}, \mathbb{R}^k)$, $\mathfrak{S}$ a sub-σ-algebra which is sufficient for $\mathscr{P}$, and $s \in \mathscr{S}_U^V(g)$. We know that under these assumptions there exists a version $s^{\mathfrak{S}}$ of $E_P^{\mathfrak{S}}(s)$ independent of the individual $P \in \mathscr{P}$. Then

(i) $s^{\mathfrak{S}} \in \mathscr{S}_U^V(g)$.

(ii) $R_{s^{\mathfrak{S}}}^V \leq R_s^V$.

(iii) For every $P \in \mathscr{P}$ we have $R_{s^{\mathfrak{S}}}^V(P) = R_s^V(P)$ iff $s = s^{\mathfrak{S}}[P]$.

Proof: 1. For every $P \in \mathscr{P}$ we obtain

$$E_P(s^{\mathfrak{S}}) = E_P(E_P^{\mathfrak{S}}(s)) = E_P(s) = g(P),$$

whence $s^{\mathfrak{S}} \in \mathscr{S}_U(g)$.

2. Let $P \in \mathscr{P}$. Then by Jensen's inequality applied to the convex function $x \to V(x,P)$ on $\mathbb{R}^k$ we get

$$\begin{aligned} R_{s^{\mathfrak{S}}}^V(P) &= \int V(P,s^{\mathfrak{S}})dP = \int V(P,E_P^{\mathfrak{S}}(s))dP \\ &\leq \int E_P^{\mathfrak{S}} V(P,s)dP \\ &= \int V(P,s)dP = R_s^V(P). \end{aligned}$$

This implies (ii) and also $s^{\mathfrak{S}} \in \mathscr{S}_U^V(g)$, which is (i).

3. By the chain of inequalities in 2 we obtain equality in (ii) iff

$$E_P^{\mathfrak{S}} V(P,s) = V(P,E_P^{\mathfrak{S}}(s)) = V(P,s^{\mathfrak{S}})[P].$$

Since the function $x \to V(x,P)$ was assumed to be strictly convex, the converse of Jensen's inequality implies that

$$E_P^{\mathfrak{S}}(P,s) = V(P,s^{\mathfrak{S}})[P]$$

iff

$$s = s^{\mathfrak{S}}[P]. \quad □$$

Corollary 16.15. If $s \in \mathscr{S}_M^V(g)$. then $s = E_P^{\mathfrak{S}}(s)\,[P]$ for all $P \in \mathscr{P}$. In particular, s is P-a.e. $\mathfrak{S}$-measurable for all $P \in \mathscr{P}$.

Proof: Since $s \in \mathscr{S}_M^V(g)$, we have $R^V_{s^{\mathfrak{S}}}(P) = R^V_s(P)$ for all $P \in \mathscr{P}$. Thus (iii) of the theorem implies $s = s^{\mathfrak{S}} = E_P^{\mathfrak{S}}(s)\,[P]$ for all $P \in \mathscr{P}$, and s is P-a.e. $\mathfrak{S}$-measurable for all $P \in \mathscr{P}$. □

Example 16.16. Let $(\Omega, \mathbf{A}, \mathscr{P}, \mathbb{R}, g)$ be the estimation experiment defined by $\Omega := \mathbb{R}^n$, $\mathbf{A} := \mathbb{B}^n$, $\mathscr{P} := \{\nu_{a,1}^{\otimes n} : a \in \mathbb{R}\}$ and $g : \mathscr{P} \to \mathbb{R}$ given by $g(P_a) = a^2 + 1$ for all $P_a := \nu_{a,1}^{\otimes n}$ with $a \in \mathbb{R}$.

In the language of samples we are dealing with a sample $X = (X_1, \ldots, X_n)$ of size n with distribution $P_{X_k} = \nu_{a,1}$ for all $k = 1, \ldots, n$, and some $a \in \mathbb{R}$. This means that the sample X has "known" variance 1 and "unknown" mean a.

We are looking for unbiased estimators for the parameter g corresponding to $(\Omega, \mathbf{A}, \mathscr{P}, \mathbb{R}, g)$ or equivalently for the parameter value $a^2 + 1 = E(X)^2 + 1 = E(X^2)$.

First we note that the function

$$x \to s(x) := s(x_1, \ldots, x_n) := \frac{1}{n} \sum_{k=1}^{n} x_k^2$$

on $\mathbb{R}^n$ is an unbiased estimator for $a^2 + 1$.

Let $T : \mathbb{R}^n \to \mathbb{R}$ be the statistic $\overline{X}$ defined by

$$\overline{X}(x) := \frac{1}{n} \sum_{k=1}^{n} x_k$$

for all $x = (x_1, \ldots, x_n) \in \mathbb{R}^n$. It has been shown in Example 8.9 that T is sufficient for $\mathscr{P}$.

We put $s^T := E_P^T(s)$. Using the equality

$$s(x) = \frac{1}{n} \sum_{k=1}^{n} (x_k - \overline{x})^2 + \overline{x}^2,$$

valid for all $x = (x_1, \ldots, x_n) \in \mathbb{R}^n$, we get

$$E_P(s|T) = \frac{1}{n} E_P\left(\sum_{k=1}^{n} (X_k - \overline{X})^2 \,\Big|\, T\right) + E_P(\overline{X}^2|T)\,[P].$$

But $E_P(\overline{X}^2|T) = E_P(\overline{X}^2|\overline{X}) = \overline{X}^2\,[P]$.

Moreover, the quadratic form $\Sigma_{k=1}^n (X_k - \overline{X})^2$ and the sample mean $\overline{X}$ are independent (by Cochran's theorem) for all $P \in \mathscr{P}$, thus

$$E_P\left(\sum_{k=1}^{n} (X_k - \overline{X})^2 | \overline{X}\right) = E_P\left(\sum_{k=1}^{n} (X_k - \overline{X})^2\right)[P].$$

Then, combining the results of this computation, we obtain

$$E_P(s|T) = \overline{X}^2 + \frac{1}{n} E_P\left(\sum_{k=1}^{n} (X_k - \overline{X})^2\right)$$
$$= \overline{X}^2 + \frac{n-1}{n},$$

since

$$E_P\left(\frac{1}{n-1} \sum_{k=1}^{n} (X_k - \overline{X})^2\right) = 1.$$

Now, Theorem 16.14 tells us that the function s^T defined by

$$s^T(x) := \frac{n-1}{n} + \overline{X}^2(x)$$

for all $x \in \mathbb{R}^n$ is an unbiased estimator for $a^2 + 1$ with $R^V_{s^T} \leq R^V_s$ for any $V \in \mathscr{V}(\mathscr{P}, \mathbb{R})$.

Example 16.17. Let $(\Omega, \mathfrak{A}, \mathscr{P}, \mathbb{R}, g)$ be the estimation experiment defined by $\Omega := \mathbb{R}^n$, $\mathfrak{A} := \mathfrak{B}^n$, $\mathscr{P} := \{P := \mu^{\otimes n} : \mu \in \mathscr{M}^1(\mathbb{R}, \mathfrak{B}), \int |\xi| \mu(d\xi) < \infty\}$ and $g: \mathscr{P} \to \mathbb{R}$ given by $g(P) = g(\mu^{\otimes n}) := \overline{\mu} := \int \xi \mu(d\mu)$ for all $P \in \mathscr{P}$.

We are searching for unbiased estimators for the parameter g corresponding to $(\Omega, \mathfrak{A}, \mathscr{P}, \mathbb{R}, g)$. The function $x \to s(x) := x_1$ for all $x = (x_1, \ldots, x_n) \in \mathbb{R}^n$ is an unbiased estimator for g. In fact, we have

$$E_P(s) = \int \cdot \ldots \cdot \int x_1 \mu(dx_1) \cdot \ldots \cdot \mu(dx_n) = g(P) = \overline{\mu}$$

for all $P \in \mathscr{P}$. We now consider the *order statistic* $T: (\mathbb{R}^n, \mathfrak{B}^n) \to (\mathbb{R}^n, \mathfrak{B}^n)$ defined by

$$T(x_1, \ldots, x_n) := (x_{i_1}, \ldots, x_{i_n})$$

for all $(x_1, \ldots, x_n) \in \mathbb{R}^n$ with $x_{i_1} \leq x_{i_2} \leq \ldots \leq x_{i_n}$.

Clearly the order statistic is a special case of the *permutation statistic* introduced in Example 4.8. From this very example we infer that T is sufficient for $\mathscr{P}$. Moreover, we have for all $x = (x_1, \ldots, x_n)$,

$$E_P^T(s)(x) = \frac{1}{n!} \sum s(x_{i_1}, \ldots, x_{i_n}) = \frac{(n-1)!}{n!}(x_1 + \ldots + x_n) = \overline{x},$$

whatever $P \in \mathscr{P}$. That is to say, $s^T = \overline{X}$, and by Theorem 16.14 we get $R^V_{s^T} \leq R^V_s$ for any $V \in \mathscr{V}(\mathscr{P}, \mathbb{R})$.

Theorem 16.18. (E. L. Lehmann, H. Scheffé). Let $(\Omega, \mathbf{A}, \mathscr{P}, \mathbb{R}^k, g)$ be an estimation experiment such that $\mathscr{S}_U(g) \neq \emptyset$, and let $\mathfrak{S}$ be a sub-σ-algebra of $\mathbf{A}$ which is sufficient and complete for $\mathscr{P}$. Then

(i) There exists an estimator s_0 which is MVU for all $V \in \mathscr{V}(\mathscr{P}, \mathbb{R}^k)$ such that $\mathscr{S}_U^V(g) \neq \emptyset$.

(ii) s_0 is uniquely determined $[\mathscr{P}]$ and $\mathfrak{S}$-measurable.

(iii) Every $\mathfrak{S}$-measurable estimator in $_U(g)$ coincides with $s_0[\mathscr{P}]$.

Proof: 1. Since $\mathfrak{S}$ is sufficient for $\mathscr{P}$, there exists for every estimator $s \in \mathscr{S}_U(g)$ a function $s^{\mathfrak{S}}$ satisfying

$$s^{\mathfrak{S}} = E_P^{\mathfrak{S}}(s)\,[\mathscr{P}].$$

By Theorem 16.14(i) we obtain $s^{\mathfrak{S}} \in \mathscr{S}_U(g)$.

2. Let s_1 be another estimator in $\mathscr{S}_U^V(g)$. Then for all $P \in \mathscr{P}$ we obtain

$$E_P(s^{\mathfrak{S}} - s_1^{\mathfrak{S}}) = E_P(s^{\mathfrak{S}}) - E_P(s_1^{\mathfrak{S}}) = g(P) - g(P) = 0,$$

whence by the completeness of $\mathfrak{S}$ for g,

$$s^{\mathfrak{S}} = s_1^{\mathfrak{S}}[\mathscr{P}].$$

3. Let $s_0 := s^{\mathfrak{S}}$. If $V \in \mathscr{V}(\mathscr{P}, \mathbb{R}^k)$ and $s_1 \in \mathscr{S}_U^V(g)$, then we conclude from Theorem 16.14(ii) that

$$R^V_{s_1^{\mathfrak{S}}} \leq R^V_{s_1}.$$

Taking together all the facts established we get

$$R^V_{s_0} = R^V_{s^{\mathfrak{S}}} = R^V_{s_1^{\mathfrak{S}}} \leq R^V_{s_1},$$

which completes the proof. □

Example 16.19. Let $(\Omega, \mathbf{A}, \mathscr{P}, \mathbb{R}, g)$ be the estimation experiment defined by $\Omega := \mathbb{R}^n$, $\mathbf{A} := \mathfrak{B}^n$, $\mathscr{P} := \{\nu_{a,1}^{\otimes n} : a \in \mathbb{R}\}$ and $g: \mathscr{P} \to \mathbb{R}$ given by $g(P_a) = a$ for all $P_a := \nu_{a,1}^{\otimes n}$ with $a \in \mathbb{R}$. From Example 16.5 we infer that the real-valued statistic $T := \overline{X}$ on $(\mathbb{R}^n, \mathfrak{B}^n)$ is an unbiased estimator for g, and from Examples 6.6 and 8.9 we know that T is sufficient and complete for $\mathscr{P}$.

By the Lehmann-Scheffé theorem T is the MVU estimator for g.

Example 16.20. Let $(\Omega, \mathbf{A}, \mathscr{P}, \mathbb{R}, g)$ be the estimation experiment defined by $\Omega := \mathbb{R}^n$, $\mathbf{A} := \mathfrak{B}^n$, $\mathscr{P} := \{\nu_{a,1}^{\otimes n} : a \in \mathbb{R}\}$ and $g : \mathscr{P} \to \mathbb{R}$ given by $g(P_a) = a^2 + 1$ for all $P_a := \nu_{a,1}^{\otimes n}$ with $a \in \mathbb{R}$.

In Example 16.16 we showed that the real-valued statistic

$$W := \frac{n-1}{n} + \overline{X}^2$$

on $(\mathbb{R}^n, \mathfrak{B}^n)$ is an unbiased estimator for g. Since $\overline{X}$ is sufficient and complete for $\mathscr{P}$, the Lehmann-Scheffé theorem implies that W is in fact the MVU estimator for g.

§17. p-MINIMALITY

In this section we continue the discussion of MVU estimators in the direction of analytic characterizations. We choose special loss functions in describing the minimality of estimators by reduction to the trivial parametrization and in terms of Gateau differentials. These characterizations of minimality yield a desirable additional insight into the structure of the set of all MVU estimators.

Definition 17.1. Let $(\Omega, \mathbf{A}, \mathscr{P}, g : \mathscr{P} \to \mathbb{R})$ be an estimation experiment. For any $p \in \mathbb{R}$, $p > 1$ we define the loss function $V_p : \mathbb{R} \times \mathscr{P} \to \mathbb{R}_+$ by

$$V_p(x,P) := |x - g(P)|^p$$

for all $x \in \mathbb{R}$, $P \in \mathscr{P}$. We furthermore introduce the sets

$$\mathscr{S}_U(p,g) := \left\{ s \in \bigcap_{P \in \mathscr{P}} \mathscr{L}^p(\Omega, \mathbf{A}, P) : \int s\, dP = g(P) \quad \text{for all} \quad p \in \mathscr{P} \right\}$$

and

$$\mathscr{S}_M(p,g) := \left\{ s \in \mathscr{S}_U(p,g) : \int V_p(s,P)dP = \inf_{s' \in \mathscr{S}_U(p,g)} \int V_p(s',P)dP \quad \text{for all} \ ,P \in \mathscr{P} \right\}.$$

The elements of $\mathscr{S}_M(p,g)$ are called p-*minimal estimators* for $(\Omega, \mathbf{A}, \mathscr{P}, g : \mathscr{P} \to \mathbb{R})$ (or for g).

Remark 17.2. p-minimal estimators are MVU estimators in the sense of Definition 16.12 for $k = 1$ and $V = V_p$. Thus, in this case, $\mathscr{S}_M(p,g) = \mathscr{S}_M^V(g)$.

Theorem 17.3. For any $s \in \mathscr{S}_U(p,g)$ $(p \geq 2)$ the following statements are equivalent:

(i) $s \in \mathscr{S}_M(p,g)$.

(ii) For every $v \in \mathscr{S}_U(p,0)$ and all $P \in \mathscr{P}$ we have

$$E_P(v|s - g(P)|^{p-1}\operatorname{sgn}(s - g(P))) = 0.$$

Proof: 1. (i) $\Rightarrow$ (ii). Let $s_0 \in \mathscr{S}_M(p,g)$ and $v \in \mathscr{S}_U(p,0)$. Since $1/p + 1/(p/p-1) = 1$, Hölder's inequality implies

$$\int |v||s_0 - g(P)|^{p-1}dP \leq \left(\int |v|^p dP\right)^{1/p}\left(\int |s_0 - g(P)|^{(p-1)\frac{p}{p-1}}dP\right)^{\frac{p-1}{p}}.$$

This inequality particularly yields the existence of the integrals appearing in the equality of (ii). The equality itself will be established in an indirect manner. Suppose there exists a $P_1 \in \mathscr{P}$ such that

$$E_{P_1}(v|s_0 - g(P_1)|^{p-1}\operatorname{sgn}(s_0 - g(P_1))) \neq 0.$$

We shall apply the fact that the function $x \to |x|^p$ on $\mathbb{R}$ is n-times differentiable for every $n < p$. Then by Taylor's theorem we obtain for all $x,h \in \mathbb{R}$

$$|x + h|^p = \sum_{\nu=0}^{[p]-1} \binom{p}{\nu}|x|^{p-\nu}(\operatorname{sgn} x)^\nu h^\nu$$
$$+ \binom{p}{[p]}|x + \upsilon h|^{p-[p]}(\operatorname{sgn}(x + \upsilon h))^{[p]}h^{[p]},$$

where υ denotes a real number satisfying $|\upsilon| \leq 1$. From

$$|x + \upsilon h|^{p-[p]} \leq |x|^{p-[p]} + |\upsilon h|^{p-[p]}$$

follows

$$|x + h|^p \leq |x|^p + p|x|^{p-1}(\operatorname{sgn} x)h$$
$$+ \sum_{\nu=2}^{[p]} \binom{p}{\nu}|x|^{p-\nu}|h|^\nu + \binom{p}{[p]}|h|^p \quad \text{for } p \geq 2.$$

This property of the function $x \to |x|^p$ implies the following inequality valid for all $\lambda \in \mathbb{R}$:

$$E_{P_1}(|(s_0 - g(P_1)) + \lambda v|^p)$$
$$\leq E_{P_1}(|s_0 - g(P_1)|^p) + \lambda p E_{P_1}(|s_0 - g(P_1)|^{p-1}\operatorname{sgn}(s_0 - g(P_1))v)$$
$$+ \sum_{\nu=2}^{[p]} \lambda^\nu \binom{p}{\nu} E_{P_1}(|s_0 - g(P_1)|^{p-\nu}|v|^\nu) + |\lambda|^p E_{P_1}\binom{p}{[p]}(|v|^p).$$

By assumption the coefficient of λ in the second summand is different from 0, whence the absolute value of this summand will be larger than the remaining sum (of higher powers of λ) if λ gets small. In fact, λ can be chosen such that the second summand becomes negative. Therefore there exists a $\lambda_1 \in \mathbb{R}$ satisfying the strict inequality

$$E_{P_1}(|(s_0 + \lambda_1 v) - g(P_1)|^p) < E_{P_1}(|s_0 - g(P_1)|^p).$$

But this contradicts the minimality of s_0. Indeed, for $v \in \mathcal{S}_U(p,0)$ and $s_0 \in \mathcal{S}_U(p,g)$ we get $s_0 + \lambda_1 v \in \mathcal{L}^p(\Omega,\mathfrak{A},P)$ and

$$E_P(s_0 + \lambda_1 v) = E_P(s_0) + E_P(\lambda_1 v) = g(P) + 0 = g(P)$$

for all $P \in \mathcal{P}$, which implies $s_0 + \lambda_1 v \in \mathcal{S}_U(p,g)$.

2. (ii) $\Rightarrow$ (i). Let $s_0 \in \mathcal{S}_U(p,g)$, and let

$$E_P(v|s_0 - g(P)|^{p-1}\operatorname{sgn}(s_0 - g(P))) = 0$$

be satisfied for all $v \in \mathcal{S}_U(p,0)$ and $P \in \mathcal{P}$. Then for any given $s \in \mathcal{S}_U(p,g)$, $s - s_0 \in \mathcal{S}_U(p,0)$, and thus we conclude for every $P \in \mathcal{P}$ from the identity $s - s_0 = (s-g(P)) - (s_0-g(P))$, the equality

$$E_P((s_0 - g(P))|s_0 - g(P)|^{p-1}\operatorname{sgn}(s_0 - g(P)))$$
$$= E_P((s - g(P))|s_0 - g(P)|^{p-1}\operatorname{sgn}(s_0 - g(P))).$$

The left hand side of this equality equals $E_P(|s_0 - g(P)|^p)$, the right hand side can be subject to Hölder's inequality. Consequently we obtain

$$E_P(|s_0 - g(P)|^p) \leq E_P(|s - g(P)|^p)^{1/p} E_P(|s_0 - g(P)|^p)^{p-1/p}.$$

If $E_P(|s_0 - g(P)|^p) = 0$, then our assertion holds trivially. If, however, $E_P(|s_0 - g(P)|^p) \neq 0$, then the above inequality yields

$$E_P(|s_0 - g(P)|^p)^{1/p} \leq E_P(|s - g(P)|^p)^{1/p}.$$

But then, for all $P \in \mathcal{P}$,

$$E_P(|s_0 - g(P)|^p) \leq E_P(|s - g(P)|^p),$$

whence $s_0 \in \mathcal{S}_M(p,g)$. □

Corollary 17.4.

$\mathscr{S}_M(2,g) = \{s \in \mathscr{S}_U(2,g)\colon E_P(vs) = 0$ for all $v \in \mathscr{S}_U(2,0)$ and $P \in \mathscr{P}\}$.

Proof: For every $x \in \mathbb{R}$ we have $x = |x| \operatorname{sgn} x$. Thus by the theorem,

$\mathscr{S}_M(2,g) = \{s \in \mathscr{S}_U(2,g)\colon E_P[v(s-g(P))] = 0$ for all $v \in \mathscr{S}_U(2,0)$ and $P \in \mathscr{P}\}$.

But for $v \in \mathscr{S}_U(2,0)$, $E_P[vg(P)] = g(P)E_P(v) = 0$ whatever $P \in \mathscr{P}$. This implies the assertion. □

Corollary 17.5. Let $(\Omega,\mathbf{A},\mathscr{P})$ be an experiment, $(g_n)_{n\geq 1}$ a sequence of parametrizations $g_n\colon \mathscr{P} \to \mathbb{R}$, $(s_n)_{n\geq 1}$ a sequence of 2-minimal estimators s_n for g_n, and $s \in \bigcap_{P\in\mathscr{P}} \mathscr{L}^2(\Omega,\mathbf{A},P)$. If for every $P \in \mathscr{P}$ we have

$$\lim_{n\to\infty} \int (s_n - s)^2 dP = 0,$$

then $g\colon = \lim_{n\to\infty} g_n$ exists, and s is a 2-minimal estimator for g.

Proof: The mean square convergence of the sequence $(s_n)_{n\geq 1}$ towards a square integrable function s implies the convergence of $(s_n)_{n\geq 1}$ in the mean. For every $n \geq 1$ and $P \in \mathscr{P}$ we have

$$\int s_n \, dP = g_n(P),$$

whence for every $P \in \mathscr{P}$,

$$\int s \, dP = \lim_{n\to\infty} \int s_n dP = \lim_{n\to\infty} g_n(P) =: g(P),$$

i.e., the existence of g and the unbiasedness of s for g. For every $n \geq 1$, s_n is 2-minimal which by Corollary 17.4 means that

$$E_P(vs_n) = 0 \quad \text{for all } v \in \mathscr{S}_U(2,0) \text{ and } P \in \mathscr{P}.$$

Thus by hypothesis we get $E_P(vs) = 0$ for all $v \in \mathscr{S}_U(2,0)$ and $P \in \mathscr{P}$, and another application of Corollary 17.4 yields the result. □

Example 17.6. Let $(\Omega,\mathbf{A},\mathscr{P},g\colon \mathscr{P} \to \mathbb{R}^2)$ be an estimation experiment with $\Omega\colon = \mathbb{R}^n$, $\mathbf{A}\colon = \mathbb{B}^n$ $(n \geq 2)$, $\mathscr{P}\colon = \{P_{a,\sigma^2}\colon (a,\sigma^2) \in \mathbb{R} \times \mathbb{R}_+^*\}$ where $P_{a,\sigma^2}\colon = \nu_{a,\sigma^2}^{\otimes n}$ for all $(a,\sigma^2) \in \mathbb{R} \times \mathbb{R}_+^*$, and $g\colon \mathscr{P} \to \mathbb{R}^2$ defined by $g(P_{a,\sigma^2})\colon = (a,\sigma^2)$ for all $(a,\sigma^2) \in \mathbb{R} \times \mathbb{R}_+^*$. Moreover, we consider the loss function $V\colon \mathbb{R}^2 \times \mathscr{P} \to \mathbb{R}_+$ given by $V(y,P)\colon = ||y - g(P)||^2$ for all $y \in \mathbb{R}^2$, $P \in \mathscr{P}$. We shall show that the function $s\colon \mathbb{R}^n \to \mathbb{R}^2$ defined by

$$s(x)\colon = \left(\overline{x}, \frac{1}{n-1} \sum_{k=1}^{n} (x_k - \overline{x})^2\right)$$

is an MVU estimator in $\mathscr{S}_M^V(g)$. We put $s := (s_1, s_2)$ with $s_1 := \overline{X}$ and $s_2 := \frac{1}{n-1} \Sigma_{k=1}^n (X_k - \overline{X})^2$. From Examples 16.5 and 16.6 we know that s is an unbiased estimator for $g := (g_1, g_2)$. In order to see that s is in fact an MVU estimator for g, it suffices to prove that for $i = 1,2$ the component function s_i is 2-minimal for g_i.

Indeed, if $s' = (s_1', s_2')$ is an unbiased estimator for $g = (g_1, g_2)$, then s_i' is an unbiased estimator for g_i whenever $i = 1,2$. From $R_{s'}^V < \infty$ we conclude $s_i' \in \mathscr{L}^2(\Omega, \mathfrak{A}, P)$ for all $P \in \mathscr{P}$. If therefore s_i is 2-minimal, then

$$\int |s_i - g_i(P)|^2 dP \leq \int |s_i' - g_i(P)|^2 dP \quad \text{for} \quad i = 1,2,$$

whence

$$\begin{aligned} R_s^V(P) &= \int |s_1 - g_1(P)|^2 dP + \int |s_2 - g_2(P)|^2 dP \\ &\leq \int |s_1' - g_1(P)|^2 dP + \int |s_2' - g_2(P)|^2 dP \\ &\leq R_{s'}^V(P) \end{aligned}$$

for all $P \in \mathscr{P}$.

For the proof of the 2-minimality of the functions s_i $(i = 1,2)$ we shall apply Corollary 17.4. Let $v \in \mathscr{S}_U(2,0)$, i.e.,

$$0 = \int v \, dP_{a,\sigma^2} = \int v(x) \frac{1}{(2\pi\sigma^2)^{n/2}} e^{-\frac{1}{2\sigma^2} \sum_{k=1}^n (x_k - a)^2} \lambda^n(dx) \qquad (*)$$

for all $(a, \sigma^2) \in \mathbb{R} \times \mathbb{R}_+^*$. Differentiation with respect to a on both sides of the equality yields

$$\begin{aligned} 0 &= \int v(x) \frac{1}{(2\pi\sigma^2)^{n/2}} \frac{1}{\sigma^2} \sum_{k=1}^n (x_k - a) e^{-\frac{1}{2\sigma^2} \sum_{k=1}^n (x_k - a)^2} \lambda^n(dx) \\ &= \frac{1}{\sigma^2} \sum_{k=1}^n \int v(x) x_k \, P_{a,\sigma^2}(dx) - \frac{na}{\sigma^2} \int v \, dP_{a,\sigma^2} \\ &= \frac{n}{\sigma^2} \int v(x) \, \overline{x} \, P_{a,\sigma^2}(dx). \end{aligned}$$

This implies that

$$0 = \int v(x) \, \overline{x} \, P(dx) = \int v s_1 \, dP$$

for all $P \in \mathscr{P}$, i.e., $s_1 := \overline{X} \in \mathscr{S}_M(2, g_1)$. We now differentiate both sides of equality (*) with respect to σ^2 and obtain

$$0 = \int v(x)\left(-\frac{n}{2}\right) \frac{1}{(2\pi)^{n/2}} \frac{1}{(\sigma^2)^{\frac{n}{2}+1}} e^{-\frac{1}{2\sigma^2}\sum_{k=1}^{n}(x_k-a)^2} \lambda^n(dx)$$

$$+ \int v(x) \frac{1}{(2\pi\sigma^2)^{n/2}} \frac{1}{2\sigma^4} \sum_{k=1}^{n} (x_k-a)^2 e^{-\frac{1}{2\sigma^2}\sum_{k=1}^{n}(x_k-a)^2} \lambda^n(dx)$$

$$= -\frac{n}{2\sigma^2}\int v\, dP_{a,\sigma^2} + \frac{1}{2\sigma^4}\int v(x) \sum_{k=1}^{n}(x_k-a)^2\, P_{a,\sigma^2}(dx).$$

This implies

$$\int v(x) \sum_{k=1}^{n} (x_k-a)^2 P(dx) = 0$$

for all $P \in \mathscr{P}$. But differentiating (*) twice with respect to a yields

$$0 = \int v(x)\, \overline{x}^{\,2} P(dx)$$

for all $P \in \mathscr{P}$. Thus an application of the identities

$$\sum_{k=1}^{n} (x_k-a)^2 = n(\overline{x}-a)^2 + (n-1)s_2$$

and

$$\int v(x)(\overline{x}-a)^2 P(dx)$$

$$= \int v(x)\overline{x}^2 P(dx) + 2a \int v(x)\overline{x}\, P(dx) + a^2 \int v(x) P(dx) = 0$$

implies

$$\int v\, s_2\, dP = 0 \quad \text{for all} \quad P \in \mathscr{P},$$

i.e., $s_2 \in \mathscr{S}_M(2, g_2)$.

Theorem 17.7. Let $(\Omega, \mathbb{A}, \mathscr{P})$ be an experiment. For $p \in]1,\infty[$ we put

$$\mathscr{S}_M(p) := \bigcup_{g \in \mathbb{R}^{\mathscr{P}}} \mathscr{S}_M(p,g)$$

and

$$\mathscr{S}_M^b(p) := \mathscr{S}_M(p) \cap \mathbb{M}^b(\Omega, \mathbb{A}).$$

Then

(i) For every $S \in \mathscr{S}_M(p)$ there exists exactly one $g_s \in \mathbb{R}^{\mathscr{P}}$ with $s \in \mathscr{S}_M(p, g_s)$.

(ii) Given $s \in \mathscr{S}_M(p)$ and $\lambda \in \mathbb{R}$, λs and $s + \lambda$ are elements of $\mathscr{S}_M(p)$, and we have

$$g_{\lambda s} = \lambda g_s$$

and

$$g_{s+\lambda} = g_s + \lambda.$$

(iii) $\mathscr{S}_M^b(2)$ is an algebra, and $\mathscr{S}_M(2)$ is a vector space and an $\mathscr{S}_M^b(2)$-module.

Proof: 1. Every element s of $\mathscr{S}_M(p)$ is unbiased by definition, thus we obtain for s the parametrization $g_s: \mathscr{P} \to \mathbb{R}$ defined by $g_s(P) := \int s\, dP$ for all $P \in \mathscr{P}$, and this proves (i).

2. Let $s \in \mathscr{S}_M(p)$. Then for every $\lambda \in \mathbb{R}$, all $v \in \mathscr{S}_U(p,0)$ and all $P \in \mathscr{P}$ we have by Theorem 17.3

$$\int v|\lambda s - \lambda g_s(P)|^{p-1} \operatorname{sgn}(\lambda s - \lambda g_s(P))dP$$
$$= |\lambda|^{p-1} \operatorname{sgn} \lambda \cdot \int v|s - g_s(P)|^{p-1} \operatorname{sgn}(s - g_s(P))dP = 0$$

and

$$\int v|(s+\lambda) - (g_s(P) + \lambda)|^{p-1} \operatorname{sgn}((s+\lambda) - (g_s(P) + \lambda))dP = 0,$$

whence, again by Theorem 17.3, $\lambda s \in \mathscr{S}_M(p, \lambda g_s)$ and $s + \lambda \in \mathscr{S}_M(p, g_s + \lambda)$, which implies (ii).

3. By Corollary 17.4 we have

$$\mathscr{S}_M(2) = \left\{ s \in \bigcap_{P \in \mathscr{P}} \mathscr{L}^2(\Omega, \mathbf{A}, P) : \int v\, s\, dP = 0 \text{ for all } v \in \mathscr{S}_U(2,0) \text{ and all } P \in \mathscr{P} \right\}.$$

This shows that $\mathscr{S}_M(2)$ is a vector space. Let now $s^b \in \mathscr{S}_M^b(2)$. Then by Corollary 17.4, for every $v \in \mathscr{S}_U(2,0)$ we get $s^b \cdot v \in \mathscr{S}_U(2,0)$. This implies

$$\int (s \cdot s^b) v\, dP = \int s(s^b \cdot v) dP = 0$$

for all $s \in \mathscr{S}_M(2)$ and $P \in \mathscr{P}$, and therefore that $s^b \cdot s \in \mathscr{S}_M(2)$ whenever $s^b \in \mathscr{S}_M^b(2)$ and $s \in \mathscr{S}_M(2)$. Since $\mathfrak{M}^b(\Omega, \mathbf{A})$ is an algebra, $\mathscr{S}_M^b(2)$ also is, and $\mathscr{S}_M(2)$ is an $\mathscr{S}_M^b(2)$-module, as asserted in (iii). □

Theorem 17.8. Let $(\Omega, \mathbf{A}, \mathscr{P})$ be an experiment. For every $f \in \bigcap_{P \in \mathscr{P}} \mathscr{L}^1(\Omega, \mathbf{A}, P)$ we define a parametrization $g_f: \mathscr{P} \to \mathbb{R}$ by

$g_f(P) := \int f\, dP$ for all $P \in \mathscr{P}$. For every $A \in \mathfrak{A}$ let $g_A := g_{1_A}$. Then

(i) $\mathfrak{S}_p := \{A \in \mathfrak{A}: 1_A \in \mathscr{S}_M(p, g_A)\}$ is a sub-σ-algebra of $\mathfrak{A}$.

(ii) $\bigcap_{P \in \mathscr{P}} \mathscr{L}^2(\Omega, \mathfrak{S}_2, P_{\mathfrak{S}_2}) \subset \mathscr{S}_M(2)$.

(iii) $\mathfrak{S}_2$ is 2-complete, and $\mathfrak{S}_2 \subset \mathfrak{T}[\mathscr{P}]$ for every sub-σ-algebra $\mathfrak{T}$ of $\mathfrak{A}$ that is sufficient for $\mathscr{P}$.

Proof: 1. For every $A \in \mathfrak{A}$, $v \in \mathscr{S}_U(p,0)$ and $P \in \mathscr{P}$ we have

$$\int_A v\, dP = -\int_{CA} v\, dP,$$

since $\int v\, dP = 0$. This implies

$$\begin{aligned}\int v|1_A - P(A)|^{p-1}\operatorname{sgn}(1_A - P(A))dP &= \int_A v(1 - P(A))^{p-1}dP - \int_{CA} v\, P(A)^{p-1}dP \\ &= \int_A v(1 - P(A))^{p-1}dP + \int_A v\, P(A)^{p-1}dP \\ &= [(1 - P(A))^{p-1} + P(A)^{p-1}] \int_A v\, dP.\end{aligned}$$

From Theorem 17.3 we thus conclude that $1_A \in \mathscr{S}_M(p, g_A)$ iff $\int 1_A v\, dP = 0$ for all $v \in \mathscr{S}_U(p,0)$.

This characterization of the elements of $\mathfrak{S}_p$ implies the following statements:

(1) $\emptyset, \Omega \in \mathfrak{S}_p$.

(2) If $A \in \mathfrak{S}_p$, then $CA \in \mathfrak{S}_p$.

(3) Let $A_n \in \mathfrak{S}_p$ and $A_n \subset A_{n+1}$ for all $n \in \mathbb{N}$, then $\bigcup_{n \in \mathbb{N}} A_n \in \mathfrak{S}_p$.

These statements follow from Lebesgue's bounded convergence theorem, since $|1_{A_n} \cdot v| \leq |v|$ for all $n \geq 1$.

(4) If $A, B \in \mathfrak{S}_p$, then $A \cap B \in \mathfrak{S}_p$.

In fact, under the hypothesis, for all $v \in \mathscr{S}_U(p,0)$ we have $1_A \cdot v \in \mathscr{S}_U(p,0)$, thus

$$\int 1_{A \cap B} v\, dP = \int 1_B(1_A \cdot v)dP = 0.$$

The properties (1) to (4) imply that $\mathfrak{S}_p$ is a σ-algebra.

2. By definition of $\mathfrak{S}_2$ we have $1_A \in \mathscr{S}_M(2, g_A) \subset \mathscr{S}_M(2)$. Therefore Theorem 17.7(iii) implies that all $\mathfrak{S}_2$-measurable step functions belong

to $\mathscr{S}_M(2)$. Let now $f \in \bigcap_{P\in\mathscr{P}} \mathscr{L}^2(\Omega,\mathfrak{S}_2,P_{\mathfrak{S}_2})$ be arbitrary and let $\varepsilon > 0$. For every $P \in \mathscr{P}$ there exists a step function $g \in \mathscr{S}_M(2)$ such that

$$\int (f - g)^2 dP < \varepsilon^2$$

holds. Then, for every $v \in \mathscr{S}_U(2,0)$ we get

$$\begin{aligned} \left|\int fv\, dP\right| &= \left|\int (f-g)v\, dP + \int gv\, dP\right| \\ &= \left|\int (f-g)v\, dP\right| \\ &\leq \left(\int (f-g)^2 dP\right)^{1/2}\left(\int v^2\, dP\right)^{1/2} \\ &< \varepsilon\left(\int v^2\, dP\right)^{1/2}, \end{aligned}$$

i.e., $\int fv\, dP = 0$ for all $v \in \mathscr{S}_U(2,0)$ and $P \in \mathscr{P}$. It follows that $\bigcap_{P\in\mathscr{P}} \mathscr{L}^2(\Omega,\mathfrak{S}_2,P_{\mathfrak{S}_2}) \subset \mathscr{S}_M(2)$, as asserted.

3. Let $f \in \bigcap_{P\in\mathscr{P}} \mathscr{L}^2(\Omega,\mathfrak{S}_2,P_{\mathfrak{S}_2})$ such that $\int f\, dP = 0$ for all $P \in \mathscr{P}$. By (ii) we conclude $f \in \mathscr{S}_M(2,0)$. Since also $1_\emptyset \in \mathscr{S}_M(2,0)$, we infer from Theorem 16.13 that $f = 1_\emptyset[\mathscr{P}]$, whence $\mathfrak{S}_2$ is 2-complete for $\mathscr{P}$.

Let now $\mathfrak{T}$ be a sub-σ-algebra of $\mathfrak{S}_2$ that is sufficient for $\mathscr{P}$. We shall show that $\mathfrak{S}_2 \subset \mathfrak{T} \vee \mathfrak{N}_{\mathscr{P}}$. For this we choose $A \in \mathfrak{S}_2$ and $Q_A \in \mathfrak{M}_{(1)}(\Omega, \mathfrak{T})$ satisfying $Q_A = \mathscr{P}^{\mathfrak{T}}(A)$. Using Jensen's inequality we obtain for all $P \in \mathscr{P}$,

$$E_P(|Q_A - P(A)|^p) \leq E_P(|1_A - P(A)|^p).$$

But since $A \in \mathfrak{S}_2$ and Q_A is an unbiased estimator for the parameter $P \to P(A)$, the above expected values are in fact equal, thus $Q_A = 1_A[\mathscr{P}]$, i.e., $A \in \mathfrak{T} \vee \mathfrak{N}_{\mathscr{P}}$. □

In the remaining part of this section we shall show that the minimization problem with respect to loss functions of type V_p can be treated with methods that are applied in calculus, in particular in the determination of extreme values of functions.

Definition 17.9. Let E be a real vector space and q any real function on E.

q is said to be *weakly differentiable* at $x_0 \in E$ if for every $y \in E$ the limit

$$q'(x_0,y) := \lim_{t\to 0} \frac{1}{t}(q(x_0 + ty) - q(x_0))$$

exists.

In the case of its existence $q'(x_0,y)$ is called the *weak derivative of* q *at (the point)* x_0 *in the direction of* y.

Lemma 17.10. Let in the setting of the above definition $E := L^p(\Omega,\mathfrak{A},P)$ $(p \in]1,\infty[)$ and $q\colon E \to \mathbb{R}_+$ the p-norm. Then q is weakly differentiable at every $x_0 \in E$, $x_0 \neq 0$, and we have

$$q'(x_0,y) = q(x_0)^{1-p} \int |x_0(\omega)|^{p-1}(\operatorname{sgn} x_0(\omega))y(\omega)P(d\omega)$$

for all $y \in E$.

Proof: For every $x_0 \in E$, $x_0 \neq 0$ and all $y \in E$,

$$\begin{aligned}
q'(x_0,y) &= \lim_{t\to 0} \frac{1}{t}(q(x_0 + ty) - q(x_0)) \\
&= \lim_{t\to 0} \frac{1}{t}\left[\left(\int |x_0(\omega)+ty(\omega)|^p P(d\omega)\right)^{1/p} - \left(\int |x_0(\omega)|^p P(d\omega)\right)^{1/p}\right] \\
&= \frac{d}{dt}\left(\int |x_0(\omega) + ty(\omega)|^p P(d\omega)\right)^{1/p}\Bigg|_{t=0} \\
&= \frac{1}{p}\left(\int |x_0(\omega)|^p P(d\omega)\right)^{(1/p)-1} \frac{d}{dt}\left(\int |x_0(\omega)+ty(\omega)|^p P(d\omega)\right)\Bigg|_{t=0} \\
&= \frac{1}{p}\, q(x_0)^{1-p} \int \frac{d}{dt}|x_0(\omega) + ty(\omega)|^p\Big|_{t=0} P(d\omega) \\
&= q(x_0)^{1-p} \int |x_0(\omega)|^{p-1}(\operatorname{sgn} x_0(\omega))y(\omega)P(d\omega).
\end{aligned}$$

Here, the existence of the limit can be seen by inversing the chain of equalities. □

Theorem 17.11. Let $(\Omega,\mathfrak{A},\mathscr{P},g\colon \mathscr{P} \to \mathbb{R})$ be an estimation experiment and let $s_0\colon \Omega \to \mathbb{R}$ be an unbiased estimator for the parameter g. The following statements are equivalent:

(i) $s_0 \in \mathscr{S}_M(p,g)$.

(ii) For every $P \in \mathscr{P}$ with $P([s_0 \neq g(P)]) > 0$ and all $v \in \mathscr{S}_U(p,0)$ we have

$$\lim_{t\to 0} \frac{1}{t}(||s_0 - g(P) + tv||_p - ||s_0 - g(P)||_p) = 0,$$

where $||\cdot||_p := ||\cdot||_p^{(P)}$ denotes the p-norm with respect to P.

Proof: By Theorem 17.3 the above statement (i) is equivalent to the validity of the equation

$$E_P(v|s_0 - g(P)|^{p-1}\operatorname{sgn}(s_0 - g(P))) = 0 \qquad (*)$$

for every $v \in \mathscr{S}_U(p,0)$ and $P \in \mathscr{P}$. On the other hand we obtain from the Lemma that for all $P \in \mathscr{P}$ satisfying $P([s_0 - g(P) \neq 0]) > 0$ and all $v \in \mathscr{S}_U(p,0)$,

$$\lim_{t\to 0} \frac{1}{t}(||s_0 - g(P) + tv||_p - ||s_0 - g(P)||_p)$$
$$= ||s_0 - g(P)||_p^{1-p}\, E_P(v|s_0 - g(P)|^{p-1}\operatorname{sgn}(s_0 - g(P))).$$

Since in the case $P([s_0 - g(P) \neq 0]) = 0$ the equality (*) is always satisfied, we achieved the equivalence of (i) and (ii). □

§18. ESTIMATION VIA THE ORDER STATISTIC

Let $(\Omega',\mathbf{A}')$ be a measurable space and $(\Omega,\mathbf{A}) := (\Omega'^n,\mathbf{A}'^{\otimes n})$ for some $n \geq 1$. By Σ_n we shall denote as usual the group of permutations of the set $\{1,\ldots,n\}$. For every $\pi \in \Sigma_n$ let the mapping $T_\pi: (\Omega,\mathbf{A}) \to (\Omega,\mathbf{A})$ be defined by

$$T_\pi((\omega'_1,\ldots,\omega'_n)) := (\omega'_{\pi(1)},\ldots,\omega'_{\pi(n)})$$

for all $(\omega'_1,\ldots,\omega'_n) \in \Omega$. Finally we introduce the σ-algebra

$$\mathbf{A}'^{\odot n} := \{A \in \mathbf{A}'^{\otimes n}: T_\pi^{-1}(A) = A \quad \text{for all} \quad \pi \in \Sigma_n\}$$

of *permutation invariant sets* in $\mathbf{A}'^{\otimes n}$. We know from Example 4.8 that $\mathbf{A}'^{\odot n}$ is sufficient for every set of product measures $P'^{\otimes n}$ with $P' \in \mathscr{M}^1(\Omega',\mathbf{A}')$. The aim of this section is to discuss the question under what conditions the σ-algebra $\mathbf{A}'^{\odot n}$ is in fact complete for a given set of product measures on $(\Omega,\mathbf{A})$. In the special case $(\Omega',\mathbf{A}') := (\mathbb{R},\mathbb{B})$ the σ-algebra $\mathbf{A}'^{\odot n}$ is generated by the *order statistic* $o_n: (\mathbb{R}^n,\mathbb{B}^n) \to (\mathbb{R}^n,\mathbb{B}^n)$. We shall see that within this general framework the Lehmann-Scheffé theorem yields for a variety of important classes of experiments that some of the most useful estimators are in fact MVU estimators and can be factorized via the order statistic.

<u>Definition 18.1</u>. Let

$$\mathbb{R}^n_{mon} := \{x = (x_1,\ldots,x_n) \in \mathbb{R}^n: x_1 \leq x_2 \leq \ldots \leq x_n\}.$$

For every $x \in \mathbb{R}^n$ we put

$$\Sigma_x^{mon} := \{\pi \in \Sigma_n: T_\pi(x) \in \mathbb{R}^n_{mon}\}.$$

The mapping $o_n: \mathbb{R}^n \to \mathbb{R}^n_{mon}$ defined by

$$\{o_n(x)\} := \{T_\pi(x) : \pi \in \Sigma_x^{mon}\}$$

for all $x \in \mathbb{R}^n$ is called the *order statistic* on $\mathbb{R}^n$.

Remark 18.2. The definition of o_n given above is based on the facts that for all $x \in \mathbb{R}^n$, $\Sigma_x^{mon} \neq \emptyset$ and that the set $\{T_\pi(x) : \pi \in \Sigma_x^{mon}\}$ is a singleton. Let $\mathfrak{B}^n_{mon} := \operatorname{Res}_{\mathbb{R}^n_{mon}} \mathfrak{B}^n$. Then obviously $\mathfrak{B}^{\odot n} = o_n^{-1}(\mathfrak{B}^n_{mon})$. In this sense $\mathfrak{B}^{\odot n}$ is generated by the order statistic o_n.

Theorem 18.3. Let $\mathcal{M}_r$ be the set of all rectangular probability distributions μ on $(\mathbb{R}, \mathfrak{B})$ of the form

$$\mu = \left(\sum_{i=1}^{k} \frac{\alpha_i}{\lambda(I_i)} 1_{I_i}\right) \cdot \lambda,$$

where $\alpha_1, \ldots, \alpha_k \geq 0$ with $\Sigma_{i=1}^k \alpha_i = 1$ and $I_1, \ldots, I_k$ are pairwise disjoint intervals in $\mathbb{R}$. We consider the experiment $(\mathbb{R}^n, \mathfrak{B}^n, \mathcal{M}_r^{\otimes n})$ where

$$\mathcal{M}_r^{\otimes n} := \{\mu^{\otimes n} : \mu \in \mathcal{M}_r\}.$$

Then o_n is complete for $\mathcal{M}_r^{\otimes n}$.

The proof of the theorem will be preceded by the following

Lemma 18.4. Let p be a homogeneous polynomial of k variables over $\mathbb{R}$ of the form

$$p(\alpha_1, \ldots, \alpha_k) = \sum_{\substack{(i_1,\ldots,i_k) \in \mathbb{Z}_+^k \\ i_1+\ldots+i_k = n}} c_{i_1,\ldots,i_k} \alpha_1^{i_1} \cdot \ldots \cdot \alpha_k^{i_k}$$

for some $n \in \mathbb{Z}_+$. If $p(\alpha_1, \ldots, \alpha_k) = 0$ for all $(\alpha_1, \ldots, \alpha_k) \in \mathbb{R}_+^k$ with $\Sigma_{i=1}^k \alpha_i = 1$, then $p(\alpha_1, \ldots, \alpha_k) = 0$ for all $(\alpha_1, \ldots, \alpha_k) \in \mathbb{R}^k$.

Proof: Since p is assumed to be homogeneous, we have $p(r\alpha) = r^n p(\alpha)$ for all $\alpha \in \mathbb{R}^k$ and all $r \in \mathbb{R}$. Every $\beta \in \mathbb{R}_+^k$ admits a representation $\beta = r\alpha$ with $r \in \mathbb{R}$ and $\alpha = (\alpha_1, \ldots, \alpha_k) \in \mathbb{R}_+^k$ such that $\Sigma_{i=1}^k \alpha_i = 1$. Therefore $p(\beta) = r^n p(\alpha) = 0$ for all $\beta \in \mathbb{R}_+^k$. We note that the element $\mathbb{1} := (1, \ldots, 1) \in \mathbb{R}_+^k$ is an interior point of $\mathbb{R}_+^k$. Then for every $(i_1, \ldots, i_k) \in \mathbb{Z}_+^k$ such that $i_1 + \ldots + i_k = n$ we obtain

$$i_1 \cdot \ldots \cdot i_k \cdot c_{i_1,\ldots,i_k} = \frac{\partial^n p}{\partial \alpha_1^{i_1} \cdot \ldots \cdot \partial \alpha_n^{i_n}}(\mathbb{1}) = 0,$$

thus $p \equiv 0$. □

We now proceed to the proof of the theorem.

Let $f \in \mathfrak{M}(\mathbb{R}^n, \mathfrak{A}(o_n)) \cap \bigcap_{P \in \mathcal{M}_r^{\otimes n}} \mathcal{L}^1(\mathbb{R}^n, \mathfrak{B}^n, P)$ satisfying $\int f\, dP = 0$ for all $P \in \mathcal{M}_r^{\otimes n}$. We recall that $\mathfrak{A}(o_n)$ is just the σ-algebra of permutation invariant sets in $\mathfrak{B}^n$. Therefore we have for every measure $P \in \mathcal{M}_r^{\otimes n}$ of the form

$$P := \left[\left(\sum_{i=1}^{k} \frac{\alpha_i}{\lambda(I_i)} 1_{I_i}\right) \cdot \lambda\right]^{\otimes n}$$

with pairwise disjoint intervals $I_1,\ldots,I_k$ in $\mathbb{R}$ and the equalities

$$\int f\, dP = \sum_{(j_1,\ldots,j_n)=(1,\ldots,1)}^{(k,\ldots,k)} \left(\frac{1}{\lambda(I_{j_1})\cdot\ldots\cdot\lambda(I_{j_n})} \cdot \int_{I_{j_1}\times\ldots\times I_{j_n}} f\, d\lambda^n\right)\alpha_{j_1}\cdot\ldots\cdot\alpha_{j_n}$$

$$= \sum_{1\le j_1\le j_2\le\ldots\le j_n\le k} \left(\frac{n!}{i_1!\cdot\ldots\cdot i_k!}\,\frac{1}{\lambda(I_{j_1})\cdot\ldots\cdot\lambda(I_{j_n})} \cdot \int_{I_{j_1}\times\ldots\times I_{j_n}} f\, d\lambda^n\right)\alpha_{j_1}\cdot\ldots\cdot\alpha_{j_n},$$

where for all $j_1,\ldots,j_n$ the symbol i_ℓ denotes the number of j's equal to ℓ. The latter sum can be rewritten as

$$\sum_{\substack{(i_1,\ldots,i_k)\in\mathbb{Z}_+^k \\ i_1+\ldots+i_k = n}} c_{i_1,\ldots,i_k}\,\alpha_1^{i_1}\cdot\ldots\cdot\alpha_n^{i_n},$$

where

$$c_{i_1,\ldots,i_k} := \frac{n!}{i_1!\cdot\ldots\cdot i_k!}\,\frac{1}{\lambda(I_1)^{i_1}\cdot\ldots\cdot\lambda(I_k)^{i_k}} \cdot \int_{I_1^{i_1}\times\ldots\times I_k^{i_k}} f\, d\lambda^n.$$

If we fix the intervals $I_1,\ldots,I_k$ and let the α_i vary, we obtain a homogeneous polynomial p over $\mathbb{R}$ of degree k, which by assumption on f vanishes for all $(\alpha_1,\ldots,\alpha_k) \in \mathbb{R}_+^k$ with $\Sigma_{i=1}^k \alpha_i = 1$. Application of the lemma yields that p vanishes identically. This implies that the measure $f \cdot \lambda^n$ equals the zero measure on all rectangles and hence on the sets of a $\cap$-stable generator of $\mathbb{B}^n$. Consequently $f \cdot \lambda^n = 0$, whence $f = o[\lambda^n]$. Since $\mathscr{M}_r^{\otimes n} \sim \lambda^n$, we get $f = o[\mathscr{M}_r^{\otimes n}]$ which is the desired result. $\square$

The following result which serves as an important tool in various applications will be presented as an example of the many far-reaching generalizations of Theorem 18.3.

Theorem 18.5. (D. A. S. Fraser). Let $(\Omega',\mathbb{A}')$ be a measurable space, $\mathbb{H}$ a semiring generating $\mathbb{A}'$ and $\mathbb{R}$ the ring generated by $\mathbb{H}$. Furthermore, let ν be a measure in $\mathscr{M}_+^\sigma(\Omega',\mathbb{A}')$ which has no atoms, and let

$$\mathscr{N}: = \{\frac{1}{\nu(B)} 1_B \cdot \nu: B \in \mathbb{R},\ 0 < \nu(B) < \infty\}.$$

Then the σ-algebra $\mathbb{A}'^{\odot n}$ is complete for $\mathscr{N}^{\otimes n}$.

Proof: Without loss of generality we may assume that the measure ν is bounded. For, if the theorem has been proved for every finite measure $\nu \in \mathscr{M}_+(\Omega',\mathbb{A}')$ and if $f \in \mathbb{M}(\Omega,\mathbb{A}'^{\odot n})$ satisfies $\int f\, d\mu^{\otimes n} = 0$ for all $\mu \in \mathscr{N}$, then for all $R \in \mathbb{R}$ with $\nu(R) < \infty$ we obtain $f = o[1_{R^n} \cdot \mu^{\otimes n}]$ for all $\mu \in \mathscr{N}$, which implies $f \cdot 1_{R^n} = o[\mu^{\otimes n}]$ for all $\mu \in \mathscr{N}$, whence $f = o[\mathscr{N}^{\otimes n}]$, and the result has been established. Now let $f \in \mathbb{M}(\Omega,\mathbb{A}'^{\odot n})$ be such that

$$\int f\, d\mu^{\otimes n} = 0 \quad \text{for all} \quad \mu \in \mathscr{N}$$

holds. We shall show, that for all $A_1',\ldots,A_n' \in \mathbb{H}$ we have

$$\int_{A_1'\times\ldots\times A_n'} f\, d\nu^{\otimes n} = 0.$$

In this case the measures $f^+ \cdot \nu^{\otimes n}$ and $f^- \cdot \nu^{\otimes n}$ coincide on $\underbrace{\mathbb{H} \times\ldots\times \mathbb{H}}_{n\text{-times}}$ and thus on $\mathbb{A}'^{\otimes n}$. It follows, that $f = o[\nu^{\otimes n}]$, whence that $f = o[\mathscr{N}^{\otimes n}]$. Let, therefore, $A_1',\ldots,A_n' \in \mathbb{H}$ and $\varepsilon > 0$ such that $\varepsilon < \nu(\Omega)$. Since ν has no atoms, there exists a measurable partition $\mathbb{Z}: = \{Z_1',\ldots,Z_k'\}$ of Ω' with $k \geq 2$ and $\nu(Z_i') = \varepsilon$ for all $i = 1,\ldots, k-1$, $\nu(Z_k') \leq \varepsilon$. Without loss of generality we assume that $\mathbb{Z} \subset \mathbb{H}$. If this cannot be achieved, then for every $i = 1,\ldots,k$ the set Z_i' has to be approximated with respect to ν by a finite disjoint union of sets in $\mathbb{H}$. This approximation, however, is possible, since $\mathbb{H}$ is a generator of $\mathbb{A}'$. We now conclude

$$\left| \int_{A_1'\times\ldots\times A_n'} f \, d\nu^{\otimes n} \right|$$

$$= \left| \sum_{(i_1,\ldots,i_n)=(1,\ldots,1)}^{(k,\ldots,k)} \int_{(A_1'\cap Z_{i_1}')\times\ldots\times(A_n'\cap Z_{i_n}')} f \, d\nu^{\otimes n} \right|$$

$$\leq \sum_{m=1}^{k} \sum_{1\leq i<j\leq n} \int_{A_1'\times\ldots\times(A_i'\cap Z_m')\times\ldots\times(A_j'\cap Z_m')\times\ldots\times A_n'} |f| d\nu^{\otimes n}.$$

The last inequality results from the fact that all summands for which the n numbers $i_1,\ldots,i_n$ are pairwise different, vanish. This fact can be proved by induction, where the hypothesis that

$$\int f \, d\left(\frac{1}{\nu(Z_k')} 1_{Z_k'} \cdot \nu\right)^{\otimes n} = 0$$

holds, is used essentially. We now proceed to majorize the above double sum by

$$\sum_{m=1}^{k} \sum_{1\leq i<j\leq n} \int_{\Omega'\times\ldots\times(A_i'\cap Z_m')\times\ldots\times(A_j'\cap Z_m')\times\ldots\times\Omega'} |f| d\nu^{\otimes n}.$$

Since

$$\nu^{\otimes n}(\Omega'\times\ldots\times(A_i' \cap Z_m')\times\ldots\times(A_j' \cap Z_m')\times\ldots\times\Omega') \leq \nu(\Omega')^{n-2}\varepsilon^2$$

and $k\varepsilon \leq 2\nu(\Omega')$, we obtain

$$\sum_{1\leq m\leq k} \sum_{1\leq i<j\leq n} \nu(\Omega')^{n-2}\varepsilon^2 \leq \binom{n}{2}\nu(\Omega')^{n-2} 2\nu(\Omega')\varepsilon.$$

But $|f| \cdot \nu^{\otimes n} \ll \nu^{\otimes n}$ then implies that

$$\left| \int_{A_1'\times\ldots\times A_n'} f d\nu^{\otimes n} \right| \leq \int_{\Omega'\times\ldots\times(A_i'\cap Z_m')\times\ldots\times(A_j'\cap Z_m')\times\ldots\times\Omega'} |f| d\nu^{\otimes n}$$

becomes arbitrarily small for suitably chosen $\varepsilon > 0$, whence

$$\int_{A_1'\times\ldots\times A_n'} f \, d\nu^{\otimes n} = 0,$$

which had to be proved. □

In contrast to Fraser's theorem where the measures involved are assumed to be free of atoms we now present the following

Theorem 18.6. Let B be a subset of $\mathbb{R}$ and let $\mathcal{M}_B$ be the convex hull of the set $\{\varepsilon_b : b \in B\}$. Then the order statistic o_n is complete for the set $\mathcal{M}_B^{\otimes n} := \{\mu^{\otimes n} : \mu \in \mathcal{M}_B\}$.

Proof: Let $P := (\Sigma_{i=1}^k \alpha_i \varepsilon_{b_i})^{\otimes n}$ with $\alpha_1,\ldots,\alpha_k \in \mathbb{R}_+$, $\Sigma_{i=1}^k \alpha_i = 1$, and $b_1,\ldots,b_k \in B$, and let $f \in \mathfrak{M}(\mathbb{R}^n, \mathfrak{B}^{\odot n})$. Then

$$\int f\,dP = \sum_{(j_1,\ldots,j_n)=(1,\ldots,1)}^{(k,\ldots,k)} \left(\int fd(\varepsilon_{b_{j_1}} \otimes \ldots \otimes \varepsilon_{b_{j_n}})\right)\alpha_{j_1}\cdot\ldots\cdot\alpha_{j_n}$$

$$= \sum_{1\le j_1\le\ldots\le j_n\le k} \frac{n!}{i_1!\cdot\ldots\cdot i_k!} f(b_{j_1},\ldots,b_{j_n})\alpha_{j_1}\cdot\ldots\cdot\alpha_{j_n}$$

$$= \sum_{\substack{(i_1,\ldots,i_k)\in \mathbb{Z}_+^k \\ i_1+\ldots+i_k = n}} c_{i_1,\ldots,i_k} \alpha_1^{i_1}\cdot\ldots\cdot\alpha_k^{i_k},$$

where

$$c_{i_1,\ldots,i_k} := \frac{n!}{i_1!\cdot\ldots\cdot i_k!} f(\underbrace{b_{i_1},\ldots,b_{i_1}}_{i_1\text{-times}},\underbrace{b_{i_2},\ldots,b_{i_2}}_{i_2\text{-times}},\ldots,\underbrace{b_{i_k},\ldots,b_{i_k}}_{i_k\text{-times}}).$$

Now, let $\int f\,dP = 0$ for all $P \in \mathcal{M}_B^{\otimes n}$. Then Lemma 18.4 yields $f = 0$ on B^n and therefore $f = o[\mathcal{M}_B^{\otimes n}]$ which proves the theorem. □

Example 18.7. Let $k \geq 1$, $\mathcal{M}^{(k)} := \{\mu \in \mathcal{M}^1(\mathbb{R},\mathfrak{B}) : \int|\xi|^k\mu(d\xi) < \infty\}$, and $\mathcal{N} \subset \mathcal{M}^{(k)}$ such that $\mathcal{N}$ is a set of λ-continuous measures with $\mathcal{N} \supset \mathcal{M}_r$ or such that $\mathcal{N}$ is a set of measures concentrated on B, with $\mathcal{N} \supset \mathcal{M}_B$ for some $B \subset \mathbb{R}$. We define an estimation experiment $(\Omega, \mathfrak{A}, \mathcal{P}, g: \mathcal{P} \to \mathbb{R})$ with $\Omega := \mathbb{R}^n$, $\mathfrak{A} := \mathfrak{B}^n$, $\mathcal{P} := \{\mu^{\otimes n} : \mu \in \mathcal{N}\}$ and

$$g(P) := \int |\xi|^k \mu(d\xi) \quad \text{for all} \quad P := \mu^{\otimes n} \in \mathcal{P}.$$

The estimator $s_k := \frac{1}{n}\Sigma_{i=1}^n |X_i|^k$ is MVU for g.

In fact, $s_k \in \mathcal{S}_U(g)$ as follows from the equalities

$$\int s_k\,dP = \int \frac{1}{n}\sum_{i=1}^n |X_i|^k d\mu^{\otimes n} = \frac{1}{n}\sum_{i=1}^n \int |X_i|^k d\mu^{\otimes n} = \frac{1}{n}\sum_{i=1}^n \int |\xi|^k d\mu = g(P),$$

valid for all $P = \mu^{\otimes n}$ with $\mu \in \mathcal{N}$. By Theorem 18.6 and 18.3 the σ-algebra $\mathfrak{B}^{\odot n}$ is complete for $\mathcal{P} = \mathcal{N}^{\otimes n}$ for both choices of $\mathcal{N}$. Clearly s_k is sufficient for $\mathcal{P}$. Since s_k is also $\mathfrak{B}^{\odot n}$-measurable, it is by the Lehmann-Scheffé theorem the unique MVU estimator for g.

Let $(\Omega,\mathbf{A},\mathscr{P},\chi:\mathscr{P}\to\Theta)$ be an m-dimensional exponential experiment, where for every $\theta\in\Theta$ the measure $P_\theta := \chi^{-1}(\theta)$ is defined by

$$P_\theta = C(\theta)\exp\Big(\sum_{k=1}^{m}\xi_k(\theta)T_k\Big)h\cdot\mu$$

with mappings $C,\zeta_k:\Theta\to\mathbb{R}$ and measurable mappings $h,T_k:(\Omega,\mathbf{A})\to(\mathbb{R},\mathbf{B})$ $(k = 1,\ldots,m)$. We specify $\Theta := \mathbb{R}^m$ and define $Z := \chi(\mathscr{P})$. Then the measures P_θ can be rewritten as

$$P_\zeta = C(\zeta)\exp\Big(\sum_{k=1}^{m}\zeta_k T_k\Big)h\cdot\mu.$$

The following result will be helpful in handling further estimation problems.

Theorem 18.8. The statistic $T = (T_1,\ldots,T_m):(\Omega,\mathbf{A})\to(\mathbb{R}^m,\mathbf{B}^m)$ is complete for the family $\{P_\zeta:\zeta\in Z\}$ if Z contains a nondegenerate m-dimensional rectangle.

Proof: Since the measures P_ζ are invariant with respect to linear transformations, the given nondegenerate m-dimensional rectangle R can be chosen of the form

$$R := \{\zeta = (\zeta_1,\ldots,\zeta_m)\in\mathbb{R}^m : \zeta_k\in\,]-a,a[\ \text{ for } k = 1,\ldots,m\}$$

for some $a\in\mathbb{R}_+^*$.

Let $\nu := h\cdot\mu$. Then we have to show that for every $f\in\mathbf{M}(\mathbb{R}^m,\mathbf{B}^m)$ the equality

$$\int f(t)\exp\Big(\sum_{k=1}^{m}\zeta_k t_k\Big)T(\nu)(dt) = 0$$

for all $\zeta\in R$ implies $f = o[T(\nu)]$. The hypothesis of this implication is equivalent to

$$\int f^+(t)\exp\Big(\sum_{k=1}^{m}\zeta_k t_k\Big)T(\nu)(dt) = \int f^-(t)\exp\Big(\sum_{k=1}^{m}\zeta_k t_k\Big)T(\nu)(dt)$$

for all $\zeta = (\zeta_1,\ldots,\zeta_m)\in R$. Since $\zeta_0 = (0,\ldots,0)\in R$, we obtain

$$\int f^+(t)T(\nu)(dt) = \int f^-(t)T(\nu)(dt).$$

We may suppose that both integrals in this equality are different from zero, and hence after an eventual norming of f even equal to 1. Thus the set functions

$$B \to \rho^{\pm}(B) := \int_B f^{\pm}(t)T(\nu)(dt)$$

are probability measures on $(\mathbb{R}^m, \mathfrak{B}^m)$. We therefore have that

$$\int \exp\left(\sum_{k=1}^{m} \zeta_k t_k\right)\rho^{+}(dt) = \int \exp\left(\sum_{k=1}^{m} \zeta_k t_k\right)\rho^{-}(dt)$$

for all $\zeta \in R$, and it remains to show that $\rho^{+} = \rho^{-}$. To this end it suffices to prove that $\widehat{\rho^{+}} = \widehat{\rho^{-}}$ or that

$$\int \exp\left(i \sum_{k=1}^{m} \eta_k t_k\right)\rho^{+}(dt) = \int \exp\left(i \sum_{k=1}^{m} \eta_k t_k\right)\rho^{-}(dt)$$

for all $\eta = (\eta_1,\ldots,\eta_m) \in \mathbb{R}^m$. We put $\xi := \zeta + i\eta \in \mathbb{C}^m$. The functions

$$\xi \to \beta^{\pm}(\xi) := \int \exp\left(\sum_{k=1}^{m} \xi_k t_k\right)\rho^{\pm}(dt)$$

are holomorphic in each component variable ξ_k within the band $-a < \zeta_k < a$, $-\infty < \eta_k < \infty$ whenever the variables $\xi_j (j \neq k,\ k = 1,\ldots,m)$ are fixed. For fixed $(\zeta_2,\ldots,\zeta_m) \in]-a,a[^{m-1}$ the functions $\xi_1 \to \beta^{+}(\xi_1,\zeta_2,\ldots,\zeta_m)$ and $\xi_1 \to \beta^{-}(\xi_1,\zeta_2,\ldots,\zeta_m)$ coincide for all real $\xi_1 = \zeta_1 \in]-a,a[$. By the uniqueness of analytic continuation these functions coincide also for all purely imaginary $\xi_1 = i\eta_1$ with $\eta_1 \in]-\infty,\infty[$. This procedure can be repeated for ξ_2 and fixed $\xi_1,\zeta_3,\ldots,\zeta_k$ and so on. We end up with $\widehat{\rho^{+}} = \widehat{\rho^{-}}$ as desired. □

CHAPTER VII

Information and Sufficiency

§19. COMPARISON OF CLASSICAL EXPERIMENTS

In this section we introduce the decision theoretic basis to our approach to mathematical statistics. We restrict ourselves to the purely measure theoretic version of decision theory. The section contains the fundamental definitions, some interpretations, and some theorems of an introductory nature designed to give a first insight into the theory of comparison of experiments.

Definition 19.1. A *decision problem* is a triplet $\underline{\overline{D}} = (I,D,\mathcal{V})$ where $I = (\Omega_I,\mathfrak{A}_I)$ and $D = (\Omega_D,\mathfrak{A}_D)$ are measurable spaces and $\mathcal{V}$ denotes a set of separately measurable real-valued functions on $\Omega_I \times \Omega_D$.

We remark that if $\mathfrak{A}_I$ and $\mathcal{V}$ are not specified any further, then $\mathfrak{A}_I$ will be the σ-algebra $\mathfrak{P}(\Omega_I)$ and $\mathcal{V}$ the set of all bounded separately measurable real-valued functions on $\Omega_I \times \Omega_D$.

Definition 19.2. Let $\underline{\overline{D}} = (I,D,\mathcal{V})$ be a decision problem.

(a) An *experiment corresponding to* $\underline{\overline{D}}$ is any pair $X = (X_X,N_X)$, where $X_X = (\Omega_X,\mathfrak{A}_X)$ is a measurable space and N_X is a kernel in $\mathrm{Stoch}(I,X_X)$.

The class of all experiments corresponding to $\underline{\overline{D}}$ will be abbreviated by $\mathscr{E}(\underline{\overline{D}})$ or exchangeably by $\mathscr{E}(I)$.

(b) A *decision function corresponding to* $X \in \mathscr{E}(\underline{\overline{D}})$ is a kernel $\delta \in \mathrm{Stoch}(X_X,D)$.

The totality of all decision functions corresponding to X will be abbreviated by $\mathscr{D}(X_X,D)$ or simply by $\mathscr{D}(X)$.

Interpretation 19.3. As in §3 we interpret I as the (measurable) space of states of nature, D as the (measurable) space of decisions and $\mathcal{V}$ as the set of loss functions corresponding to the given decision problem $\underline{D}$. A set up of this kind is familiar in the theory of games, where I and $\mathcal{D}(X)$ are interpreted as spaces of strategies. Within this framework the gambler often replaces the space D by its mixed extension $\mathcal{M}^1(D)$. In contrast to the gambler the statistician prefers to consider situations in which the states of nature are realized in an explicit way. There is a close analogy between this interpretation and the basic model of information theory; in the latter case signals are sent along a channel and in the process are distorted by noise.

Remark 19.4. The notion $(\Omega,\mathbf{A},\mathcal{P})$ of an experiment as we used it in the preceding sections can be obtained from the more general definition above by putting $I := (\mathcal{P},\Sigma)$ with a σ-algebra Σ in $\mathcal{P}$ satisfying $\Sigma_{\mathbf{A}} \subset \Sigma \subset \mathfrak{P}(\mathcal{P})$ and choosing for $\mathcal{V}$ the set of all bounded separately measurable functions V from $\mathcal{P}\times D$ into $\mathbb{R}$. Here we recall that $\Sigma_{\mathbf{A}}$ denotes the σ-algebra in $\mathcal{P}$ generated by the mappings $P \to P(A)$ from $\mathcal{P}$ into $\mathbb{R}$, where $A \in \mathbf{A}$.

Given an indexed experiment $(\Omega,\mathbf{A},(P_i)_{i\in I})$ one can interpret the index set as the space of states of nature in a more suggestive way. If one considers the parametrized experiment $(\Omega,\mathbf{A},\mathcal{P},\chi: \mathcal{P}\to\Theta)$, then evidently Θ will be taken as the decision space and χ will signify the mapping which to every $P \in \mathcal{P}$ assigns the "correct" decision $\chi(P)$.

Definition 19.5. Let $\underline{D} = (I,D,\mathcal{V})$ be a decision problem.

(a) Any mapping $\varepsilon: \Omega_I \to \mathbb{R}_+$ will be called a *tolerance function corresponding to* $\underline{D}$.

(b) Let $X \in \mathcal{E}(\underline{D})$, $\delta \in \mathcal{D}(X)$ and $V \in \mathcal{V}$ be such that for every $i \in \Omega_I$ the function $V(i,\cdot)$ is integrable with respect to the measure $(N_X\delta)(i,\cdot) = \int \delta(x,\cdot)N_X(i,dx)$. Then the function $R_\delta^V: \Omega_I \to \mathbb{R}$ defined by

$$R_\delta^V(i) := \langle N_X\delta(i,\cdot), V(i,\cdot)\rangle \quad \text{for all} \quad i \in \Omega_I$$

is called the *risk function corresponding to* δ *and* V.

Clearly $R_\delta^V(i)$ is describing the *risk* occuring with the choices δ and V if $i \in \Omega_I$ is obtained.

The following definition contains the basic comparison relation for experiments. Its consequences will occupy us for much of what follows.

Definition 19.6. Let $\underline{D} = (I,D,\mathcal{V})$ be a decision problem, $X,Y \in \mathcal{E}(\underline{D})$ two experiments and $\varepsilon: \Omega_I \to \mathbb{R}_+$ any tolerance function corresponding to $\underline{\bar{D}}$. X is said to be *more informative than* Y *at level* ε, in symbols $X \overset{\bar{D}}{>}_{\varepsilon} Y$, if for every $V \in \mathscr{V}$ with

$$||V||: = \sup_{(i,d)\in\Omega_I\times\Omega_D} |V(i,d)| < \infty$$

and every $\delta_Y \in \mathscr{D}(Y)$ there exists a $\delta_X \in \mathscr{D}(X)$ satisfying

$$R^V_{\delta_X} \leq R^V_{\delta_Y} + \varepsilon||V||.$$

Remark 19.7. It is easily seen that the comparison relation $X \overset{\bar{D}}{>}_{\varepsilon} Y$ is satisfied if for every $V \in \mathscr{V}$ with $||V|| < \infty$ and every $\delta_Y \in \mathscr{D}(Y)$ there exists a $\delta_X \in \mathscr{D}(X)$ satisfying

$$R^V_{\delta_X}(i) \leq R^V_{\delta_Y}(i) + \varepsilon(i)||V(i,\cdot)||$$

for all $i \in \Omega_I$.

Theorem 19.8. For any two experiments $X,Y \in \mathscr{E}(\underline{\bar{D}})$ of the form $X: = (X_X,N_X)$ and $Y: = (X_Y,N_Y)$ respectively, and for any tolerance function $\varepsilon \geq 2$ corresponding to $\underline{\bar{D}}$ we have $X \overset{\bar{D}}{>}_{\varepsilon} Y$.

Proof: For every loss function $V \in \mathscr{V}$ and decision functions $\delta_Y \in \mathscr{D}(Y)$, $\delta_X \in \mathscr{D}(X)$ one obtains the following chain of inequalities valid for all $i \in \Omega_I$:

$$\begin{aligned}
&|R^V_{\delta_X}(i) - R^V_{\delta_Y}(i)| \\
&\leq |N_X\delta_X(i,\cdot)(V(i,\cdot)) - N_Y\delta_Y(i,\cdot)(V(i,\cdot))| \\
&\leq ||N_X\delta_X(i,\cdot)||\ ||V(i,\cdot)|| + ||N_Y\delta_Y(i,\cdot)||\ ||V(i,\cdot)|| \\
&= 2||V(i,\cdot)||.
\end{aligned}$$

By the above remark the assertion follows. □

Definition 19.9. For two given experiments $X,Y \in \mathscr{E}(\underline{\bar{D}})$ we introduce the number

$$\rho_{\underline{\bar{D}}}(X,Y): = \inf\{\varepsilon \in \mathbb{R}_+ : X \overset{\bar{D}}{>}_{\varepsilon 1_{\Omega_I}} Y\},$$

and, since in general $\rho_{\underline{\bar{D}}}(X,Y) \neq \rho_{\underline{\bar{D}}}(Y,X)$, also

$$\Delta_{\underline{\bar{D}}}(X,Y): = \rho_{\underline{\bar{D}}}(X,Y) \vee \rho_{\underline{\bar{D}}}(Y,X).$$

$\rho_{\underline{\bar{D}}}(X,Y)$ is called the *deficiency of* X *relative to* Y.

Interpretation 19.10. The number $\rho_{\underline{\bar{D}}}(X,Y)$ can be interpreted as a measure for the *maximal loss of information* occuring by observing X instead of Y.

Properties 19.11. Let $X,Y,Z \in \mathscr{E}(\underline{\bar{D}})$.

19.11.1. $0 \leq \rho_{\underline{\bar{D}}}(X,Y) \leq 2$.

19.11.2. $\rho_{\underline{\bar{D}}}(X,X) = 0$.

19.11.3. $\rho_{\underline{\bar{D}}}(X,Z) \leq \rho_{\underline{\bar{D}}}(X,Y) + \rho_{\underline{\bar{D}}}(Y,Z)$.

All of these properties extend to $\Delta_{\underline{\bar{D}}}$.

19.11.4. $0 \leq \Delta_{\underline{\bar{D}}}(X,Y) \leq 2$.

19.11.5. $\Delta_{\underline{\bar{D}}}(X,X) = 0$.

19.11.6. $\Delta_{\underline{\bar{D}}}(X,Y) = \Delta_{\underline{\bar{D}}}(Y,X)$.

19.11.7. $\Delta_{\underline{\bar{D}}}(X,Z) \leq \Delta_{\underline{\bar{D}}}(X,Y) + \Delta_{\underline{\bar{D}}}(Y,Z)$.

Properties 19.11.4 to 19.11.7 show that $\Delta_{\underline{\bar{D}}}$ behaves like a pseudometric on $\mathscr{E}(\underline{\bar{D}})$.

Definition 19.12. Two experiments $X,Y \in \mathscr{E}(\underline{\bar{D}})$ are said to be *equivalent with respect to* $\underline{\bar{D}}$, in symbols $X \overset{\bar{D}}{\sim} Y$, if $X \overset{\bar{D}}{>}_0 Y$ and $Y \overset{\bar{D}}{>}_0 X$.

Obviously, $X \overset{\bar{D}}{\sim} Y$ implies $\Delta_{\underline{\bar{D}}}(X,Y) = 0$. The converse of this implication will be discussed in §30.

Definition 19.13. Let $\underline{\bar{D}} = (I,D,\mathscr{V})$ be a decision problem and $\varepsilon: \Omega_I \to \mathbb{R}_+$ a tolerance function.

(a) We put

$$\mathscr{V}^b := \{V \in \mathscr{V}: ||V|| = \sup_{(i,d)\in\Omega_I\times\Omega_D} |V(i,d)| < \infty\},$$

$$\mathscr{V}^I := \{V \in \mathscr{V}: ||V(i,\cdot)|| = \sup_{d\in\Omega_D} |V(i,d)| < \infty \text{ for all } i \in \Omega_I\},$$

$$\mathscr{V}_+ := \{V \in \mathscr{V}: V \geq 0\},$$

$$\mathscr{V}_1 := \{V \in \mathscr{V}: ||V|| \leq 1\},$$

and finally

$$\mathscr{V}_{(1)} := \mathscr{V}_1 \cap \mathscr{V}_+.$$

(b) The decision problems corresponding to the sets of loss functions introduced in (a) will be denoted by $\underline{\bar{D}}^b$, $\underline{\bar{D}}^I$, $\underline{\bar{D}}_+$, $\underline{\bar{D}}_1$ and $\underline{\bar{D}}_{(1)}$ respectively. It will be clear what is meant by $\underline{\bar{D}}^{bI}$, $\underline{\bar{D}}^I_+$ and $\underline{\bar{D}}^I_1$.

(c) For a given tolerance function $\varepsilon: \Omega_I \times \mathscr{V} \to \overline{\mathbb{R}}_+$ and every $k = 0,1,2$ we define functions $\tilde{\varepsilon}^k: \Omega_I \times \overline{\mathbb{R}}_+$ by

$$\tilde{\varepsilon}^0(i,V) := \varepsilon(i),$$

$$\tilde{\varepsilon}^1(i,V) := \varepsilon(i)||V||$$

and

$$\tilde{\varepsilon}^2(i,V) := \varepsilon(i)||V(i,\cdot)|| \quad \text{for all } i \in \Omega_I,\ V \in \mathscr{V}.$$

(d) Given $X,Y \in \mathscr{E}(\underline{\bar{D}})$ and $\tilde{\varepsilon}: \Omega_I \times \mathscr{V} \to \overline{\mathbb{R}}_+$ we define $X \overset{\underline{\bar{D}}^I}{>_{\tilde{\varepsilon}}} Y$ if to every $V \in \mathscr{V}^I$ and every $\delta_Y \in \mathscr{D}(Y)$ there exists a $\delta_X \in \mathscr{D}(X)$ satisfying

$$R^V_{\delta_X} \leq R^V_{\delta_Y} + \tilde{\varepsilon}(\cdot,V).$$

Clearly this definition can be extended to all the decision problems appearing in (b) if one chooses the set of loss functions appropriately.

Remark 19.14. 1. Plainly $\underline{\bar{D}}^{bI} = \underline{\bar{D}}^b$ and $\underline{\bar{D}}^I_1 = \underline{\bar{D}}_1$.

2. Definition 19.6 (d) contains Definition 19.4 in the sense that $X \overset{\underline{\bar{D}}}{>_{\varepsilon}} Y$ is equivalent to $X \overset{\underline{\bar{D}}^b}{>_{\tilde{\varepsilon}^1}} Y$.

Theorem 19.15. Let $\underline{\bar{D}} = (I,D,\mathscr{V})$ be a decision problem, $X,Y \in \mathscr{E}(\underline{\bar{D}})$ and $\varepsilon: \Omega_I \times \mathbb{R}_+$ a tolerance function corresponding to $\underline{\bar{D}}$. We consider the following properties of $\mathscr{V}$:

(1) If $V \in \mathscr{V}$, $c \in \mathbb{R}$ and $n \in \mathbb{N}$, then $V + c,\ nV \in \mathscr{V}$.

(2) If $V \in \mathscr{V}$ and $W: \Omega_I \times \Omega_D \to \mathbb{R}$ is defined by

$$W(i,d) := \frac{V(i,d)}{||V(i,\cdot)||} 1_{\{j \in \Omega_I:\ ||V(j,\cdot)||>0\}}(i)$$

for all $(i,d) \in \Omega_I \times \Omega_D$, then $W \in \mathscr{V}$.

Moreover, we consider the subsequent assertions:

(a) $X \overset{\underline{\bar{D}}^I}{>_{\tilde{\varepsilon}^2}} Y$ (b) $X \overset{\underline{\bar{D}}}{>_{\varepsilon}} Y$

(c) $X \overset{\underline{\bar{D}}_1}{>_{\tilde{\varepsilon}^0}} Y$ (d) $X \overset{\underline{\bar{D}}^I_+}{>_{\frac{\tilde{\varepsilon}^2}{2}}} Y$

(e) $X \underset{\frac{\varepsilon}{2}}{\overset{\bar{D}_+}{>}} Y$ (f) $X \underset{\frac{\tilde{\varepsilon}^o}{2}}{\overset{\bar{D}(1)}{>}} Y$

Then we have (a) ⇒ (b) ⇒ (c) and (d) ⇒ (e) ⇒ (f).

If, moreover, (1) is satisfied, then also (b) ⟺ (e) and (c) ⟺ (f).

If, finally, both (1) and (2) are satisfied, then all assertions (a) to (f) are equivalent.

Proof: The implications (a) ⇒ (b) ⇒ (c) and (d) ⇒ (e) ⇒ (f) are clear, since the inequalities involved are weakened.

Now let (1) be satisfied. We shall show (b) ⟺ (e). The implication (c) ⟺ (f) is proved analoguously.

To this end take $V \in \mathscr{V}^b_+$ and $\delta_Y \in \mathscr{D}(Y)$. Since by (1), $2V - ||V|| \in \mathscr{V}^b$, (b) implies the existence of $\delta_X \in \mathscr{D}(X)$ satisfying

$$<N_X\delta_X(i,\cdot),\ 2V(i,\cdot) - ||V||>$$

$$\leq <N_Y\delta_Y(i,\cdot),\ 2V(i,\cdot) - ||V||> + \varepsilon(i)||2V - ||V||\ ||,$$

whence

$$2<N_X\delta_X(i,\cdot),\ V(i,\cdot)> - ||V||$$

$$\leq 2<N_Y\delta_Y(i,\cdot),\ V(i,\cdot)> - ||V|| + \varepsilon(i)||2V - ||V||\ ||.$$

But now $||2V|| \leq ||2V - ||V||\ || + ||V||$ implies $||2V - ||V||\ || \geq ||V||$. For $V \geq 0$ we have furthermore $-||V|| \leq 2V - ||V|| \leq ||V||$ and therefore $||2V - ||V||\ || \leq ||V||$, thus altogether $||2V - ||V||\ || = ||V||$. Consequently we obtain for all $i \in \Omega_I$ the inequality

$$<N_X\delta_X(i,\cdot),V(i,\cdot)> \leq <N_Y\delta_Y(i,\cdot),V(i,\cdot)> + \frac{1}{2}\varepsilon(i)||V||,$$

i.e., (e).

Conversely let $V \in \mathscr{V}^b$ and $\delta_Y \in \mathscr{D}(Y)$ be given and suppose that (e) holds. From $V + ||V|| \in \mathscr{V}^b_+$ we conclude the existence of $\delta_X \in \mathscr{D}(X)$ satisfying

$$<N_X\delta_X(i,\cdot),V(i,\cdot) + ||V||> \leq <N_Y\delta_Y(i,\cdot),V(i,\cdot) + ||V||>$$
$$+ \frac{1}{2}\varepsilon(i)||V + ||V||\ ||,$$

whence

$$<N_X\delta_X(i,\cdot),V(i,\cdot)> + ||V||$$

$$\leq <N_Y\delta_Y(i,\cdot),V(i,\cdot)> + ||V|| + \frac{1}{2}\varepsilon(i)(||V|| + ||V||)$$

for all $i \in \Omega_I$. But this is (b).

Now suppose (1) and (2) to be satisfied. We shall prove (c) $\Rightarrow$ (a). Assume (c) and let $V \in \mathscr{V}^I$, $\delta_Y \in \mathscr{D}(Y)$ be given. Writing

$$W(i,d) := \frac{V(i,d)}{||V(i,\cdot)||} 1_{\{j \in \Omega_I:\ ||V(j,\cdot)||>0\}}(i)$$

for all $(i,d) \in \Omega_I \times \Omega_D$ we define a function $W \in \mathscr{V}_1$. By (c) we choose $\delta_X \in \mathscr{D}(X)$ with

$$<N_X\delta_X(i,\cdot),W(i,\cdot)> \leq <N_Y\delta_Y(i,\cdot),W(i,\cdot)> + \varepsilon(i)$$

for all $i \in \Omega_I$. Then on the set $\{j \in \Omega_I: ||V(j,\cdot)|| > 0\}$ we obtain the inequality

$$\frac{1}{||V(i,\cdot)||} <N_X\delta_X(i,\cdot),V(i,\cdot)>$$

$$\leq \frac{1}{||V(i,\cdot)||} <N_Y\delta_Y(i,\cdot),V(i,\cdot)> + \varepsilon(i),$$

whence

$$<N_X\delta_X(i,\cdot),V(i,\cdot)> \leq <N_Y\delta_Y(i,\cdot),V(i,\cdot)> + \varepsilon(i)||V(i,\cdot)||.$$

But the last inequality is true also for all $i \in \Omega_I$ with $||V(i,\cdot)|| = 0$, i.e., (a) is verified.

The implication (f) $\Rightarrow$ (d) is proved by applying the reasoning above to $\mathscr{V}_+$ instead of $\mathscr{V}$, since the property (2) is also satisfied by $\mathscr{V}_+$.

Thus the properties (a) to (c) and (d) to (f) are equivalent, and from (b) $\Leftrightarrow$ (e) we deduce the full equivalence of properties. □

<u>Definition 19.16</u>. Let $\bar{D} = (I,D,\mathscr{V})$ be a decision problem, $X,Y \in \mathscr{E}(\bar{D})$ and $\varepsilon: \Omega_I \to \mathbb{R}_+$ a tolerance function. X is called *Blackwell more informative than* Y *at level* ε, in symbols $X >^B_\varepsilon Y$, if there exists a kernel $N \in \text{Stoch}(X_X,X_Y)$ with

$$||N_XN - N_Y|| \leq \varepsilon.$$

<u>Theorem 19.17</u>. Let $\bar{D} = (I,D,\mathscr{V})$ be a decision problem, $X,Y \in \mathscr{E}(\bar{D})$ and $\varepsilon: \Omega_I \to \mathbb{R}_+$ a tolerance function. Then $X >^B_\varepsilon Y$ implies $X >^{\bar{D}}_\varepsilon Y$.

<u>Proof</u>: Let $V \in \mathscr{V}^b$ and $\delta_Y \in \mathscr{D}(Y)$ be given. We choose $N \in \text{Stoch}(X_X,X_Y)$ with $||N_XN - N_Y|| \leq \varepsilon$ and put $\delta_X := N\delta_Y$. Then clearly $\delta_X \in \text{Stoch}(X_X,D) = \mathscr{D}(X)$, and we obtain

$$\begin{aligned}
\langle N_X\delta_X(i,\cdot),V(i,\cdot)\rangle &= \langle N_X N\delta_Y(i,\cdot),V(i,\cdot)\rangle \\
&= \langle N_X N(i,\cdot),\delta_Y V(i,\cdot)\rangle \\
&= \langle N_Y(i,\cdot),\delta_Y V(i,\cdot)\rangle + \langle (N_X N - N_Y)(i,\cdot),\delta_Y V(i,\cdot)\rangle \\
&\leq \langle N_Y(i,\cdot),\delta_Y V(i,\cdot)\rangle + ||(N_X N - N_Y)(i,\cdot)||\ ||\delta_Y V(i,\cdot)|| \\
&\leq \langle N_Y(i,\cdot),\delta_Y V(i,\cdot)\rangle + \varepsilon(i)||V(i,\cdot)|| \\
&\leq \langle N_Y\delta_Y(i,\cdot),V(i,\cdot)\rangle + \varepsilon(i)||V|| \quad \text{for all } i \in \Omega_I,
\end{aligned}$$

i.e., $X >_{\varepsilon}^{\bar{D}} Y$. □

The preceding theorem shows that Blackwell informativity at level ε always implies informativity at level ε. Notice that the notion of Blackwell informativity at level ε has the advantage of fitting perfectly into the formalism of stochastic kernels with which we began.

We are now going to discuss a few operations on the set $\mathscr{E}(\bar{D})$.

Definition 19.18. Let $\bar{D} = (I,D,\mathscr{V})$ be a decision problem and $X,Y \in \mathscr{E}(\bar{D})$. Then the experiment

$$X \otimes Y := (X_X \otimes X_Y,\ N_X \otimes N_Y)$$

is called the *product* (experiment) of X and Y.

Theorem 19.19. Let $\bar{D} = (I,D,\mathscr{V})$ be a decision problem. If $X_1, Y_1, X_2, Y_2 \in \mathscr{E}(\bar{D})$ and $\varepsilon_1,\varepsilon_2\colon \Omega_I \to \mathbb{R}_+$ are tolerance functions such that $X_j >_{\varepsilon_j}^{B} Y_j$ holds for $j = 1,2$, then we have

$$X_1 \otimes X_2 >_{\varepsilon_1+\varepsilon_2}^{B} Y_1 \otimes Y_2.$$

Proof: For $j = 1,2$ let $N_j \in \mathrm{Stoch}(X_{X_j},X_{Y_j})$ satisfy $||N_{X_j}N_j - N_{Y_j}|| \leq \varepsilon_j$. We define $N := N_1 \otimes N_2 \in \mathrm{Stoch}(X_{X_1\otimes X_2},X_{Y_1\otimes Y_2})$. Then we obtain for all $i \in \Omega_I$ the inequalities

$$\begin{aligned}
&||N_{X_1\otimes X_2}N(i,\cdot) - N_{Y_1\otimes Y_2}(i,\cdot)|| \\
&\quad = ||[N_{X_1}N_1(i,\cdot)] \otimes [N_{X_2}N_2(i,\cdot)] - N_{Y_1}(i,\cdot) \otimes N_{Y_2}(i,\cdot)|| \\
&\quad \leq ||N_{X_1}N_1(i,\cdot) - N_{Y_1}(i,\cdot)|| + ||N_{X_2}N_2(i,\cdot) - N_{Y_2}(i,\cdot)|| \\
&\quad \leq \varepsilon_1(i) + \varepsilon_2(i),
\end{aligned}$$

where the second last inequality follows from the general inequality

$$||\mu_1 \otimes \mu_2 - \nu_1 \otimes \nu_2|| \leq ||\mu_1 - \nu_1|| + ||\mu_2 - \nu_2||$$

valid for all measures $\mu_1,\nu_1 \in \mathscr{M}^1(X_{Y_1})$ and $\mu_2,\nu_2 \in \mathscr{M}^1(X_{Y_2})$, and the last one by hypothesis. □

Definition 19.20. Let $\overline{D} = (I,D,\mathscr{V})$ be a decision problem, $X \in \mathscr{E}(\overline{D})$ an experiment corresponding to D and M any measurable space.

(a) An *M-decomposition of* X is a pair (N,μ), where N is a kernel in $\mathrm{Stoch}(M \otimes I, X_X)$ and μ is a probability measure in $\mathscr{M}^1(M)$ such that for all $i \in \Omega_I$ and $A \in \mathbf{A}_X$ one has

$$N_X(i,A) = \int N((\cdot,i),A)\,d\mu.$$

(b) Conversely, suppose that to every $\omega \in \Omega_M$ an experiment $X(\omega) \in \mathscr{E}(\overline{D})$ is given with a fixed measurable space $X := X_{X(\omega)}$ for all $\omega \in \Omega_M$. Then the family $\Gamma := (X(\omega))_{\omega\in\Omega_M}$ is called a *measurable field of experiments* when the function $N_\Gamma: (\Omega_M \times \Omega_I) \times \mathbf{A}_X \to [0,1]$ defined by

$$N_\Gamma((\omega,i),A) := N_{X(\omega)}(i,A) \quad \text{for all} \quad (\omega,i) \in \Omega_M \times \Omega_I,\ A \in \mathbf{A}_X$$

is a kernel in $\mathrm{Stoch}(M \otimes I, X_X)$.

In addition if $\mu \in \mathscr{M}^1(M)$ is given such that the pair (N_Γ,μ) is an M-decomposition of X, then X is called *the mixture of the family* Γ with respect to μ, and we write

$$X = \int^{\oplus} X(\omega)\ \mu(d\omega).$$

Remark 19.21. For every $\mu \in \mathscr{M}^1(M)$ we can define a kernel $K^\mu \in \mathrm{Stoch}(I, M \otimes I)$ by $K^\mu(i,\tilde{A}) := \mu(\tilde{A}_i)$ for all $i \in \Omega_I$ and $\tilde{A} \in \mathbf{A}_M \otimes \mathbf{A}_I$, where $\tilde{A}_i$ denotes the i-section of $\tilde{A}$. The condition that the pair (N,μ) be an M-decomposition of X is equivalent to the condition that the kernels $K^\mu \in \mathrm{Stoch}(I, M \otimes I)$ and $N \in \mathrm{Stoch}(M \otimes I, X_X)$ satisfy the relation $K^\mu N = N_X \in \mathrm{Stoch}(I, X_X)$.

§20. REPRESENTATION OF POSITIVE LINEAR OPERATORS BY STOCHASTIC KERNELS

The following discussion is of auxiliary interest to further studies of Blackwell informativity and its relationship to informativity between two experiments.

Let $(\Omega,\mathbf{A},P)$ be a probability space, $(\Omega',\mathbf{A}')$ the Borel space of a compact metrizable space Ω' and K a Stonian vector lattice with

$\mathscr{C}(\Omega') \subset K \subset \mathfrak{M}^b(\Omega',\mathfrak{A}')$. Given a substochastic kernel N from $(\Omega,\mathfrak{A})$ to $(\Omega',\mathfrak{A}')$ the definition

$$T_N(f) := [N(\cdot,f)]_P \quad \text{for all} \quad f \in K$$

yields a positive linear operator T_N from K into $L^\infty(\Omega,\mathfrak{A},P)$ satisfying $||T_N|| \leq 1$ and $N(\cdot,f) \in T_N(f)$ for all $f \in K$.

In the sequel we are interested in a converse of this implication.

Theorem 20.1. Let $(\Omega,\mathfrak{A},P)$ be a probability space, $(\Omega',\mathfrak{A}')$ the Borel space of a compact metrizable space Ω' and K a Stonian vector lattice with $\mathscr{C}(\Omega') \subset K \subset \mathfrak{M}^b(\Omega',\mathfrak{A}')$. For any positive linear operator T from K into $L^\infty(\Omega,\mathfrak{A},P)$ with $||T|| \leq 1$ (or $1_\Omega \in T(1_{\Omega'})$) the following statements are equivalent:

(i) There exists a substochastic (or stochastic) kernel N from $(\Omega,\mathfrak{A})$ to $(\Omega',\mathfrak{A}')$ with the properties

(a) $N(\cdot,f) \in T(f)$ for all $f \in K$,

(b) $N(\cdot,f) = 0$ for all $f \in \ker T$.

(ii) There exists a lifting L (or a lifting L with $L \circ T(1_{\Omega'}) = 1_\Omega$) on $T(K)$ such that $L \circ T$ is σ-continuous.

(iii) There exists a mapping $\mu: \Omega \to \mathscr{M}^{(1)}(\Omega',\mathfrak{A}')$ (or a mapping $\mu: \Omega \to \mathscr{M}^1(\Omega',\mathfrak{A}')$) of the form $\omega \to \mu_\omega$ with the properties

(a) For every $f \in K$ the mapping $\omega \to \int f\, d\mu_\omega$ is an element of $\mathfrak{M}^b(\Omega,\mathfrak{A})$.

(b) For every $f \in K$ the mapping $\omega \to \int f\, d\mu_\omega$ is a representative of $T(f)$.

(c) For every $f \in \ker T$ and $\omega \in \Omega$ one has $\int f\, d\mu_\omega = 0$.

If one of the equivalent statements holds, then T is necessarily σ-continuous.

Proof: 1. (i) $\Rightarrow$ (iii). Let N be as in (i). The mapping $\mu: \Omega \to \mathscr{M}^{(1)}(\Omega',\mathfrak{A}')$ of the form $\omega \to \mu_\omega$ defined by $\mu_\omega(A') := N(\omega,A')$ for all $\omega \in \Omega$ and $A' \in \mathfrak{A}'$ possesses the desired properties.

2. (iii) $\Rightarrow$ (ii). Let $\omega \to \mu_\omega$ be the mapping μ assumed to be given in (iii). For every $F \in T(K)$ we choose f from $T^{-1}(F)$. By assumptions (a) to (c) of (iii) the definition

$$[L(F)](\omega) := \int f\, d\mu_\omega$$

for all $F \in T(K)$ and $\omega \in \Omega$ yields a mapping from $T(K)$ into $\mathfrak{M}^b(\Omega,\mathfrak{A})$ with $L(F) \in F$ and therefore a lifting on $T(K)$. $L \circ T$ is σ-continuous, since by definition of L we have

$$L \circ T(f)(\omega) = [L(T(f))](\omega) = \int f\, d\mu_\omega$$

for all $f \in K$ and $\omega \in \Omega$.

3. (ii) ⇒ (i). Let L be the lifting on $T(K)$ assumed in (ii). For every $\omega \in \Omega$ the mapping $f \to [(L \circ T)(f)](\omega)$ from K into $\mathbb{R}$ is an abstract integral on K. Hence for every $\omega \in \Omega$ the Daniell-Stone theorem implies the existence of $\mu_\omega \in \mathscr{M}_+(\Omega',\mathbf{A}')$ satisfying

(α) $\int f\, d\mu_\omega = [(L \circ T)(f)](\omega)$ for all $f \in K$.

(β) $\mu_\omega(A') = \inf\left\{\int f\, d\mu_\omega : f \in K, f \geq 1_{A'}\right\}$ for all $A' \in \mathbf{A}'$.

We now define $N: \Omega \times \mathbf{A}' \to \mathbb{R}$ by $N(\omega,A') := \mu_\omega(A')$ for all $\omega \in \Omega$, $A' \in \mathbf{A}'$ and observe that N is a substochastic (or stochastic) kernel from $(\Omega,\mathbf{A})$ to $(\Omega',\mathbf{A}')$ with the properties (a) and (b) of (i). □

Corollary 20.2. If $K := \mathscr{C}(\Omega')$, then there exists a substochastic (or stochastic) kernel N from $(\Omega,\mathbf{A})$ to $(\Omega',\mathbf{A}')$ with

(a) $N(\cdot,f) \in T(f)$ for all $f \in \mathscr{C}(\Omega')$.

(b) $N(\cdot,f) = 0$ for all $f \in \ker T$.

Proof: By assumption on Ω' the subset $T(\mathscr{C}(\Omega'))$ of $L^\infty(\Omega,\mathbf{A},P)$ is separable. Moreover, Dini's theorem implies the σ-continuity of T. Since on a separable linear subspace of $L^\infty(\Omega,\mathbf{A},P)$ there exists always a lifting, the theorem yields the assertion. □

Corollary 20.3. If $K := \mathfrak{M}^b(\Omega',\mathbf{A}')$ and T is σ-continuous, then there exists a substochastic (or stochastic) kernel N from $(\Omega,\mathbf{A})$ to $(\Omega',\mathbf{A}')$ with the properties

(a) $N(\cdot,f) \in T(f)$ for all $f \in \mathfrak{M}^b(\Omega',\mathbf{A}')$.

(b) $N(\cdot,f) = 0$ for all $f \in (\ker T) \cap \mathscr{C}(\Omega')$.

Proof: By Corollary 20.2 there exists a substochastic (or stochastic) kernel N from $(\Omega,\mathbf{A})$ to $(\Omega',\mathbf{A}')$ with

(a') $N(\cdot,f) \in T(f)$ for all $f \in \mathscr{C}(\Omega')$ and

(b') $N(\cdot,f) = 0$ for all $f \in (\ker T) \cap \mathscr{C}(\Omega')$.

But since T is σ-continuous, we also obtain $N(\cdot,f) \in T(f)$ for all $f \in \mathfrak{M}^b(\Omega',\mathbf{A}')$, i.e., (a). □

Theorem 20.4. Let $(\Omega,\mathbf{A},\mu)$ be a finite measure space, $(\Omega',\mathbf{A}')$ the Borel space of a compact metrizable space Ω' and $(N_n)_{n\geq 1}$ a sequence in $\mathrm{Stoch}((\Omega,\mathbf{A}),(\Omega',\mathbf{A}'))$. Then there exist a stochastic kernel N in $\mathrm{Stoch}((\Omega,\mathbf{A}),(\Omega',\mathbf{A}'))$ and a subsequence $(N_{n_k})_{k\geq 1}$ of $(N_n)_{n\geq 1}$ such that for any measure $\nu \in \mathscr{M}_+^b(\Omega,\mathbf{A})$ with $\nu \ll \mu$ the sequence $(N_{n_k}(\nu))_{k\geq 1}$ converges weakly to $N(\nu)$.

The proof of the theorem will be preceded by two lemmas.

Lemma 20.5. The set

$$\{g \in L^{\infty}(\Omega,\mathbf{A},\mu): ||g|| \leq c\}$$

considered as a subset of $L^1(\Omega,\mathbf{A},\mu)$ is weakly sequentially compact with respect to the topology $\sigma(L^1,L^{\infty})$.

Proof: It is well-known that the set $\{g \in L^{\infty}: ||g|| \leq c\}$ is $\sigma(L^{\infty},L^1)$-compact, and that the injection from L^{∞} into L^1 is $\sigma(L^{\infty},L^1)$-$\sigma(L^1,L^{\infty})$-continuous. The theorem of Eberlein yields the assertion. □

Lemma 20.6. Let $(F_n)_{n\geq 1}$ be a sequence in $L^{\infty}(\Omega,\mathbf{A},\mu)$, bounded and $\sigma(L^1,L^{\infty})$-weakly convergent to $F \in L^{\infty}(\Omega,\mathbf{A},\mu)$. Then for any $\nu \in \mathcal{M}^b_+(\Omega,\mathbf{A})$ with $\nu << \mu$ the limit $\lim\limits_{n\to\infty} \int F_n d\nu$ exists, and

$$\lim_{n\to\infty} \int F_n \, d\nu = \int F \, d\nu.$$

Proof: Let $\nu \in \mathcal{M}^b_+(\Omega,\mathbf{A})$ with $\nu << \mu$. By the Radon-Nikodym theorem there exists an $h \in L^1_+$ with $\nu = h \cdot \mu$. For h there is an isotone sequence $(h_k)_{k\geq 1}$ of step functions $h_k \in L^{\infty}$ with $h = \sup\limits_{k>1} h_k$. By Legesgue's dominated convergence theorem $\lim\limits_{n\to\infty} \int|h-h_k|d\mu = \bar{0}$. The boundedness hypothesis yields the existence of $\gamma \in \mathbb{R}_+$ with $||F_n|| \leq \gamma$ for all $n \geq 1$, hence

$$\left|\int (F_n h_k - F_n h) d\mu\right| \leq \gamma \int |h_k - h| d\mu$$

for all $n \geq 1$ and $k \geq 1$, thus we obtain the existence of $\lim\limits_{k\to\infty} \int F_n h_k \, d\mu$ for all $n \geq 1$, where the convergence takes place uniformly in $n \geq 1$. This implies the existence of

$$\begin{aligned}\int F \, d\nu &= \lim_{k\to\infty} \int Fh_k \, d\mu = \lim_{k\to\infty} \lim_{n\to\infty} \int F_n h_k \, d\mu = \lim_{n\to\infty} \lim_{k\to\infty} \int F_n h_k \, d\mu \\ &= \lim_{n\to\infty} \int F_n (\lim_{k\to\infty} h_k) d\mu = \lim_{n\to\infty} \int F_n h \, d\mu = \lim_{n\to\infty} \int F_n \, d\nu,\end{aligned}$$

which shows the assertion. □

Proof of Theorem 20.4: By assumption $\mathscr{C}(\Omega')$ is separable. Hence there exists a countable dense subset H of $\mathscr{C}(\Omega')$ such that for $f,g \in H$ and $r \in \mathbb{Q}$ the functions $r\, 1_{\Omega'}$, $|f|$, $f + g$ and rf are elements of H. For every $n \geq 1$ we define positive linear operators T_n from $\mathscr{C}(\Omega')$ into L^{∞} satisfying $1_{\Omega} \in T_n(1_{\Omega'})$ by

$$T_n(f) := [N_n(\cdot,f)]_P,$$

whenever $f \in \mathscr{C}(\Omega')$. For every $f \in H$ we have

$$\{T_n(f): n \geq 1\} \subset A_f := \{g \in L^\infty: ||g|| \leq ||f||\}.$$

By Lemma 20.5 the set A_f is weakly sequentially compact in $L^1(f \in H)$. Since H is countable, a diagonal sequence argument shows the existence of a subsequence $(T_{n_k})_{k\geq 1}$ of $(T_n)_{n\geq 1}$ such that for all $f \in H$ the sequence $(T_{n_k}(f))_{k\geq 1}$ converges weakly to $T(f) \in L^\infty$, since A_f is $\sigma(L^1,L^\infty)$-closed as a weakly compact subset of L^1. As a consequence we obtain a mapping $T: H \to L^\infty$ possessing the following properties:

(a) T is positive.

(b) For $f,g \in H$ and $r,s \in \mathbb{Q}$ one gets $T(rf + sg) = rT(f) + sT(g)$.

(c) $1_\Omega \in T(1_{\Omega'})$.

(d) T is continuous on H, i.e., for every sequence $(f_n)_{n\geq 1}$ in H converging in $\mathscr{C}(\Omega')$ the sequence $(T(f_n))_{n\geq 1}$ converges in L^∞.

Thus T can be extended to a positive linear operator from $\mathscr{C}(\Omega')$ into L^∞. Using (c) we conclude from Corollary 20.2 that there exists a kernel $N \in \mathrm{Stoch}((\Omega,\mathbf{A}), (\Omega',\mathbf{A}'))$ satisfying $N(\cdot,f) \in T(f)$ for all $f \in \mathscr{C}(\Omega')$. Moreover,

$$\lim_{k\to\infty} \int N_{n_k}(\cdot,f)\, F\, d\mu = \int N(\cdot,f)\, F\, d\mu$$

holds for all $f \in H$, $F \in L^\infty$. By Lemma 20.6 we therefore get

$$\lim_{k\to\infty} \int f\, dN_{n_k}(\nu) = \lim_{k\to\infty} \int N_{n_k}(\cdot,f)d\nu$$
$$= \int N(\cdot,f)d\nu = \int f\, dN(\nu).$$

Since H is dense in $\mathscr{C}(\Omega')$, the theorem is proved. □

§21. THE STOCHASTIC KERNEL CRITERION

From now on we consider decision problems $\underline{D} := (I,D,\mathscr{V})$ with index space $I := (\Omega_I,\mathbf{A}_I)$, decision space $D := (\Omega_D,\mathbf{A}_D)$ and the set $\mathscr{V}$ of all bounded separately measurable functions on $\Omega_I \times \Omega_D$ as well as experiments $X := (\Omega,\mathbf{A},(P_i)_{i\in\Omega_I})$ corresponding to $\underline{D}$ with sample space $(\Omega,\mathbf{A})$

and parametrized family $(P_i)_{i\in\Omega_I}$ of measures in $\mathscr{M}^1(\Omega,\mathbb{A})$ such that for every $A \in \mathbb{A}$ the mapping $i \to P_i(A)$ from Ω_I into $\mathbb{R}$ is $\mathbb{A}_I$-$\mathbb{B}$-measurable. As usual the class of all such experiments corresponding to $\bar{D}$ will be abbreviated by $\mathscr{E}(\bar{D})$. Given $X \in \mathscr{E}(\bar{D})$ the set of all decision functions corresponding to X will be denoted by $\mathscr{D}(X) := \mathscr{D}(X,D)$. We make the convention that any measurable space $(C,\mathbb{C})$ with a finite set C necessarily has the σ-algebra $\mathbb{C} := \mathfrak{P}(C)$.

Theorem 21.1. Let $\bar{D} := (I,D,\mathscr{V})$ be a decision problem with $\Omega_I := \{1,\ldots,n\}$ and $\Omega_D := \{1,\ldots,k\}$, and let $X := (\Omega,\mathbb{A},(P_i)_{i\in\Omega_I})$ and $Y := (\Omega_1,\mathbb{A}_1,(Q_i)_{i\in\Omega_I})$ be two experiments in $\mathscr{E}(\bar{D})$. The following statement are equivalent:

(i) $X \overset{\bar{D}}{>-} Y$.

(ii) For every $V \in \mathscr{V}$ and $\delta_Y \in \mathscr{D}(Y)$ there exists $\delta_X \in \mathscr{D}(X)$ satisfying

$$\sum_{i=1}^{n} R^V_{\delta_X}(i) \le \sum_{i=1}^{n} R^V_{\delta_Y}(i)$$

(iii) For every $\delta_Y \in \mathscr{D}(Y)$ there exists $\delta_X \in \mathscr{D}(X)$ with $\delta_X(P_i) = \delta_Y(Q_i)$ for all $i \in \Omega_I$.

Proof: Since the implications (iii) ⇒ (i) ⇒ (ii) are obvious, we are left to show (ii) ⇒ (iii). Let $P := \frac{1}{n}\sum_{i=1}^{n} P_i$ and let $\mathscr{R}$ be the set of all positive linear operators from $\mathbb{R}^k$ into $L^\infty(\Omega,\mathbb{A},P)$ which satisfy $T(\mathbb{1}) = [1_\Omega]_P$, where $\mathbb{1}$ denotes the vector $(1,\ldots,1) \in \mathbb{R}^k$. Plainly

(1) $\mathscr{V}$ and $\mathscr{R}$ are convex sets.

For every $a := (a_1,\ldots,a_k) \in \mathbb{R}^k$ and $f \in L^1(\Omega,\mathbb{A},P)$ we define a function $F_{a,f}: \mathscr{R} \to \mathbb{R}$ by

$$F_{a,f}(T) := \int T(a)f\, dP \quad \text{for all} \quad T \in \mathscr{R}.$$

Let $\mathscr{T}$ be the topology on $\mathscr{R}$ induced by the functions $F_{a,f}(a \in \mathbb{R}^k, f \in L^1(\Omega,\mathbb{A},P))$. Then

(2) $(\mathscr{R},\mathscr{T})$ is a compact space.

In fact, $\mathscr{R}$ is a closed subset of the set

$$\prod_{a\in\mathbb{R}^k}\{g \in L^\infty(\Omega,\mathbb{A},P): \|g\|_\infty \le \|a\|_\infty\},$$

which furnished with the product topology of the $\sigma(L^\infty,L^1)$-compact factors is compact.

Now let $\delta_Y \in \mathscr{D}(Y)$ be fixed. We define a mapping $\Phi: \mathscr{V}\times\mathscr{R} \to \mathbb{R}$ by

$$\Phi(V,T) := \sum_{i=1}^{n}\left(\int T(v(i))dP_i - \int V(i,\cdot)d\delta_Y(Q_i)\right)$$

for all $V \in \mathscr{V}$ and $T \in \mathscr{R}$, where $v(i) := (V(i,1),\ldots,V(i,k)) \in \mathbb{R}^k$ for $i = 1,\ldots,n$. The function Φ enjoys the following properties:

(3) For any $T \in \mathscr{R}$ the mapping $\Phi(\cdot,T): \mathscr{V} \to \mathbb{R}$ is affine-linear.

(4) For every $V \in \mathscr{V}$ the mapping $\Phi(V,\cdot): \mathscr{R} \to \mathbb{R}$ is affine-linear and continuous on $(\mathscr{R},\mathscr{T})$.

(5) $\sup_{V\in\mathscr{V}} \inf_{T\in\mathscr{R}} \Phi(V,T) \leq 0$.

Since properties (3) and (4) are obvious, it suffices to verify (5): By assumption for $V_0 \in \mathscr{V}$ there exists $\delta_X \in \mathscr{D}(X)$ with

$$\sum_{i=1}^{n}\left(R_{\delta_X}^{V_0}(i) - R_{\delta_Y}^{V_0}(i)\right) \leq 0.$$

For every $a := (a_1,\ldots,a_k) \in \mathbb{R}^k$ we define

$$T_{V_0}(a) := \left[\sum_{j=1}^{k} a_j\delta_X(\cdot,\{j\})\right]_P$$

and obtain an element T_{V_0} of $\mathscr{R}$ satisfying $\Phi(V_0,T_{V_0}) \leq 0$. But this yields (5).

Now consider the 2-person zero sum game $\Gamma = (A,B,M)$ with $A := \mathscr{V}$, $B := \mathscr{R}$ and $M := \Phi$. We have shown that Γ is concave-convex, and the topology $\mathscr{T}$ on $\mathscr{R}$ has been constructed in such a way that for every $V \in \mathscr{V}$ the mapping $T \to \Phi(V,T)$ is continuous on $\mathscr{R}$. Moreover, we have

$$\underline{V} := \sup_{V\in\mathscr{V}} \inf_{T\in\mathscr{R}} \Phi(V,T) \leq 0,$$

thus we are left to show that $\underline{V} = \overline{V}$ and that P_{II} has a minimax strategy. This follows from Theorem 2.7, which provides us with the existence of a $T_0 \in \mathscr{R}$ satisfying

$$\sup_{V\in\mathscr{V}} \Phi(V,T_0) = \overline{V} = \underline{V} \leq 0.$$

We now realize that $(\Omega_D,\mathfrak{P}(\Omega_D))$ is a compact metrizable space and that $\mathscr{C}(\Omega_D,\mathfrak{P}(\Omega_D)) = \mathbb{R}^k = \mathfrak{M}^b(\Omega_D,\mathfrak{A}_D)$ holds. Under these assumptions Corollary 20.2 yields the existence of $\delta_X \in \mathrm{Stoch}((\Omega,\mathfrak{A}),(\Omega_D,\mathfrak{A}_D))$ with the property $\delta_X(\cdot,f) \in T_0(f)$ for all $f \in \mathfrak{M}^b(\Omega_D,\mathfrak{A}_D)$.

The above inequality implies

$$\sum_{i=1}^{n}\left(\int V(i,\cdot)d\delta_X(P_i) - \int V(i,\cdot)d\delta_Y(Q_i)\right) = \Phi(V,T_0) \leq 0$$

for all $V \in \mathscr{V}$. Given $f \in \mathfrak{M}^b(\Omega_D,\mathbf{A}_D)$ and $j \in \{1,\ldots,n\}$ the function V_j^f on $\Omega_I \times \Omega_D$ defined by $V_j^f(i,d) := \delta_{ij}\cdot f(d)$ for all $(i,d) \in \Omega_I \times \Omega_D$ lies in $\mathscr{V}$. Hence the above inequality implies

$$\int f\, d\delta_X(P_i) - \int f\, d\delta_Y(Q_i) \leq 0$$

for all $i \in \Omega_I$, $f \in \mathfrak{M}^b(\Omega_D,\mathbf{A}_D)$, which is the desired statement (iii). □

Theorem 21.2. Let $\underline{D} = (I,D,\mathscr{V})$ be a decision problem with $\Omega_D := \{1,\ldots,k\}$ and $X,Y \in \mathscr{E}(\underline{D})$ as in Theorem 21.1. We further assume that X is μ-dominated by a measure $\mu \in \mathscr{M}_+^\sigma(\Omega,\mathbf{A})$. The following statements are equivalent:

(i) $X \overset{\underline{D}}{>} Y$.

(ii) To every $\delta_Y \in \mathscr{D}(Y)$ there exists $\delta_X \in \mathscr{D}(X)$ satisfying $\delta_X(P_i) = \delta_Y(Q_i)$ for all $i \in \Omega_I$.

Proof: Again it suffices to show the implication (i) ⇒ (ii). First we observe that the system $\mathfrak{F}$ of finite subsets of Ω_I is filtering upwards with respect to inclusion. For every $A \in \mathfrak{F}$ we put $\underline{D}_A := (I_A, D, \mathscr{V}_A)$, where $I_A := (A,\mathbf{A}_I \cap A)$, $\mathscr{V}_A$ correspondingly, and $X_A := (\Omega,\mathbf{A},(P_i)_{i\in A})$, $Y_A := (\Omega_1,\mathbf{A}_1,(Q_i)_{i\in A})$. Then $X_A, Y_A \in \mathscr{E}(\underline{D}_A)$. Theorem 21.1 associates with any given kernel $\delta_Y \in \mathscr{D}(Y)$ and $A \in \mathfrak{F}$ a kernel $\delta_A \in \mathscr{D}(X_A) = \mathscr{D}(X)$ such that $\delta_A(P_i) = \delta_Y(Q_i)$ holds for all $i \in A$. With δ_A we define by

$$T_A(f) := [\delta_A(\cdot,f)]_\mu \quad \text{for all } f \in \mathscr{C}(\Omega_D)$$

a positive linear operator T_A from $\mathscr{C}(\Omega_D)$ into $L^\infty(\Omega,\mathbf{A},\mu)$ satisfying $1_\Omega \in T_A(1_{\Omega_D})$. But since the unit ball of $L^\infty(\Omega,\mathbf{A},\mu)$ is $\sigma(L^\infty,L^1)$-compact, Tychonov's theorem yields the existence of a subnet $(T_A)_{A\in\mathfrak{F}'}$ of $(T_A)_{A\in\mathfrak{F}}$ such that $(T_A(f))_{A\in\mathfrak{F}'}$ converges to $T(f) \in L^\infty(\Omega,\mathbf{A},\mu)$ with respect to $\sigma(L^\infty,L^1)$ for all $f \in \mathscr{C}(\Omega_D)$. Hence we obtain a positive linear operator T from $\mathscr{C}(\Omega_D)$ into $L^\infty(\Omega,\mathbf{A},\mu)$ with $1_\Omega \in T(1_{\Omega_D})$. The assumptions on $(\Omega_D,\mathfrak{P}(\Omega_D))$ now enable us to apply Corollary 20.2 and so there exists a kernel $\delta_X \in \mathrm{Stoch}((\Omega,\mathbf{A}),(\Omega_D,\mathbf{A}_D))$ such that $\delta_X(\cdot,f) \in T(f)$ holds for all $f \in \mathfrak{M}^b(\Omega_D,\mathbf{A}_D)$. By hypothesis for every $i \in \Omega_I$ there exists a $g_i \in L^1(\Omega,\mathbf{A},\mu)$ such that $P_i = g_i\cdot\mu$. Thus for every $f \in \mathfrak{M}^b(\Omega_D,\mathbf{A}_D)$ we obtain

$$\delta_X(P_i)(f) = \langle T(f), g_i\rangle = \lim_{A\in\mathfrak{F}'} \langle T_A(f), g_i\rangle$$

$$= \lim_{A\in\mathfrak{F}'} \delta_A(P_i)(f) = \delta_Y(Q_i)(f),$$

and $\delta_X(P_i) = \delta_Y(Q_i)$ for all $i \in \Omega_I$ has been proved. □

Theorem 21.3. Let $\underline{D}: = (I,D,\mathscr{V})$ be a decision problem and X,Y experiments in $\mathscr{E}(\underline{D})$ with the above data. We further assume that $(\Omega_D,\mathfrak{A}_D)$ is the Borel space of a compact metrizable space $(\Omega_D,\mathscr{T}_D)$ and that X is μ-dominated by a measure $\mu \in \mathscr{M}_+^\sigma(\Omega,\mathfrak{A})$. The following statements are equivalent:

(i) $X \overset{\bar{D}}{>^-} Y$.

(ii) For every $\delta_Y \in \mathscr{D}(Y)$ there exists a $\delta_X \in \mathscr{D}(X)$ satisfying $\delta_X(P_i) = \delta_Y(Q_i)$ for all $i \in \Omega_I$.

The proof of the theorem will be preceded by a

Lemma 21.4. Let $\underline{D}: = (I,D,\mathscr{V})$ be a decision problem and $X,Y \in \mathscr{E}(\underline{D})$. For $\Omega_{\tilde{D}} \in \mathfrak{A}_D$ we put $\mathfrak{A}_{\tilde{D}}: = \Omega_{\tilde{D}} \cap \mathfrak{A}_D, \tilde{D}: = (\Omega_{\tilde{D}},\mathfrak{A}_{\tilde{D}})$ and $\underline{\tilde{D}}: = (I,\tilde{D},\tilde{\mathscr{V}})$, where $\tilde{\mathscr{V}}$ is defined correspondingly. Then $\mathscr{E}(\underline{D}) = \mathscr{E}(\underline{\tilde{D}})$, and $X \overset{\bar{D}}{>^-} Y$ implies $X \overset{\bar{\tilde{D}}}{>^-} Y$.

Proof: The first statement is clear. Now let $\tilde{V} \in \tilde{\mathscr{V}}$ and $\tilde{\delta}_Y \in \mathscr{D}(Y,\tilde{D})$. Define $V \in \mathscr{V}$ by

$$V(i,d): = \begin{cases} \tilde{V}(i,d) & \text{if } (i,d) \in \Omega_I \times \Omega_{\tilde{D}} \\ ||\tilde{V}|| & \text{if } (i,d) \in \Omega_I \times (\Omega_D \setminus \Omega_{\tilde{D}}) \end{cases}$$

and $\delta_Y \in \mathscr{D}(Y,D)$ by

$$\delta_Y(\omega_1,A): = \tilde{\delta}_Y(\omega_1,A \cap \Omega_{\tilde{D}}) \quad \text{for all } \omega_1 \in \Omega_1 \text{ and } A \in \mathfrak{A}_D.$$

Clearly $\int V(i,\cdot)d\delta_Y(Q_i) = \int \tilde{V}(i,\cdot)d\tilde{\delta}_Y(Q_i)$ for all $i \in \Omega_I$. From the hypothesis we infer the existence of $\delta_X \in \mathscr{D}(X)$ with

$$\int V(i,\cdot)d\delta_X(P_i) \leq \int V(i,\cdot)d\delta_Y(Q_i) \quad \text{for all } i \in \Omega_I.$$

Let $d \in \Omega_{\tilde{D}}$. For $\tilde{\delta}_X \in \mathscr{D}(X,\tilde{D})$ defined by

$$\tilde{\delta}_X(\omega,B): = \delta_X(\omega,B) + \delta_X(\omega,\Omega_D \setminus \Omega_{\tilde{D}})\cdot\varepsilon_d$$

for all $\omega \in \Omega$, $B \in \mathfrak{A}_{\tilde{D}}$ we obtain

$$\int \tilde{V}(i,\cdot)d\tilde{\delta}_X(P_i) = \tilde{V}(i,d)\delta_X(P_i)(\Omega_D \setminus \Omega_{\tilde{D}}) + \int \tilde{V}(i,\cdot)d\ \mathrm{Res}_{\Omega_{\tilde{D}}}(\delta_X(P_i))$$

$$\leq ||\tilde{V}|| \int 1_{\Omega_D \setminus \Omega_{\tilde{D}}} d\delta_X(P_i) + \int V(i,\cdot)1_{\Omega_{\tilde{D}}} d\delta_X(P_i)$$

$$= \int V(i,\cdot)d\delta_X(P_i) \quad \text{whenever} \quad i \in \Omega_I.$$

This inequality together with the above inequality yields the assertion. □

Proof of Theorem 21.3: We are left with the proof of the implication (i) ⇒ (ii). Without loss of generality we assume μ to be finite. By hypothesis there exists a countable dense subset $\{t_k : k \geq 1\}$ of Ω_D. For every $k \geq 1$ we introduce $D_k := (\Omega_k, \mathfrak{A}_k)$ with $\Omega_k := \{t_1, \ldots, t_k\}$ and $\mathfrak{A}_k := \Omega_k \cap \mathfrak{A}_D = \mathfrak{P}(\Omega_k)$ as well as $\overline{D}_k := (I, D_k, \mathscr{V}_k)$ with $\mathscr{V}_k$ defined correspondingly. Clearly $\mathscr{B}(\overline{D}) = \mathscr{B}(\overline{D}_k)$, and from the Lemma we deduce $X \overset{\overline{D}_k}{>} Y$ for all $k \geq 1$. Let ρ be the metric inducing the topology $\mathscr{T}_D$ in Ω_D. For every $t \in \Omega_D$ and $k \geq 1$ we define the natural number

$$j(t,k) := \max\{i \in \{1,\ldots,k\} : \rho(t,t_i) = \min_{1 \leq \ell \leq k} \rho(t,t_\ell)\}$$

between 1 and k. For every $k \geq 1$ the mapping $f_k : \omega \to t_{j(\omega,k)}$ from Ω_D into Ω_k is $\mathfrak{A}_D$ - $\mathfrak{A}_k$-measurable. We take $\delta_Y \in \mathscr{D}(Y,D)$. Then for any $k \geq 1$ we have a kernel δ_Y^k in $\mathscr{D}(Y,D_k)$ defined by

$$\delta_Y^k(\omega_1,A) := \delta_Y(\omega_1, f_k^{-1}(A))$$

for all $\omega_1 \in \Omega_1$, $A \in \mathfrak{A}_k$. By Theorem 21.2 there exists for every $k \geq 1$ a kernel $\delta_X^k \in \mathscr{D}(X,D_k)$ with $\delta_X^k(P_i) = \delta_Y^k(Q_i)$ for all $i \in \Omega_I$. Consequently for $k \geq 1$ the function

$$\delta_k := \sum_{j=1}^{k} \varepsilon_{t_j} \delta_X^k(\cdot,\{t_j\}) \quad \text{on} \quad \mathscr{C}(\Omega_D)$$

defines a stochastic kernel $\delta_k \in \mathrm{Stoch}((\Omega,\mathfrak{A}),(\Omega_D,\mathfrak{A}_D))$. But now Theorem 20.4 applies, and there exist a kernel $\delta_X \in \mathscr{D}(X,D)$ and a subsequence $(\delta_{k_\ell})_{\ell \geq 1}$ of $(\delta_k)_{k \geq 1}$ such that

$$\lim_{\ell \to \infty} \int f\, d\delta_{k_\ell}(P_i) = \int f\, d\delta_X(P_i)$$

holds for all $f \in \mathscr{C}(\Omega_D)$ and $i \in \Omega_I$. For such f and i we now obtain

$$\left|\int f d\delta_X(P_i) - \int f d\delta_Y(Q_i)\right|$$

$$\leq \left|\int f d\delta_X(P_i) - \sum_{j=1}^{k_\ell} f(t_j)(\delta_X^{k_\ell}(P_i))(\{t_j\})\right|$$

$$+ \left|\sum_{j=1}^{k_\ell} f(t_j)(\delta_X^{k_\ell}(P_i))(\{t_j\}) - \sum_{j=1}^{k_\ell} f(t_j)(\delta_Y^{k_\ell}(Q_i))(\{t_j\})\right|$$

$$+ \left|\sum_{j=1}^{k_\ell} f(t_j)(\delta_Y^{k_\ell}(Q_i))(\{t_j\}) - \int f d\delta_Y(Q_i)\right|$$

$$= \left|\int f d\delta_X(P_i) - \int f d\delta_{k_\ell}(P_i)\right|$$

$$+ \left|\sum_{j=1}^{k_\ell} f(t_j)\Big((\delta_X^{k_\ell}(P_i))(\{t_j\}) - (\delta_Y^{k_\ell}(Q_i))(\{t_j\})\Big)\right|$$

$$+ \left|\int f \circ f_{k_\ell} d\delta_Y(Q_i) - \int f d\delta_Y(Q_i)\right|.$$

But $\lim_{k\to\infty} (f \circ f_k - f) = 0$ uniformly on Ω_D. Thus $\int f d\delta_X(P_i) = \int f d\delta_Y(Q_i)$ for all $f \in \mathscr{C}(\Omega_D)$ and $i \in \Omega_I$, which implies the assertion. □

Theorem 21.5. (L. LeCam). Let $\underline{\bar{D}}: = (I,D,\mathscr{V})$ be a decision problem such that $D = (\Omega_D,\mathbf{A}_D)$ is a standard Borel space. We furthermore assume to be given two experiments $X: = (\Omega,\mathbf{A},(P_i)_{i\in\Omega_I})$ and $Y: = (\Omega_1,\mathbf{A}_1,(Q_i)_{i\in\Omega_I})$ in $\mathscr{E}(\underline{\bar{D}})$ such that $(\Omega_1,\mathbf{A}_1)$ is a standard Borel space and that X is μ-dominated by a measure $\mu \in \mathscr{M}_+^\sigma(\Omega,\mathbf{A})$. The following statements are equivalent:

(i) $X \overset{\bar{D}}{>^-} Y$.

(ii) There exists a kernel $N \in \text{Stoch}((\Omega,\mathbf{A}),(\Omega_1,\mathbf{A}_1))$ satisfying $N(P_i) = Q_i$ for all $i \in \Omega_I$.

Proof: By Theorem 19.17 it suffices to show the implication (i) ⇒ (ii). We first note that the degenerate kernel $\delta \in \text{Stoch}((\Omega_1,\mathbf{A}_1),(\Omega_1,\mathbf{A}_1))$ defined by $\delta(\omega_1,A_1): = \varepsilon_{\omega_1}(A_1)$ for all $(\omega_1,A_1) \in \Omega_1 \times \mathbf{A}_1$ satisfies $\delta(Q_i) = Q_i$ for all $i \in \Omega_I$. Since $(\Omega_D,\mathbf{A}_D)$ and $(\Omega_1,\mathbf{A}_1)$ are standard Borel spaces, there exists a bimeasurable bijection f from Ω_1 onto Ω_D. Let $\delta_Y(\omega_1,B): = \delta(\omega_1,f^{-1}(B))$ for all $(\omega_1,B) \in \Omega_1 \times \mathbf{A}_D$. Then $\delta_Y \in \mathscr{D}(Y)$, and by Theorem 21.3 there exists $\delta_X \in \mathscr{D}(X)$ with $\delta_X(P_i) = \delta_Y(Q_i)$ for all $i \in \Omega_I$. We now define a kernel $N \in \text{Stoch}((\Omega,\mathbf{A}),(\Omega_1,\mathbf{A}_1))$ by $N(\omega,A_1): = \delta_X(\omega,f(A_1))$ for all $(\omega,A_1) \in \Omega \times \mathbf{A}_1$. Plainly for every $i \in \Omega_I$ and $A \in \mathfrak{A}_1$ one has

$$(N(P_i))(A) = (\delta_X(P_i))(f(A)) = (\delta_Y(Q_i))(f(A))$$
$$= (\delta(Q_i))(f^{-1}(f(A))) = Q_i(A),$$

and the theorem has been proved. □

Example 21.6. We shall solve the problem stated in Example 3.24. The probabilities concerning the characteristics A and B under the hypothesis H_0 and under the alternative H_1 respectively are given in the following schemes:

	H_0	H_1
$P(A \cap B)$	πp	ρ
$P(A \cap CB)$	$(1-\pi)p$	$p-\rho$
$P(CA \cap B)$	$\pi(1-p)$	$\pi-\rho$
$P(CA \cap CB)$	$(1-\pi)(1-p)$	$1-\pi-p+\rho$

	H_0	H_1
$P(B\|A)$	π	ρ/p
$P(A\|B)$	p	ρ/π
$P(A\|CB)$	p	$(p-\rho)/(1-\pi)$
$P(B\|CA)$	π	$(\pi-\rho)/(1-p)$

1. We consider the experiment attached to the first line of the schemes, i.e.,

$$X_{A,B} := (\Omega, \mathfrak{A}, \{P_1, P_2\})$$

with $\Omega := \{0,1\}$, $\mathfrak{A} := \mathfrak{P}(\{0,1\})$, $P_1 := B(1, p_{H_0})$ and $P_2 := B(1, p_{H_1})$, where $p_{H_0} := \pi$ and $p_{H_1} := \frac{\rho}{p}$. Thus the experiment $X_{A,B}$ is determined by the stochastic matrix

$$P^{X_{A,B}} = \begin{pmatrix} 1-\pi & \pi \\ 1-\frac{\rho}{p} & \frac{\rho}{p} \end{pmatrix}.$$

Analogously we compute the corresponding matrices of the remaining experiments $X_{B,A}$, $X_{CB,A}$ and $X_{CA,B}$. In particular we get

$$P^{X_{B,A}} = \begin{pmatrix} 1-p & p \\ 1-\frac{\rho}{\pi} & \frac{\rho}{\pi} \end{pmatrix}.$$

2. Now we apply Theorem 21.5 in order to show the comparison relation

$$X_{A,B} \overset{\bar{D}}{\geq} X_{B,A}$$

for any admissible decision problem $\bar{D} = (I,D,\mathscr{V})$ where $I = \{1,2\}$. Similar relations are established by replacing $X_{B,A}$ by $X_{CB,A}$ or $X_{CA,B}$.

In fact, let M be a kernel from $(\Omega,\mathfrak{A})$ to itself and introduce the matrix $N := (n_{ij}) \in \mathbb{M}(2,\mathbb{R})$ by $n_{ij} := M(i,\{j\})$ for all $i,j \in \{0,1\}$. Then M is a Markov kernel iff $N = (n_{ij})$ is a stochastic matrix. In order to deduce our assertion from the stochastic kernel criterion we have to show that there exists a stochastic matrix N satisfying

$$P^{X_{A,B}} N = P^{X_{B,A}}.$$

Since $P^{X_{A,B}}$ is regular with $\det P^{X_{A,B}} = \frac{\rho-\pi p}{p} \neq 0$ and

$$(P^{X_{A,B}})^{-1} = \frac{p}{\rho-\pi p}\begin{pmatrix} \frac{\rho}{p} & -\pi \\ \frac{\rho}{p}-1 & 1-\pi \end{pmatrix},$$

it suffices to show that $N := (P^{X_{A,B}})^{-1} P^{X_{B,A}}$ is a stochastic matrix. But this can easily be seen from the computation

$$N = \frac{p}{\rho-\pi p}\begin{pmatrix} \frac{\rho-\pi p}{p} & 0 \\ \frac{\rho-\pi p}{p} - \frac{\rho-\pi p}{\pi} & \frac{\rho-\pi p}{\pi} \end{pmatrix} = \begin{pmatrix} 1 & 0 \\ 1-\frac{p}{\pi} & \frac{p}{\pi} \end{pmatrix}$$

and from the assumption $p \leq \pi$.

More generally we have

<u>Example 21.7</u>. Let $X := (\Omega,\mathfrak{A},(P_i)_{i\in\Omega_I})$ and $Y := (\Omega_1,\mathfrak{A}_1,(Q_i)_{i\in\Omega_I})$ with $\Omega = \Omega_1 := \{0,1\}$, $\mathfrak{A} = \mathfrak{A}_1 := \mathfrak{P}(\{0,1\})$, $\Omega_I = \{1,2\}$ and $P_1 := B(1,p_0)$, $P_2 := B(1,p_1)$ as well as $Q_1 := B(1,q_0)$, $Q_2 := B(1,q_1)$ $(p_0,p_1,q_0,q_1 \in [0,1])$. If $p_0 < q_0$, $p_0 < p_1$, and $q_0 < q_1$, one can show that

$$X \overset{\bar{D}}{\geq} Y \text{ iff } (1-p_1)(1-q_0) \leq (1-p_0)(1-q_1).$$

Resuming the terminology of the preceding example one obtains under slightly different assumptions on p, π and ρ that $X_{CB,A} \overset{\bar{D}}{>-} X_{CA,B}$ and that the experiments $X_{B,A}$ and $X_{CB,A}$ are not comparable with respect to the comparison relation $\overset{\bar{D}}{>-}$.

§22. SUFFICIENCY IN THE SENSE OF BLACKWELL

We are now going to incorporate into the framework of informativity for experiments the classical notions of sufficiency which have been at the center of discussion in the Chapters II and III.

In order to get a general starting point we first introduce the notion of a sufficient kernel. This notion contains the notions of sufficient statistic and σ-algebra. On the other hand, the sufficiency of a kernel can also be defined through the sufficiency of some corresponding σ-algebra. The main result of this section will be a theorem which characterizes the sufficiency of an experiment with only two measures in terms of the f-divergence. Under certain regularity assumptions we obtain from this the equivalence of sufficiency, Blackwell sufficiency and informativity. It will also be shown that these assumptions cannot be removed.

Definition 22.1. Let $(\Omega,\mathfrak{A},\mathscr{P})$ be an experiment, $(\Omega_1,\mathfrak{A}_1)$ a measurable space and N a kernel in $\mathrm{Stoch}((\Omega,\mathfrak{A}),(\Omega_1,\mathfrak{A}_1))$. For every $P \in \mathscr{P}$ we define the mapping $E_P^N: \mathfrak{M}^b(\Omega,\mathfrak{A}) \to L^\infty(\Omega_1,\mathfrak{A}_1,N(P))$ by

$$E_P^N(f) := \frac{dN(f\cdot P)}{dN(P)} \quad \text{for all} \quad f \in \mathfrak{M}^b(\Omega,\mathfrak{A}).$$

(a) N is called *Blackwell sufficient* (for $\mathscr{P}$) if there exists a kernel $N' \in \mathrm{Stoch}((\Omega_1,\mathfrak{A}_1),(\Omega,\mathfrak{A}))$ such that $N'(N(P)) = P$ holds for all $P \in \mathscr{P}$.

(b) N is said to be *sufficient* (for $\mathscr{P}$) if to every $A \in \mathfrak{A}$ there exists $Q'_A \in \mathfrak{M}_{(1)}(\Omega_1,\mathfrak{A}_1)$ with

$$E_P^N(1_A) = Q'_A[N(P)] \quad \text{for all} \quad P \in \mathscr{P}.$$

Remarks 22.2. 1. The notion of a sufficient kernel generalizes that of a sufficient statistic: Let $(\Omega,\mathfrak{A},\mathscr{P})$ be an experiment and $T: (\Omega,\mathfrak{A}) \to (\Omega_1,\mathfrak{A}_1)$ a statistic. Then T is sufficient for $\mathscr{P}$ (in the Halmos-Savage sense) iff the kernel $N_T \in \mathrm{Stoch}((\Omega,\mathfrak{A}),(\Omega_1,\mathfrak{A}_1))$ defined by $N_T(\omega,A_1) := \varepsilon_{T(\omega)}(A_1)$ for all $(\omega,A_1) \in \Omega \times \mathfrak{A}_1$ is sufficient for $\mathscr{P}$.

2. The notion of a sufficient statistic had been introduced as a generalization of the notion of a sufficient σ-algebra. In fact, a

sub-σ-algebra $\mathfrak{S}$ of $\mathfrak{A}$ is sufficient for $\mathscr{P}$ iff the kernel $N \in \text{Stoch}((\Omega,\mathfrak{A}),(\Omega,\mathfrak{S}))$ defined by $N(\omega,A): = \varepsilon_\omega(A)$ for all $(\omega,A) \in \Omega \times \mathfrak{S}$ is sufficient for $\mathscr{P}$.

In the next theorem we shall show that the notion of sufficient σ-algebra suffices already in order to define sufficient kernels; one just has to admit appropriate operations on the experiments involved.

3. If $N \in \text{Stoch}((\Omega,\mathfrak{A}),(\Omega_1,\mathfrak{A}_1))$ is sufficient and if there exists another kernel $N' \in \text{Stoch}((\Omega_1,\mathfrak{A}_1),(\Omega,\mathfrak{A}))$ such that $N'(\cdot,A) = E_P^N(1_A)[N(P)]$ holds for all $P \in \mathscr{P}$, then N is Blackwell sufficient. Indeed, for every $P \in \mathscr{P}$, $A \in \mathfrak{A}$ and $N' \in \text{Stoch}((\Omega_1,\mathfrak{A}_1),(\Omega,\mathfrak{A}))$ with

$$N'(\cdot,A): = E_P^N(1_A) = \frac{dN(1_A\cdot P)}{dN(P)}\ [N(P)]$$

we obtain

$$\begin{aligned}(N'(N(P)))(A) &= \int_{\Omega_1} N'(\omega',A)(N(P))(d\omega') = (N'(\cdot,A)N(P))(\Omega_1)\\ &= \left[\frac{dN(1_A\cdot P)}{dN(P)} N(P)\right](\Omega_1) = (N(1_A\cdot P))(\Omega_1) = (1_A\cdot P)(\Omega)\\ &= P(A).\end{aligned}$$

4. The notion of Blackwell sufficiency has been introduced to contrast with the notion of Blackwell informativity already discussed in §19. It is easily verified that a kernel $N \in \text{Stoch}((\Omega,\mathfrak{A}),(\Omega_1,\mathfrak{A}_1))$ is Blackwell sufficient iff $(\Omega_1,\mathfrak{A}_1,N(\mathscr{P})) >_0^B (\Omega,\mathfrak{A},\mathscr{P})$.

<u>Theorem 22.3</u>. Let $(\Omega,\mathfrak{A},\mathscr{P})$ be an experiment, $(\Omega_1,\mathfrak{A}_1)$ a measurable space and N a kernel in $\text{Stoch}((\Omega,\mathfrak{A}),(\Omega_1,\mathfrak{A}_1))$. Moreover, let $(\tilde{\Omega},\tilde{\mathfrak{A}}): = (\Omega \times \Omega_1, \mathfrak{A} \otimes \mathfrak{A}_1)$, $\pi_1: = \tilde{\Omega} \to \Omega$ and $\pi_2: \tilde{\Omega} \to \Omega_1$ the coordinate projections and $\tilde{\mathfrak{A}}_1: = \pi_1^{-1}(\mathfrak{A})$ and $\tilde{\mathfrak{A}}_2: = \pi_2^{-1}(\mathfrak{A}_1)$. Finally let $\tilde{\mathscr{P}}$: $\{P \otimes N: P \in \mathscr{P}\}$, where $P \otimes N$ denotes the measure on $\tilde{\mathfrak{A}}: = \mathfrak{A} \otimes \mathfrak{A}_1$ defined by

$$(P \otimes N)(A \times A_1): = \int_A N(\omega,A_1)P(d\omega).$$

for all $A \in \mathfrak{A}$, $A_1 \in \mathfrak{A}_1$. The following statements are equivalent:

(i) N is sufficient.

(ii) $\tilde{\mathfrak{A}}_2$ is a sufficient σ-algebra for the experiment $(\tilde{\Omega},\tilde{\mathfrak{A}},\tilde{\mathscr{P}})$.

<u>Proof</u>: 1. For every $f \in \mathfrak{M}(\Omega,\mathfrak{A})$ and $g \in \mathfrak{M}(\Omega_1,\mathfrak{A}_1)$ we define the function $f \otimes g \in \mathfrak{M}(\tilde{\Omega},\tilde{\mathfrak{A}})$ by $f \otimes g(\omega,\omega_1): = f(\omega)g(\omega_1)$ for all

$(\omega,\omega_1) \in \Omega \times \Omega_1$. Given $g \in \mathfrak{M}(\Omega_1,\mathfrak{A}_1)$ we thus obtain $\int 1_\Omega \otimes g\,d(P \otimes N) = \int g\,dN(P)$, whenever $P \in \mathscr{P}$. Consequently we have the following chain of equalities valid for any $A_1 \in \mathfrak{A}_1$, $A \in \mathfrak{A}$ and every $P \in \mathscr{P}$:

$$\begin{aligned}\int_{\Omega\times A_1} 1_\Omega \otimes E_P^N(1_A)\,d(P \otimes N) &= \int 1_\Omega \otimes (1_{A_1} E_P^N(1_A))\,d(P \otimes N)\\ &= \int 1_{A_1} E_P^N(1_A)\,dN(P) = \int 1_{A_1} \frac{dN(1_A\cdot P)}{dN(P)}\,dN(P)\\ &= (N(1_A\cdot P))(A_1) = \int N(\cdot,A_1)\,d(1_A\cdot P)\\ &= (P \otimes N)(A \times A_1) = \int 1_{A\times\Omega_1} 1_{\Omega\times A_1}\,d(P \otimes N)\\ &= \int_{\Omega\times A_1} 1_{A\times\Omega_1}\,d(P \otimes N).\end{aligned}$$

Since $\tilde{\mathfrak{A}}_2 = \{\Omega \times A_1 : A_1 \in \mathfrak{A}_1\}$, we obtain the formula

$$E_{P\otimes N}^{\tilde{\mathfrak{A}}_2}(1_{A\times\Omega_1}) = 1_\Omega \otimes E_P^N(1_A)\,[P \otimes N] \tag{*}$$

for all $A \in \mathfrak{A}$ and $P \in \mathscr{P}$.

2. (i) ⇒ (ii). We define the system

$$\mathscr{D} := \{\tilde{A} \in \tilde{\mathfrak{A}} : \text{There exists } Q_{\tilde{A}} \in \mathfrak{M}(\tilde{\Omega},\tilde{\mathfrak{A}}_2) : Q_{\tilde{A}} = E_{P\otimes N}^{\tilde{\mathfrak{A}}_2}(1_{\tilde{A}})\,[P \otimes N] \text{ for all } P \in \mathscr{P}\}.$$

From formula (*) in 1 we conclude

$$\begin{aligned}E_{P\otimes N}^{\tilde{\mathfrak{A}}_2}(1_{A\times A_1}) &= E_{P\otimes N}^{\tilde{\mathfrak{A}}_2}(1_{A\times\Omega_1})1_{\Omega\times A_1}\\ &= (1_\Omega \otimes E_P^N(1_A))(1_\Omega \otimes 1_{A_1})\,[P \otimes N]\end{aligned}$$

for all $A \in \mathfrak{A}$, $A_1 \in \mathfrak{A}_1$ and $P \in \mathscr{P}$. This implies that $\{A \times A_1 : A \in \mathfrak{A}, A_1 \in \mathfrak{A}_1\} \subset \mathfrak{D}$, and hence $\mathfrak{D}$ contains a $\cap$-stable generating system. Since $\mathfrak{D}$ is a Dynkin system, we get $\mathfrak{D} = \tilde{\mathfrak{A}}$ and thus (ii).

3. (ii) ⇒ (i). By hypothesis to every $\tilde{A} \in \tilde{\mathfrak{A}}$ there exists a $Q_{\tilde{A}} \in \mathfrak{M}(\tilde{\Omega},\tilde{\mathfrak{A}}_2)$ satisfying

$$E_{P\otimes N}^{\tilde{\mathfrak{A}}_2}(1_{\tilde{A}}) = Q_{\tilde{A}}\,[P \otimes N]$$

for all $P \in \mathscr{P}$. In particular for every $A \in \mathfrak{A}$ there is a $Q_A \in \mathfrak{M}_{(1)}(\Omega_1,\mathfrak{A}_1)$ satisfying

$$E_{P\otimes N}^{\tilde{\mathfrak{A}}_2}(1_{A\times\Omega_1}) = 1_\Omega \otimes Q_A\,[P \otimes N]$$

for all $P \in \mathscr{P}$. Formula (*) of 1. therefore yields $E_P^N(1_A) = Q_A[N(P)]$ for all $P \in \mathscr{P}$ and $A \in \mathfrak{A}$, which shows (i). □

Definition 22.4. Let $(\Omega,\mathfrak{A})$ be a measurable space, P and Q probability measures in $\mathscr{M}^1(\Omega,\mathfrak{A})$ such that P admits the Lebesgue decomposition $P = P_1 + P_2$ with respect to Q, and let f be a convex function on $\mathbb{R}_+$. The extended real number

$$\mathscr{I}_f(P,Q) := \int \left(f \circ \frac{dP_1}{dQ}\right) dQ + P_2(\Omega) \lim_{u\to\infty} \frac{f(u)}{u}$$

is called the *f-divergence of* P with respect to Q.

Remarks 22.5. 1. The f-divergence of P with respect to Q is well-defined, since both of the defining summands are in $\mathbb{R} \setminus \{-\infty\}$. In fact, if γ denotes the right derivative of f at 1, then one has $f(u) \geq \gamma(u-1) + f(1)$ for all $u \in \mathbb{R}_+$, whence

$$\int\left(f \circ \frac{dP_1}{dQ}\right)dQ \geq \int\left(\gamma\left(\frac{dP_1}{dQ} - 1\right) + f(1)\right)dQ > -\infty.$$

Moreover, we note that for convex functions f on $\mathbb{R}_+$ one gets $\lim_{u\to\infty} \frac{f(u)}{u} > -\infty$.

2. The f-divergence of P with respect to Q is a generalization of the total variation of $P - Q$. Choosing $f(u) := |u - 1|$ for all $u \in \mathbb{R}_+$ we obtain $\mathscr{I}_f(P,Q) = ||P-Q||$.

It should be noted that for general convex functions f on $\mathbb{R}_+$ we do not even have the symmetry of $\mathscr{I}_f$ in the sense of $\mathscr{I}_f(P,Q) = \mathscr{I}_f(Q,P)$.

The subsequent discussion of the statistical implications of the notion of f-divergence will be preceded by three lemmas which are of interest in themselves.

Lemma 22.6. Let f be a convex function on $\mathbb{R}_+$. Then for any $\alpha,\beta \in \mathbb{R}_+$ we have the inequality

$$f(\alpha + \beta) \leq f(\alpha) + \beta \lim_{u\to\infty} \frac{f(u)}{u}.$$

Moreover, if f is strictly convex, then equality holds iff $\beta = 0$.

Proof: Let $D_r f$ denote the right derivative of the convex function f. Then for $\alpha,\beta \in \mathbb{R}_+$ we clearly have

$$f(\alpha + \beta) - f(\alpha) \leq \beta \sup_{t \in [\alpha,\alpha+\beta]} D_r f(t)$$

$$\leq \beta D_r f(\alpha + \beta) \leq \beta \lim_{u\to\infty} \frac{f(u)}{u} ,$$

which proves the first assertion of the Lemma. If

$$f(\alpha + \beta) = f(\alpha) + \beta \lim_{u\to\infty} \frac{f(u)}{u} ,$$

then in particular one gets

$$\beta D_r f(\alpha + \beta) = \beta \lim_{u\to\infty} \frac{f(u)}{u} \quad (\alpha,\beta \in \mathbb{R}_+).$$

But for strictly convex f this equality can only be true if $\beta = 0$. □

The following two Lemmas show that the concept of f-divergence can be extended to arbitrary bounded (positive) measures.

Lemma 22.7. Let $(\Omega,\mathfrak{A})$ be a measurable space, $\mu,\nu \in \mathscr{M}_+^b(\Omega,\mathfrak{A})$ two measures with $\nu \neq 0$ and f a convex function on $\mathbb{R}_+$. By analogy with Definition 22.4 we put

$$\mathscr{I}_f(\mu,\nu) := \int \left(f \circ \frac{d\mu_1}{d\nu}\right) d\nu + \mu_2(\Omega) \lim_{u\to\infty} \frac{f(u)}{u} ,$$

where $\mu := \mu_1 + \mu_2$ is the Lebesgue decomposition of μ with respect to ν. Then

$$\mathscr{I}_f(\mu,\nu) \geq \nu(\Omega) f(\rho) \quad \text{with} \quad \rho := \frac{\mu(\Omega)}{\nu(\Omega)} .$$

In addition, if f is strictly convex, then we have equality in the above inequality iff $\mu = \rho\nu$.

Proof: We suppose $\mu \neq 0$, since for $\mu = 0$ the assertion becomes trivial. Let γ be the arithmetic mean of the left- and right derivatives of f in ρ. Then for all $u \in \mathbb{R}_+$ we have $f(u) \geq \gamma(u - \rho) + f(\rho)$, whence

$$\mathscr{I}_f(\mu,\nu) = \int \left(f \circ \frac{d\mu_1}{d\nu}\right) d\nu + \mu_2(\Omega) \lim_{u\to\infty} \frac{f(u)}{u}$$

$$\geq \int \left[\gamma\left(\frac{d\mu_1}{d\nu} - \rho\right) + f(\rho)\right] d\nu + \mu_2(\Omega) \lim_{u\to\infty} \frac{f(u)}{u}$$

$$= \mu_2(\Omega)\left(\lim_{u\to\infty} \frac{f(u)}{u} - \gamma\right) + \nu(\Omega) f(\rho) \geq \nu(\Omega) f(\rho) .$$

When f is strictly convex, equality holds for both of the preceding inequalities iff $d\mu_1/d\nu = \rho\,[\nu]$ and $\mu_2(\Omega) = 0$, that is, iff $\mu = \rho\nu$. Here we use the fact that for a strictly convex function f on $\mathbb{R}_+$ we have $\lim\limits_{u\to\infty} \frac{f(u)}{u} > \gamma$ and $f(u) > \gamma(u-\rho) + f(\rho)$ for $u \neq \rho$. □

Lemma 22.8. With the notation of the preceding lemma let us suppose that $\lim\limits_{u\to\infty} \frac{f(u)}{u} < \infty$ holds. Then

$$\lim_{n\to\infty} \mathscr{I}_f(\mu, \frac{n-1}{n} \nu + \frac{1}{n} \mu) = \mathscr{I}_f(\mu,\nu).$$

Proof: For every $n \in \mathbb{N}$ we put $\nu_n := \frac{n-1}{n} \nu + \frac{1}{n} \mu$. Since $\nu, \mu \ll \nu_n$ and $d\mu_2/d\nu_n = 0[\nu]$, $d\mu_2/d\nu_n = 0[\mu_1]$ as well as $d\mu_1/d\nu_n = 0[\mu_2]$ for all $n \in \mathbb{N}$ we obtain

$$\begin{aligned}\mathscr{I}_f(\mu,\nu_n) &= \int (f \circ \frac{d\mu}{d\nu_n}) d\nu_n \\ &= \int \frac{n-1}{n} (f \circ \frac{d\mu_1}{d\nu_n}) d\nu + \int \frac{1}{n}(f \circ \frac{d\mu_1}{d\nu_n}) d\mu_1 + \int \frac{1}{n}(f \circ \frac{d\mu_2}{d\nu_n}) d\mu_2,\end{aligned}$$

where $\mu := \mu_1 + \mu_2$ is the Lebesgue decomposition of μ with respect to ν. By assumption we can choose numbers $a,b \in \mathbb{R}_+$ such that for all $u \in \mathbb{R}_+$ we have $|f(u)| \leq au + b$. But then we obtain the estimates

$$\left|\frac{n-1}{n}(f \circ \frac{d\mu_1}{d\nu_n})\right| \leq \frac{n-1}{n}(a \frac{d\mu_1}{d\nu_n} + b) \leq a \frac{d\mu_1}{d\nu_n} + b,$$

$$\left|\frac{1}{n}(f \circ \frac{d\mu_1}{d\nu_n})\right| \leq \frac{1}{n}(a \frac{d\mu_1}{d\nu_n} + b) \leq a + b[\mu_1]$$

and

$$\frac{1}{n}(f \circ \frac{d\mu_2}{d\nu_n}) = \frac{1}{n} f(n)[\mu_2],$$

the last two following from $d\mu_1/d\nu_n \leq n[\mu_1]$ and $d\mu_2/d\nu_n = n[\mu_2]$ respectively. Lebesgue's dominated convergence theorem enables us to compute the limits for $n \to \infty$ of the summands of $\mathscr{I}_f(\mu,\nu_n)$:

$$\lim_{n\to\infty} \int \frac{n-1}{n} (f \circ \frac{d\mu_1}{d\nu_n}) d\nu = \int (f \circ \frac{d\mu_1}{d\nu}) d\nu,$$

$$\lim_{n\to\infty} \int \frac{1}{n}(f \circ \frac{d\mu_1}{d\nu_n}) d\mu_1 = \int \lim_{n\to\infty} \frac{1}{n}(f \circ \frac{d\mu_1}{d\nu_n}) d\mu_1 = 0,$$

and

$$\lim_{n\to\infty} \int \frac{1}{n}(f \circ \frac{d\mu_2}{d\nu_n}) d\mu_2 = \lim_{n\to\infty} \int \frac{1}{n} f(n) d\mu_2 = \mu_2(\Omega) \lim_{u\to\infty} \frac{f(u)}{u}.$$

But this implies the assertion. □

The importance of the f-divergence for the theory of sufficiency becomes evident from the following result.

Theorem 22.9. (S. Kullback, R. A. Leibler, I. Csiszár). Let $(\Omega,\mathbf{A})$, $(\Omega_1,\mathbf{A}_1)$ be measurable spaces, N a kernel in $\mathrm{Stoch}((\Omega,\mathbf{A}),(\Omega_1,\mathbf{A}_1))$ and $(\Omega,\mathbf{A},\mathscr{P})$ an experiment with $\mathscr{P}: = \{P,Q\} \subset \mathscr{M}^1(\Omega,\mathbf{A})$. Then

(i) For any convex function f on $\mathbb{R}_+$ one has

$$\mathscr{I}_f(N(P),N(Q)) \leq \mathscr{I}_f(P,Q).$$

(ii) If f is a strictly convex function on $\mathbb{R}_+$ such that

$$\mathscr{I}_f(N(P),N(Q)) = \mathscr{I}_f(P,Q) < \infty$$

holds, then N is sufficient for $\mathscr{P}$.

Proof of Theorem 22.9, part (i): Let $N(P_2): = \tau' + \sigma'$ be the Lebesgue-decomposition of $N(P_2)$ with respect to $N(Q)$. Then clearly $N(P) = (N(P_1) + \tau') + \sigma'$ is the Lebesgue decomposition of $N(P)$ with respect to $N(Q)$. With reference to Theorem 22.3 we put $(\tilde{\Omega},\tilde{\mathbf{A}}): = (\Omega,\mathbf{A}) \otimes (\Omega_1,\mathbf{A}_1)$ and $\tilde{\mathbf{A}}': = \{\emptyset,\Omega\} \otimes \mathbf{A}_1$. Then

$$E^{\tilde{\mathbf{A}}'}_{Q\otimes N}\left(\frac{dP_1}{dQ} \otimes 1_{\Omega_1}\right) = 1_\Omega \otimes \frac{dN(P_1)}{dN(Q)} \tag{**}$$

as can be derived in a manner analogous to that giving formula (*) of the proof of Theorem 22.3 or directly by passing from indicator functions to general Q-integrable functions. The desired inequality is now implied by the following chain of inequalities:

$$\mathscr{I}_f(N(P),N(Q))$$

$$= \int\left(f \circ \left[\frac{dN(P_1)}{dN(Q)} + \frac{d\tau'}{dN(Q)}\right]\right)dN(Q) + \sigma'(\Omega_1) \lim_{u\to\infty} \frac{f(u)}{u}$$

$$\leq \int\left(f \circ \frac{dN(P_1)}{dN(Q)}\right)dN(Q) + (\tau' + \sigma')(\Omega_1) \lim_{u\to\infty} \frac{f(u)}{u}$$

(by Lemma 22.6)

$$= \int f \circ \left(1_\Omega \otimes \frac{dN(P_1)}{dN(Q)}\right)d(Q \otimes N) + (N(P_2))(\Omega_1) \lim_{u\to\infty} \frac{f(u)}{u}$$

$$= \int f \circ \left(E^{\tilde{\mathbf{A}}'}_{Q\otimes N}\left[\frac{dP_1}{dQ} \otimes 1_{\Omega_1}\right]\right)d(Q \otimes N) + P_2(\Omega) \lim_{u\to\infty} \frac{f(u)}{u}$$

(by (**))

$$\leq \int E^{\tilde{\mathbf{A}}'}_{Q\otimes N}\left(f \circ \left[\frac{dP_1}{dQ} \otimes 1_{\Omega_1}\right]\right)d(Q \otimes N) + P_2(\Omega) \lim_{u\to\infty} \frac{f(u)}{u}$$

(by Jensen's inequality)

$$= \int f \circ \left(\frac{dP_1}{dQ} \otimes 1_{\Omega_1}\right) d(Q \otimes N) + P_2(\Omega) \lim_{u\to\infty} \frac{f(u)}{u}$$

$$= \int f \circ \frac{dP_1}{dQ}\, dQ + P_2(\Omega) \lim_{u\to\infty} \frac{f(u)}{u} = \mathcal{I}_f(P,Q).$$

This proves part (i) of Theorem 22.9.

For the proof of part (ii) of Theorem 22.9 we need the following

Lemma 22.10. We keep the notation from above. Let $\lambda' := N(P_1) + \tau'$ be the absolutely continuous part of $N(P)$ with respect to $N(Q)$. If f is strictly convex, then $\mathcal{I}_f(N(P),N(Q)) = \mathcal{I}_f(P,Q)$ holds iff either $\mathcal{I}_f(N(P),N(Q)) = \infty$ or

$$1_\Omega \otimes \frac{d\lambda'}{dN(Q)} = \frac{dP_1}{dQ} \otimes 1_{\Omega_1} \; [Q \otimes N]$$

is satisfied.

Proof: Let $\mathcal{I}_f(P,Q) = \mathcal{I}_f(N(P),N(Q)) < \infty$. Then for both of the inequalities in the proof of part (i) of Theorem 22.9 we actually have equality. Lemma 22.6 together with the first equality then implies that $d\tau'/dN(Q) = 0\,[N(Q)]$, whence $1_\Omega \otimes d\tau'/dN(Q) = 0\,[Q \otimes N]$, and from the equality in Jensen's inequality we get

$$\frac{dP_1}{dQ} \otimes 1_{\Omega_1} = E^{\tilde{\mathbf{A}}'}_{Q\otimes N}\left(\frac{dP_1}{dQ} \otimes 1_{\Omega_1}\right) [Q \otimes N].$$

But this implies via formula (**) of part (i) of the proof of Theorem 22.9 the equalities

$$1_\Omega \otimes \frac{d\lambda'}{dN(Q)} = 1_\Omega \otimes \frac{d(N(P_1)+\tau')}{dN(Q)} = 1_\Omega \otimes \frac{dN(P_1)}{dN(Q)}$$
$$= E^{\tilde{\mathbf{A}}'}_{Q\otimes N}\left(\frac{dP_1}{dQ} \otimes 1_{\Omega_1}\right) = \frac{dP_1}{dQ} \otimes 1_{\Omega_1} \; [Q \otimes N],$$

which yield the one implication of the assertion in the Lemma.

The remaining implication is checked by reversing the arguments. □

Proof of Theorem 22.9, part (ii): By Theorem 22.3 it suffices to show that $\tilde{\mathbf{A}}'$ is sufficient for $\{P \otimes N, Q \otimes N\}$, which by Theorem 8.4 means that the measures $P \otimes N$ and $Q \otimes N$ admit $\tilde{\mathbf{A}}'$-measurable densities with respect to $\frac{1}{2}(P+Q) \otimes N$. From Lemma 22.10 we obtain

$$(N(P_1))(\Omega_1) + \tau'(\Omega_1) = \lambda'(\Omega_1) = \int \frac{d\lambda'}{dN(Q)}\, dN(Q)$$

$$= \int 1_\Omega \otimes \frac{d\lambda'}{dN(Q)}\, d(Q \otimes N) = \int \frac{dP_1}{dQ} \otimes 1_{\Omega_1}\, d(Q \otimes N) = P_1(\Omega)$$

$$= (N(P_1))(\Omega_1),$$

whence $\tau' = 0$. Therefore there exists a set $A_1 \in \mathbb{A}_1$ satisfying $(N(P_2))(CA_1) = 0 = (N(Q))(A_1)$. It follows that

$$(P_2 \otimes N)(C(\Omega \times A_1)) = 0 = (Q \otimes N)(\Omega \times A_1)$$

and using the notation $R := \frac{1}{2}(P + Q)$ we get $P_2 \otimes N = 2\cdot 1_{\Omega\times A_1}(R \otimes N)$ as well as

$$(P_1 + Q) \otimes N = (1 - 2\cdot 1_{\Omega\times A_1})(R \otimes N).$$

From Lemma 22.10 and $\tau' = 0$ we infer that

$$\frac{dP_1}{dQ} \otimes 1_{\Omega_1} = 1_\Omega \otimes \frac{dN(P_1)}{dN(Q)}\ [Q \otimes N],$$

whence that dP_1/dQ is Q-a.s. constant and thus that there exists an $\alpha \in [0,1]$ with $P_1 = \alpha Q$. This implies

$$P_1 \otimes N = \frac{\alpha}{\alpha+1}(P_1 + Q) \otimes N = \frac{\alpha}{\alpha+1}(1 - 2\cdot 1_{\Omega\times A_1})(R \otimes N)$$

and

$$\frac{d(P \otimes N)}{d(R \otimes N)} = \frac{\alpha}{\alpha+1} + \frac{2}{\alpha+1} 1_{\Omega\times A_1}\ [R \otimes N],$$

which shows that $P \otimes N$ and therefore also $Q \otimes N$ admits an $\tilde{\mathbb{A}}'$-measurable density with respect to $R \otimes N$. □

Theorem 22.11. Let $(\Omega,\mathbb{A},\mathscr{P})$ be an experiment, $(\Omega_1,\mathbb{A}_1)$ a measurable space and N a kernel in $\mathrm{Stoch}((\Omega,\mathbb{A}),(\Omega_1,\mathbb{A}_1))$.

(i) If $(\Omega,\mathbb{A},\mathscr{P})$ is μ-dominated by a measure $\mu \in \mathscr{M}_+^\sigma(\Omega,\mathbb{A})$ and N is Blackwell sufficient, then N is sufficient for $\mathscr{P}$.

(ii) If $(\Omega_1,\mathbb{A}_1,N(\mathscr{P}))$ is μ_1-dominated by a measure $\mu_1 \in \mathscr{M}_+^\sigma(\Omega_1,\mathbb{A}_1)$, $(\Omega,\mathbb{A})$ is a standard Borel space and N sufficient for $\mathscr{P}$, then N is also Blackwell sufficient.

Proof: (i) By Theorem 22.3 it suffices to show that the σ-subalgebra $\{\emptyset,\Omega\} \otimes \mathbb{A}_1$ is sufficient for the experiment $(\Omega \times \Omega_1, \mathbb{A} \otimes \mathbb{A}_1, \{P \otimes N: P \in \mathscr{P}\})$. Since $\mathscr{P} \ll \mu$, we have $\mathscr{P} \otimes N \ll \mu \otimes N$. It therefore suffices by Theorem 8.18 to prove the sufficiency of $\{\emptyset,\Omega\} \otimes \mathbb{A}_1$ for every two element subset of $\{P \otimes N: P \in \mathscr{P}\}$. Again by Theorem 22.3 it even suffices to show the sufficiency of N for any two element subset $\mathcal{P}_0$ of $\mathcal{P}$. We may thus take $\mathcal{P} = \{P_1, P_2\}$. By assumption there exists a kernel $M \in \mathrm{Stoch}((\Omega_1, \mathfrak{A}_1), (\Omega, \mathfrak{A}))$ with $M(N(P_i)) = P_i$ for $i = 1,2$. Thus for every strictly convex function f on $\mathbb{R}_+$ we get

$$\mathcal{I}_f(M(N(P_1)), M(N(P_2))) = \mathcal{I}_f(P_1, P_2).$$

Theorem 22.9 yields the equality

$$\mathcal{I}_f(N(P_1), N(P_2)) = \mathcal{I}_f(P_1, P_2),$$

whence that N is sufficient for $\mathcal{P}$.

(ii) We assumed $N(\mathcal{P}) \ll \mu_1$ for a measure $\mu_1 \in \mathcal{M}_+^\sigma(\Omega_1, \mathfrak{A}_1)$. Since N is sufficient for $\mathcal{P}$, with every $f \in \mathfrak{M}^b(\Omega, \mathfrak{A})$ we can associate a $Q_f \in \bigcap_{P \in \mathcal{P}} E_P^N(f)$. The mapping $f \to T(f) := [Q_f]_{\mu_1}$ defines a σ-continuous positive linear operator on $\mathfrak{M}^b(\Omega, \mathfrak{A})$ satisfying $1_{\Omega_1} \in T(1_\Omega)$. But then Corollary 20.3 implies the existence of a kernel $M \in \mathrm{Stoch}((\Omega_1, \mathfrak{A}_1), (\Omega, \mathfrak{A}))$ with the property

$$M(\cdot, f) \in E_P^N(f) = \frac{dN(f \cdot P)}{dN(P)}$$

for all $f \in \mathfrak{M}^b(\Omega, \mathfrak{A})$. By Remark 3 of 22.2 N is Blackwell sufficient. □

<u>Theorem 22.12</u>. Let $(\Omega, \mathfrak{A})$ be a standard Borel space, $\mathcal{P}$ a dominated family of measures in $\mathcal{M}^1(\Omega, \mathfrak{A})$, $(\Omega_1, \mathfrak{A}_1)$ another measurable space and N a kernel in $\mathrm{Stoch}((\Omega, \mathfrak{A}), (\Omega_1, \mathfrak{A}_1))$. By $\bar{D} := (I, D, \mathcal{V})$ we denote the decision problem with standard Borel space $D := (\Omega_D, \mathfrak{A}_D)$ and the set $\mathcal{V}$ of all bounded separately measurable functions on $\Omega_I \times \Omega_D$. Then $X := (\Omega, \mathfrak{A}, \mathcal{P})$ and $Y := (\Omega_1, \mathfrak{A}_1, N(\mathcal{P}))$ are experiments in $\mathcal{E}(\bar{D})$, and the following statements are equivalent:

(i) N is sufficient for $\mathcal{P}$.

(ii) N is Blackwell sufficient for $\mathcal{P}$.

(iii) $Y \overset{B}{\underset{0}{>}} X$.

(iv) $Y \overset{\bar{D}}{\underset{-}{>}} X$.

The <u>proof</u> of (i) $\Leftrightarrow$ (ii) follows from Theorem 22.11, that of (ii) $\Leftrightarrow$ (iii) is Remark 4 of 22.2, that of (iii) $\Leftrightarrow$ (iv) is a direct consequence of the stochastic kernel criterion 21.5. □

The assumptions made in Theorem 22.12 cannot be dropped. In order to describe the domain of validity of the theorem we present two counter-examples.

Example 22.13. Without the domination hypothesis the statement of Theorem 22.12 is in general false. To see this we consider Example 9.3. Let $\Omega = \Omega_1 := \mathbb{R}$, $\mathbb{A} := \mathbb{B}$, $\mathbb{A}_1 := \{A_1 \cup A_2 : A_i \in \mathbb{A}\ (i = 1,2),\ A_1 = -A_1, A_2 \subset M\}$, where $M \subset \mathbb{R}$ is a non-$\mathbb{A}$-measurable set with $0 \in M$ and $M = -M$, and $\mathscr{P} := \{\frac{1}{2}\varepsilon_x + \frac{1}{2}\varepsilon_{-x} : x \in \mathbb{R}\}$. Then by the discussion of §9, $\mathbb{A}_1$ is a σ-algebra. Let N be the kernel in $\mathrm{Stoch}((\Omega,\mathbb{A}),(\Omega_1,\mathbb{A}_1))$ corresponding to the mapping $x \to x$ from Ω onto Ω_1. Again from §9 we know that $\mathbb{A}_1 \subset \mathbb{A}$ is not sufficient for $\mathscr{P}$, i.e., N is not sufficient. On the other hand we have a kernel $M \in \mathrm{Stoch}((\Omega_1,\mathbb{A}_1),(\Omega,\mathbb{A}))$ defined by

$$M(\cdot,A) := \frac{1}{2}(1_A + 1_{-A}) = 1_{A\cap(-A)} + \frac{1}{2}\,1_{A\Delta(-A)}$$

for all $A \in \mathbb{A}$. For this kernel, however, $M(N(P)) = P$ holds for all $P \in \mathscr{P}$.

Example 22.14. Without the hypothesis that $(\Omega,\mathbb{A})$ is a standard Borel space the statement of Theorem 22.12 is in general false. Let $(\Omega,\mathbb{A},\nu)$ be a probability space and $\mathbb{A}_1 \subset \mathbb{A}$ a sub-σ-algebra for which the conditional expectation with respect to ν cannot be described by a stochastic kernel. Such probability spaces and sub-σ-algebras exist as is well-known (see the argument in Example 9.8). Now let

$$\mathscr{P} := \left\{\frac{1}{\nu(A_1)}\,1_{A_1}\cdot\nu : A_1 \in \mathbb{A}_1,\ \nu(A_1) > 0\right\}.$$

Then for all $P \in \mathscr{P}$ and $A \in \mathbb{A}$ we have

$$E_P^{\mathbb{A}_1}(1_A) = E_\nu^{\mathbb{A}_1}(1_A)\,[P],$$

i.e., $\mathbb{A}_1$ is sufficient for $\mathscr{P}$. Assuming that there exists a kernel $M \in \mathrm{Stoch}((\Omega,\mathbb{A}_1),(\Omega,\mathbb{A}))$ satisfying $M(P_{\mathbb{A}_1}) = P$ for all $P \in \mathscr{P}$, we obtain for all $A \in \mathbb{A}$ and $A_1 \in \mathbb{A}_1$ the equality

$$(M(1_{A_1}\cdot\nu))(A) = \int 1_{A_1}(\omega)\nu(d\omega)M(\omega,A) = \int_{A_1} M(\omega,A)\nu(d\omega).$$

On the other hand we have

$$(M(1_{A_1}\cdot\nu))(A) = (1_{A_1}\cdot\nu)(A) = \nu(A \cap A_1).$$

But then M would be an expectation kernel contrary to the hypothesis above.

CHAPTER VIII

Invariance and the Comparison of Experiments

§23. EXISTENCE OF INVARIANT STOCHASTIC KERNELS

Invariant Markov kernels can be used with success whenever the general theory of comparison of experiments is applied to special classes of experiments like those classical experiments involving location parameters. Our first aim in this section is a strengthening of LeCam's Markov kernel criterion in the case of invariant experiments.

Let $\underline{\overline{D}} = (I, D, \mathcal{V})$ be a decision problem with a standard Borel decision space $D = (\Omega_D, \mathfrak{A}_D)$ and the set $\mathcal{V}$ of all bounded, separately measurable functions on $\Omega_I \times \Omega_D$ as the set of loss functions corresponding to $\underline{\overline{D}}$. We assume given two experiments $X = (\Omega, \mathfrak{A}, (P_i)_{i \in \Omega_I})$ and $Y = (\Omega_1, \mathfrak{A}_1, (Q_i)_{i \in \Omega_I})$ corresponding to $\underline{\overline{D}}$ satisfying the following hypotheses:

(1) X is μ-dominated by a measure $\mu \in \mathcal{M}_+^\sigma(\Omega, \mathfrak{A})$.

(2) $(\Omega_1, \mathfrak{A}_1)$ is a standard Borel space (generated by a compact metrizable topology on Ω_1).

We recall that a pair $(G, \mathfrak{B})$ consisting of an abstract group G and a σ-algebra $\mathfrak{B}$ in G is said to be a *measurable group* if for every $g_0 \in G$ the mapping $g \to g g_0^{-1}$ from G into G is measurable. $(G, \mathfrak{B})$ is called *amenable* if G is amenable in the sense that there exists a (left) invariant mean on $\mathscr{B}(G)$. It is known that any Abelian (measurable) group is amenable. For a measurable group $(G, \mathfrak{B})$ we define a measure $\lambda \in \mathcal{M}_+^\sigma(G, \mathfrak{B})$ with $\lambda \neq 0$ to be *quasi-invariant* if for all $B \in \mathfrak{B}$ and $g \in G$ the relation $\lambda(Bg) = 0$ holds iff $\lambda(B) = 0$.

Now, let $(G, \mathfrak{B})$ be a measurable group having the following properties: G operates on Ω_I, Ω and Ω_1 via the bijections $g_I: \Omega_I \to \Omega_I$,

$g_\Omega: \Omega \to \Omega$ and $g_{\Omega_1}: \Omega_1 \to \Omega_1$ respectively such that

(α) The mappings $(\omega,g) \to g_\Omega(\omega)$ from $\Omega \times G$ into Ω, $(\omega_1,g) \to g_{\Omega_1}(\omega_1)$ from $\Omega_1 \times G$ into Ω_1 are measurable.

(β) For every $g \in G$ the mapping g_Ω is bimeasurable and the mapping g_{Ω_1} is a homeomorphism and hence bimeasurable.

(γ) For all $i \in \Omega_I$ and $g \in G$ we have $g_\Omega(P_i) = P_{g_I(i)}$ and $g_{\Omega_1}(Q_i) = Q_{g_I(i)}$ as well as $g_\Omega(\mu) = \mu$.

Definition 23.1. Let $N \in \mathrm{Stoch}((\Omega,\mathbf{A}),(\Omega_1,\mathbf{A}_1))$. For every $g \in G$ we denote by N_g the mapping from $\Omega \times \mathbf{A}_1$ into $\mathbb{R}$ defined by

$$N_g(\omega,A_1) := N(g_\Omega(\omega), g_{\Omega_1}(A_1))$$

for all $(\omega,A_1) \in \Omega \times \mathbf{A}_1$.

Clearly $N_g \in \mathrm{Stoch}((\Omega,\mathbf{A}),(\Omega_1,\mathbf{A}_1))$.

(a) N is called *almost* (surely) *invariant* if for every $g \in G$ there exists a $C_g \in \mathbf{A}$ such that $\mu(C_g) = 0$ and

$$N_g(\omega,A_1) = N(\omega,A_1)$$

for all $\omega \in CC_g$ and all $A_1 \in \mathbf{A}_1$.

(b) N is called *invariant* if there exists a set $C \in \mathbf{A}$ such that $\mu(C) = 0$, $g_\Omega(C) = C$ for all $g \in G$ and

$$N_g(\omega,A_1) = N(\omega,A_1)$$

holds for all $\omega \in CC$ and all $A_1 \in \mathbf{A}_1$.

Theorem 23.2. Let $(G,\mathfrak{B})$ be a measurable group which is assumed to be amenable, and let there exist a quasi-invariant measure λ on $(G,\mathfrak{B})$. Then the following statements are equivalent:

(i) $X \overset{\bar{D}}{\geq} Y$.

(ii) There exists an invariant kernel $N \in \mathrm{Stoch}((\Omega,\mathbf{A}),(\Omega_1,\mathbf{A}_1))$ such that $N(P_i) = Q_i$ for all $i \in \Omega_1$.

Proof: By Theorem 21.5 it remains to prove the implication (i) ⇒ (ii). Let $\mathscr{R}$ be the set of all positive linear operators $T: \mathscr{C}(\Omega_1) \to L^\infty(\Omega,\mathbf{A},\mu)$ such that $1_\Omega \in T(1_{\Omega_1})$ holds. Then $\mathscr{R}$ is a closed and therefore compact subsets of the compact space

$$K := \prod_{f \in \mathscr{C}(\Omega_1)} \{g \in L^\infty(\Omega,\mathbf{A},\mu): ||g||_\infty \leq ||f||\},$$

where K is furnished with the product topology with respect to the topology $\sigma(L^\infty, L^1)$.

For every $T \in \mathscr{R}$ and $f \in \mathscr{L}(\Omega_1)$ we denote a representative of the class $T(f)$ by $T^*(f)$. Given $T \in \mathscr{R}$, $g \in G$ and $f \in \mathscr{L}(\Omega_1)$ we define

$$T^g(f) := [(T^*(f \circ g_{\Omega_1}^{-1})) \circ g_\Omega]_\mu \in L^\infty(\Omega, \mathbf{A}, \mu).$$

For $T \in \mathscr{R}$ and $g \in G$ we put

$$\Phi_g(T) := T^g.$$

The following properties are easily verified:

(1) $T^g \in \mathscr{R}$ for all $g \in G$.

(2) For every $g \in G$ the mapping $\Phi_g: \mathscr{R} \to \mathscr{R}$ is linear.

(3) Φ_g is continuous for every $g \in G$.

The latter property can be seen as follows.

Let $(T_\alpha)_{\alpha \in A}$ be a net in $\mathscr{R}$ which converges to T in the topology of $\mathscr{R}$. We want to show that

$$\lim_{\alpha \in A} \Phi_g(T_\alpha)(f) = \Phi_g(T)(f)$$

for all $f \in \mathscr{L}(\Omega_1)$ with respect to the topology $\sigma(L^\infty, L^1)$. Let $f \in \mathscr{L}(\Omega_1)$. Then $f \circ g_{\Omega_1}^{-1} \in \mathscr{L}(\Omega_1)$, whence

$$\sigma(L^\infty, L^1) - \lim_{\alpha \in A} T_\alpha(f \circ g_{\Omega_1}^{-1}) = T(f \circ g_{\Omega_1}^{-1}).$$

Since $g_\Omega(\mu) = \mu$ has been assumed, we get that $[h \circ g_\Omega^{-1}]_\mu \in L^1(\Omega, \mathbf{A}, \mu)$ for every $[h]_\mu \in L^1(\Omega, \mathbf{A}, \mu)$. Consequently

$$\begin{aligned} &\int T_\alpha^*(f \circ g_{\Omega_1}^{-1}) \circ g_\Omega \cdot h \, d\mu \\ &\qquad = \int T_\alpha^*(f \circ g_{\Omega_1}^{-1}) \circ g_\Omega \cdot h \circ g_\Omega^{-1} \circ g_\Omega \, d\mu \\ &\qquad = \int T_\alpha^*(f \circ g_{\Omega_1}^{-1}) \cdot h \circ g_\Omega^{-1} \, d\mu \end{aligned}$$

converges to

$$\int T^*(f \circ g_{\Omega_1}^{-1}) \cdot h \circ g_\Omega^{-1} \, d\mu = \int T^*(f \circ g_{\Omega_1}^{-1}) \circ g_\Omega \cdot h \, d\mu$$

which is equivalent to the assertion.

Now let

$$\mathscr{R}_0 := \{T \in \mathscr{R}: \int T(f)dP_i = \int f\, dQ_i \quad \text{for all} \quad i \in \Omega_I,\ f \in \mathscr{C}(\Omega_1)\}.$$

Clearly $\mathscr{R}_0$ is a closed and thus compact convex subset of $\mathscr{R}$.

Since by hypothesis $X \overset{\bar{D}}{>-} Y$, Theorem 21.5 implies that $\mathscr{R}_0 \neq \emptyset$. Moreover, we have $\Phi_g(\mathscr{R}_0) \subset \mathscr{R}_0$ for all $g \in G$, and the set $S := \{\Phi_g : g \in G\}$ forms a semigroup of mappings with respect to composition, as follows from the relation $(T^{g_1})^{g_2} = T^{(g_1 g_2)}$ for all $g_1, g_2 \in G$. The mapping $g \to \Phi_g$ is a semigroup homomorphism from G onto S. Since G is assumed to be amenable, also S is amenable, and hence the Markov-Kakutani-Day fixed point theorem can be applied in order to provide us with a kernel $T_0 \in \mathscr{R}_0$ satisfying $T_0^g = T_0$ for all $g \in G$.

By Corollary 20.2 there exists a kernel $N \in \text{Stoch}((\Omega, \mathbf{A}), (\Omega_1, \mathbf{A}_1))$ such that

$$N(\cdot, f) \in T_0(f)$$

for all $f \in \mathscr{C}(\Omega_1)$. N is almost invariant, since

$$[N_g(\cdot, f)]_\mu = T_0^g(f) = T_0(f) = [N(\cdot, f)]_\mu$$

for all $g \in G$ and $f \in \mathscr{C}(\Omega_1)$. It remains to be shown that to N there exists an invariant kernel $\hat{N} \in \text{Stoch}((\Omega, \mathbf{A}), (\Omega_1, \mathbf{A}_1))$ which is μ-equivalent to N in the sense that

$$\hat{N}(\cdot, f) = N(\cdot, f)\,[\mu]$$

for every $f \in \mathscr{C}(\Omega_1)$.

The proof of this statement is based on the separability of $\mathscr{C}(\Omega_1)$, the existence of a quasi-invariant measure $\lambda \in \mathscr{M}_+^\sigma(G, \mathbf{B})$ and the fact that the mapping $(\omega, g) \to N_g(\omega, f)$ from $\Omega \times G$ into $\mathbb{R}$ is $\mathbf{A} \otimes \mathbf{B}$-measurable.

Indeed, suppose that $\mathscr{A}$ is a countable dense subset of $\mathscr{C}(\Omega_1)$. We consider the set

$$V := \{(\omega, g) \in \Omega \times G: \sum_{h \in \mathscr{A}} |(N_g h)(\omega) - (Nh)(\omega)| > 0\} \in \mathbf{A} \otimes \mathbf{B}.$$

By V_g and V_ω we denote the g- and ω-sections of V respectively. We note that without loss of generality the quasi-invariant measure λ on $(G, \mathbf{B})$ can be chosen as a probability measure. By assumption $\mu(V_g) = 0$ for all $g \in G$, whence $\mu(M) = 0$ for the set

$$M := \{\omega \in \Omega: \lambda(V_\omega) > 0\}.$$

We consider the set

$$A: = \{\omega \in \Omega: (N_g h)(\omega) = \text{const}[\lambda] \quad \text{for all} \quad h \in \mathscr{A}\} \in \mathbf{A}.$$

The quasi-invariance of λ yields $g_\Omega(A) = A$ for all $g \in G$. Since $CA \subset M$, we obtain $\mu(A) = 1$. Now we define the kernel $\hat{N}$ by

$$(\hat{N}f)(\omega): = \int (N_g f)(\omega)\lambda(dg)$$

for all $f \in \mathscr{C}(\Omega_1)$, $\omega \in \Omega$. Then $\hat{N}f = Nf[\mu]$, and $\hat{N}$ is invariant, since for all $\bar{g} \in G$, $f \in \mathscr{C}(\Omega_1)$ and $\omega \in A$ we have

$$\begin{aligned}(\hat{N}_{\bar{g}}f)(\omega) &= \int N_g(f \circ \bar{g}_{\Omega_1}^{-1})(\bar{g}_\Omega(\omega))\lambda(dg) \\ &= \int (N_{g\bar{g}}f)(\omega)\lambda(dg) \\ &= \int (N_g f)(\omega)\lambda(dg) \\ &= (\hat{N}f)(\omega). \quad \square\end{aligned}$$

§24. COMPARISON OF TRANSLATION EXPERIMENTS

We shall now specialize the situation described in the preceding section by replacing the general measurable group $(G,\mathfrak{B})$ by the Borel group $(G,\mathfrak{B}(G))$ of a locally compact group G with a countable basis of its topology together with its Borel-σ-algebra $\mathfrak{B}(G)$. We recall that on any locally compact group G there exists a (left-invariant) Haar measure $\lambda \in \mathscr{M}_+^\sigma(G,\mathfrak{B}(G))$ which is unique up to a positive multiplicative constant. Amenability of the group G means the existence of a (left-) invariant mean on $\mathscr{B}(G)$. Examples of amenable groups are all Abelian groups and all solvable groups.

Let $\bar{D} = (I,D,\mathscr{V})$ be a decision problem with $I: = (G,\mathfrak{B}(G))$, $D: = (G,\mathfrak{B}(G))$ and the set $\mathscr{V}$ of all bounded, separately measurable functions on $G \times G$. Under these specialized hypotheses we shall write $X > Y$ instead of $X \overset{\bar{D}}{>} Y$.

<u>Definition 24.1</u>. For any measure $\mu \in \mathscr{M}^1(G,\mathfrak{B}(G))$ the experiment

$$X(\mu): = (G,\mathfrak{B}(G), \{\mu * \varepsilon_x: x \in G\})$$

corresponding to the decision problem $\bar{D}$ is called the *translation experiment* with defining measure μ.

<u>Remark 24.2</u>. Translation experiments arise in connection with testing or estimation of *location parameters*. In these cases $\Omega: = \Omega_1: = I: =$]

$\mathfrak{X}: = \mathfrak{X}(\mu)$ for some λ^1-absolutely continuous measure $\mu \in \mathscr{M}^1(\mathbb{R},\mathfrak{B}^1)$ and the group $\mathbb{R}$ operates via right or left translations. For any number $x \in \mathbb{R}$ the measure $P_x: = \mu * \varepsilon_x$ is interpreted as the distribution of $x + E$ where the "error" E is distributed according to μ.

Definition 24.3. A kernel $N \in \mathrm{Stoch}(G,\mathfrak{B}(G)): = \mathrm{Stoch}((G,\mathfrak{B}(G)),(G,\mathfrak{B}(G)))$ is said to be *translation invariant* if

$$N(xy,By) = N(x,B)$$

for all $x,y \in G$, $B \in \mathfrak{B}(G)$.

Theorem 24.4. Let G be an amenable locally compact group with a countable basis of its topology and let $\mu,\nu \in \mathscr{M}^1(G,\mathfrak{B}(G))$ such that $\mu,\nu \ll \lambda$. Then the following statements are equivalent:

(i) $\mathfrak{X}(\mu) > \mathfrak{X}(\nu)$.

(ii) There exists a translation invariant kernel $N \in \mathrm{Stoch}(G,\mathfrak{B}(G))$ satisfying

$$N(\mu * \varepsilon_x) = \nu * \varepsilon_x \quad \text{for all } x \in G.$$

(iii) There exists a measure $\rho \in \mathscr{M}^1(G,\mathfrak{B}(G))$ such that

$$\rho * \mu = \nu.$$

(iv) For all $f \in \mathscr{C}^b(G)$ we have

$$\int f\, d\nu \leq \sup_{x \in G} \int f\, d(\varepsilon_x * \mu)$$

Proof: 1. (i) $\Rightarrow$ (ii). In order to apply Theorem 23.2 we choose $\Omega: = G$, $\Omega_1: = G \cup \{\infty\}$ (the one-point compactification of G), $\Omega_I: = G$ and define

$$g_I(g'): = g_\Omega(g'): = g'g^{-1}$$

as well as

$$g_{\Omega_1}(g'): = \begin{cases} g'g^{-1} & \text{if } g' \in G \\ \infty & \text{if } g' = \infty. \end{cases}$$

Clearly g_{Ω_1} is bicontinuous, and $g \to g_{\Omega_1}^{-1}(\omega_1)$ is measurable for all $\omega_1 \in \Omega_1$.

Let $P_x: = \mu * \varepsilon_x$ and $Q_x(B): = \nu * \varepsilon_x(B \cap G)$ for all $x \in G = \Omega_I$ and $B \in \mathfrak{A}_1$. Then Theorem 23.2 yields the existence of an invariant kernel $\overline{N} \in \mathrm{Stoch}((\Omega,\mathfrak{A}),(\Omega_1,\mathfrak{A}_1))$ such that $\overline{N}(P_x) = Q_x$ holds for all

$x \in G$. Since the exceptional set C of $\overline{N}$ is G-invariant of λ-measure zero, we obtain $C = \emptyset$, in particular that

$$\overline{N}(g,\{\infty\}) = \overline{N}(e,\{\infty\})$$

for all $g \in G$. This implies that

$$\begin{aligned}\overline{N}(e,\{\infty\}) &= \int \overline{N}(g,\{\infty\})P_e(dg)\\ &= \overline{N}(P_e)(\{\infty\})\\ &= Q_e(\{\infty\}) = 0,\end{aligned}$$

i.e.,

$$N(g,B) := \overline{N}(g,B)$$

for all $g \in G$ and $B \in \mathfrak{B}(G)$ defines a translation invariant kernel $N \in \mathrm{Stoch}(G,\mathfrak{B}(G))$ having the desired property

$$N(\mu * \varepsilon_x) = \nu * \varepsilon_x$$

for all $g \in G$.

2. (ii) ⇒ (iii). Let N be chosen as in (ii). The measure $\rho := N(e,\cdot)$ satisfies the convolution equation of (iii).

3. (iii) ⇒ (ii). Let ρ be as in (iii). The kernel $N \in \mathrm{Stoch}(G,\mathfrak{B}(G))$ defined by

$$N(x,B) := \rho(Bx^{-1}) \quad \text{for all} \quad (x,B) \in G \times \mathfrak{B}(G)$$

satisfies the equation stated in (ii).

4. (ii) ⇒ (i) is a straightforward consequence of Theorem 23.2.

5. (iii) ⇒ (iv). Let $\rho * \mu = \nu$. Then for every $f \in \mathscr{C}^b(G)$ we have

$$\begin{aligned}\int f\,d\nu = \int f\,d(\rho * \mu) &= \iint f(xy)\rho(dx)\mu(dy)\\ &= \int\left(\int f\,d(\varepsilon_x * \mu)\right)\rho(dx)\\ &\le \sup_{x\in G}\int f\,d(\varepsilon_x * \mu).\end{aligned}$$

6. (iv) ⇒ (iii). For every $f \in \mathscr{C}^b(G)$ we define a real-valued function g_f on G by

$$g_f(x) := \int f\,d(\varepsilon_x * \mu)$$

for all $x \in G$ and put

$$\mathscr{W}_\mu := \{g \in \mathscr{C}^b(G): \text{There is an } f \in \mathscr{C}^b(G) \text{ such that } g = g_f\}.$$

Clearly $\mathscr{W}_\mu$ is a linear subspace of $\mathscr{C}^b(G)$ containing 1_G. On $\mathscr{W}_\mu$ we consider the positive linear functional T_0 given by

$$T_0(g) := \int f \, d\nu$$

for all $g \in \mathscr{W}_\mu$ of the form $g = g_f$ for $f \in \mathscr{C}^b(G)$.

In order to justify this definition we observe that $g_{f_1} = g_{f_2}$ for $f_1, f_2 \in \mathscr{C}^b(G)$ implies $\int (f_1 - f_2) d(\varepsilon_x * \mu) = 0$ for all $x \in G$ and thus

$$\left| \int f_1 d\nu - \int f_2 d\nu \right| \leq \left| \int (f_1 - f_2) d\nu \right|$$

$$\leq \sup_{\substack{x \in G \\ g \in \{f_1 - f_2, f_2 - f_1\}}} \int g \, d(\varepsilon_x * \mu) = 0.$$

We now apply the Hahn-Banach extension theorem to extend T_0 from $\mathscr{W}_\mu$ to a linear functional T on $\mathscr{C}^b(G)$ satisfying $||T|| = ||T_0||$. Since $T(1_G) = T_0(1_G) = ||T_0|| = ||T||$, T is positive. Thus the Riesz representation theorem yields the existence of a measure $\rho \in \mathscr{M}^1(G, \mathfrak{B}(G))$ such that

$$T(g) = \int g \, d\rho \quad \text{for all } g \in \mathscr{C}^b(G).$$

This implies for all $f \in \mathscr{C}^b(G)$

$$\int f \, d\nu = T(g_f) = \int g_f \, d\rho = \int \left(\int f \, d(\varepsilon_x * \mu) \right) \rho(dx)$$

$$= \iint f(xy) \rho(dx) \mu(dy),$$

i.e., $\nu = \rho * \mu$. □

§25. COMPARISON OF LINEAR NORMAL EXPERIMENTS

In specializing the theory of translation experiments to linear experiments involving the normal distribution we obtain an additional insight into the decision theoretical comparison which has become the dominant aspect of our exposition. We shall discuss in some detail translation experiments that are invariant with respect to measurable groups of the type $(\mathbb{R}^n, \mathfrak{B}^n)$ or $(\mathbb{R}^n \times \mathbb{R}_+^*, \mathfrak{B}(\mathbb{R}^n \times \mathbb{R}_+^*))$ for $n \geq 1$. Here we need to explain how we are going to make $\mathbb{R}^n \times \mathbb{R}_+^*$ a locally compact group. In order to achieve this we introduce for elements

$(x_1,\ldots,x_n,x^2)$ and $(\beta_1,\ldots,\beta_n,\sigma^2) \in \mathbb{R}^n \times \mathbb{R}_+^*$ the sum

$$(x_1,\ldots,x_n,x^2) \oplus (\beta_1,\ldots,\beta_n,\sigma^2) := (\beta_1 + |\sigma|x_1,\ldots,\beta_n + |\sigma|x_n,\sigma^2x^2).$$

The so defined operation $\oplus$ admits a neutral element $(0,\ldots,0,1)$, and for every $(x_1,\ldots,x_n,x^2) \in \mathbb{R}^n \times \mathbb{R}_+^*$ an inverse with respect to $(0,\ldots,0,1)$ is given by

$$\ominus(x_1,\ldots,x_n,x^2) := \left(-\frac{x_1}{|x|},\ldots,-\frac{x_n}{|x|},\frac{1}{x^2}\right).$$

In this fashion $\mathbb{R}^n \times \mathbb{R}_+^*$ becomes a group, which together with the natural topology in $\mathbb{R}^n \times \mathbb{R}_+^*$ is a locally compact group with a countable basis of its topology. Since the closed normal subgroup $\mathbb{R}^n \times \{1\}$ of $\mathbb{R}^n \times \mathbb{R}_+^*$ and the quotient group $\mathbb{R}^n \times \mathbb{R}_+^* / \mathbb{R}^n \times \{1\}$ are amenable as Abelian groups, $\mathbb{R}^n \times \mathbb{R}_+^*$ itself is amenable and the theorems of the preceding two sections can be applied.

<u>Preparations 25.1</u> on *linear normal experiments*. Let $n \geq 1$ be fixed. For any $k \geq n$ and $c := (c_1,\ldots,c_n) \in \mathbb{R}^n$ with $c_i \neq 0$ for all $i = 1,\ldots,n$ we consider the experiment

$$X(k;c,n) := \left(\mathbb{R}^k,\mathfrak{B}^k,\left\{\left(\bigotimes_{i=1}^{n} \nu_{\beta_i,\sigma^2/c_i^2}\right) \otimes \left(\bigotimes_{n+1}^{k} \nu_{0,\sigma^2}\right): (\beta_1,\ldots,\beta_n) \in \mathbb{R}^n,\ \sigma^2 \in \mathbb{R}_+^*\right\}\right)$$

and if additionally $\sigma^2 \in \mathbb{R}_+^*$ is given, the experiment

$$X(k;\sigma^2,c,n) := \left(\mathbb{R}^k,\mathfrak{B}^k,\left\{\left(\bigotimes_{i=1}^{n} \nu_{\beta_i,\sigma^2/c_i^2}\right) \otimes \left(\bigotimes_{n+1}^{k} \nu_{0,\sigma^2}\right): (\beta_1,\ldots,\beta_n) \in \mathbb{R}^n\right\}\right).$$

We note that the groups $\mathbb{R}^n$ and $\mathbb{R}^n \times \mathbb{R}_+^*$ act only on the first n components of $\mathbb{R}^k$.

In order that the comparison relationships

$$X(k;c,n) > X(\ell;c',n)$$

and

$$X(k;\sigma^2,c,n) > X(\ell;\sigma^2,c'n)$$

make sense also for $c := (c_1,\ldots,c_n) \in \mathbb{R}^n \setminus \{0\}$ and $c' := (c_1',\ldots,c_m') \in \mathbb{R}^m \setminus \{0\}$ with $n > m$, we extend the above definitions by putting

$$X(\ell,c;,n):=\left(\mathbb{R}^\ell,\mathfrak{B}^\ell,\left\{\left(\bigotimes_{i=1}^m \nu_{\beta_i,\sigma^2/c_i'^2}\right)\otimes\left(\bigotimes_{m+1}^\ell \nu_{0,\sigma^2}\right): (\beta_1,\ldots,\beta_n)\in\mathbb{R}^n,\ \sigma^2\in\mathbb{R}_+^*\right\}\right)$$

and

$$X(\ell;\sigma^2,c',n):=\left(\mathbb{R}^\ell,\mathfrak{B}^\ell,\left\{\left(\bigotimes_{i=1}^m \nu_{\beta_i,\sigma^2/c_i'^2}\right)\otimes\left(\bigotimes_{m+1}^\ell \nu_{0,\sigma^2}\right): (\beta_1,\ldots,\beta_n)\in\mathbb{R}^n\right\}\right)$$

respectively.

If $c:=(1,\ldots,1)\in\mathbb{R}^n$, the corresponding experiments will be abbreviated by $X(k;1,n)$ and $X(k;\sigma^2,1,n)$ respectively.

Theorem 25.2. Let $m,\ell,p\in\mathbb{N}$ with $m,\ell\geq p$ and $c:=(c_1,\ldots,c_p)\in\mathbb{R}^p$ with $c_i\neq 0$ for all $i=1,\ldots,p$. For every $\sigma^2\in\mathbb{R}_+^*$ the following statements are equivalent:

(i) $X(m;\sigma^2,1,p) > X(\ell;\sigma^2,c,p)$.

(ii) $X\left(\bigotimes_1^p \nu_{0,\sigma^2}\right) > X\left(\bigotimes_{i=1}^p \nu_{0,\sigma^2/c_i^2}\right)$.

(iii) $|c_i|\leq 1$ for all $i=1,\ldots,p$.

Proof: 1. (i) $\Leftrightarrow$ (ii). We consider the mappings $T:\mathbb{R}^m\to\mathbb{R}^p$ and $\overline{T}:\mathbb{R}^\ell\to\mathbb{R}^p$ defined by

$$T(x_1,\ldots,x_m):=(x_1,\ldots,x_p)\quad\text{for all}\quad (x_1,\ldots,x_m)\in\mathbb{R}^m$$

and

$$\overline{T}(x_1,\ldots,x_\ell):=(x_1,\ldots,x_p)\quad\text{for all}\quad (x_1,\ldots,x_\ell)\in\mathbb{R}^\ell.$$

From the Neyman criterion (Theorem 8.7) we conclude that T and $\overline{T}$ are sufficient statistics for the experiments $X(m;\sigma^2,1,p)$ and $X(\ell;\sigma^2,c,p)$ respectively (with $\sigma^2\in\mathbb{R}_+^*$). But then Theorems 22.12 and 21.5 imply the equivalences

$$X(m;\sigma^2,1,p)\sim\left(\mathbb{R}^p,\mathfrak{B}^p,\left\{T\left(\left(\bigotimes_{i=1}^p \nu_{\beta_i,\sigma^2}\right)\otimes\left(\bigotimes_{p+1}^m \nu_{0,\sigma^2}\right)\right): (\beta_1,\ldots,\beta_p)\in\mathbb{R}^p\right\}\right)$$

and

$$X(\ell;\sigma^2,c,p)\sim\left(\mathbb{R}^p,\mathfrak{B}^p,\left\{\overline{T}\left(\left(\bigotimes_{i=1}^p \nu_{\beta_i,\sigma^2/c_i^2}\right)\otimes\left(\bigotimes_{p+1}^\ell \nu_{0,\sigma^2}\right)\right): (\beta_1,\ldots,\beta_p)\in\mathbb{R}^p\right\}\right).$$

Since the experiments on the right side of these equivalences are identical with the experiments $X\left(\bigotimes_1^p \nu_{0,\sigma^2}\right)$ and $X\left(\bigotimes_{i=1}^p \nu_{\beta_i,\sigma^2/c_i^2}\right)$ respectively, the proof of 1 is complete.

2. (ii) ⇒ (iii). We apply Theorem 24.4 to obtain the existence of a measure $\rho \in \mathcal{M}^1(\mathbb{R}^p,\mathfrak{B}^p)$ satisfying

$$\rho * \bigotimes_1^p \nu_{0,\sigma^2} = \bigotimes_{i=1}^p \nu_{0,\sigma^2/c_i^2}.$$

For every $i = 1,\ldots,p$ let X_i denote the projection from $\mathbb{R}^p$ onto its i-th component. Then for every $i = 1,\ldots,p$ we have

$$X_i(\rho) * \nu_{0,\sigma^2} = \nu_{0,\sigma^2/c_i^2},$$

whence by Cramér's characterization theorem that $X_i(\rho)$ is either a normal distribution or a Dirac measure on $\mathbb{R}$. We conclude $\sigma^2 \le \sigma^2/c_i^2$ and thus $|c_i| \le 1$ for all $i = 1,\ldots,p$.

3. (iii) ⇒ (ii). Let $\sigma^2 \in \mathbb{R}_+^*$. Given the assumption $|c_i| \le 1$ for all $i = 1,\ldots,p$ we can define measures $\rho_i \in \mathcal{M}^1(\mathbb{R},\mathfrak{B})$ by

$$\rho_i := \begin{cases} \nu_{0,(1-c_i^2)\sigma^2/c_i^2} & \text{if } |c_i| < 1 \\ \varepsilon_0 & \text{if } |c_i| = 1. \end{cases}$$

Obviously the measure $\rho := \bigotimes_{i=1}^p \rho_i \in \mathcal{M}^1(\mathbb{R}^p,\mathfrak{B}^p)$ satisfies the equation

$$\rho * \bigotimes_1^p \nu_{0,\sigma^2} = \bigotimes_{i=1}^p \nu_{0,\sigma^2/c_i^2}.$$

Then the assertion follows from Theorem 24.4. □

<u>Thoerem 25.3</u>. Let $m,\ell,p \in \mathbb{N}$ and $c := (c_1,\ldots,c_p) \in \mathbb{R}^p$ with $c_i \neq 0$ for all $i = 1,\ldots,p$. The following statements are equivalent:

(i) $X(p + m;1,p) > X(p + \ell;c,p)$.

(ii) $X\left(\left(\bigotimes_1^p \nu_{0,1}\right) \otimes \chi_m^2\right) > X\left(\left(\bigotimes_{i=1}^p \nu_{0,1/c_i^2}\right) \otimes \chi_\ell^2\right)$.

(iii) $|c_i| \leq 1$ for all $i = 1,\ldots,p$, and

$$m \geq \ell + \mathrm{card}(\{i = 1,\ldots,p: |c_i| < 1\}).$$

Proof: 1. (i) ⇒ (ii). For all $\beta = (\beta_1,\ldots,\beta_p) \in \mathbb{R}^p$ and $\sigma^2 \in \mathbb{R}_+^*$ we define the measures

$$P_{\beta,\sigma^2} := \left(\bigotimes_{i=1}^{p} \nu_{\beta_i,\sigma^2}\right) \otimes \left(\bigotimes_{p+1}^{p+m} \nu_{0,\sigma^2}\right)$$

and

$$Q_{\beta,\sigma^2} := \left(\bigotimes_{i=1}^{p} \nu_{\beta_i,\sigma^2/c_i^2}\right) \otimes \left(\bigotimes_{p+1}^{p+\ell} \nu_{0,\sigma^2}\right).$$

Clearly $X(p + m;1,p) > X(p + \ell;c,p)$ iff

$$\left(\hat{\mathbb{R}}^{p+m},\hat{\mathfrak{B}}^{p+m}, \left\{P_{\beta,\sigma^2}: \beta \in \mathbb{R}^p, \sigma^2 \in \mathbb{R}_+^*\right\}\right) > \left(\hat{\mathbb{R}}^{p+\ell},\hat{\mathfrak{B}}^{p+\ell}, \left\{Q_{\beta,\sigma^2}: \beta \in \mathbb{R}^p, \sigma^2 \in \mathbb{R}_+^*\right\}\right),$$

where $\hat{\mathbb{R}} := \mathbb{R} \setminus \{0\}$ and $\hat{\mathfrak{B}} := \mathfrak{B}(\hat{\mathbb{R}})$.

The mappings $T: \hat{\mathbb{R}}^{p+m} \to \mathbb{R}^p \times \mathbb{R}_+^*$ and $\overline{T}: \hat{\mathbb{R}}^{p+\ell} \to \mathbb{R}^p \times \mathbb{R}_+^*$ defined by

$$T(x_1,\ldots,x_{p+m}) := (x_1,\ldots,x_p, \sum_{i=p+1}^{p+m} x_i^2)$$

for all $(x_1,\ldots,x_{p+m}) \in \hat{\mathbb{R}}^{p+m}$ and

$$\overline{T}(x_1,\ldots,x_{p+\ell}) := (x_1,\ldots,x_p, \sum_{i=p+1}^{p+\ell} x_i^2)$$

for all $(x_1,\ldots,x_{p+\ell}) \in \hat{\mathbb{R}}^{p+\ell}$ are sufficient statistics for the experiments

$$\left(\hat{\mathbb{R}}^{p+m},\hat{\mathfrak{B}}^{p+m},\left\{P_{\beta,\sigma^2}: \beta \in \mathbb{R}^p, \sigma^2 \in \mathbb{R}_+^*\right\}\right)$$

and

$$\left(\hat{\mathbb{R}}^{p+\ell},\hat{\mathfrak{B}}^{p+\ell},\left\{Q_{\beta,\sigma^2}: \beta \in \mathbb{R}^p, \sigma^2 \in \mathbb{R}_+^*\right\}\right)$$

respectively. By Theorems 21.5 and 22.12 these experiments are equivalent to the experiments

$$\left(\mathbb{R}^p \times \mathbb{R}_+^*, \mathfrak{B}(\mathbb{R}^p \times \mathbb{R}_+^*),\left\{T(P_{\beta,\sigma^2}): \beta \in \mathbb{R}^p, \sigma^2 \in \mathbb{R}_+^*\right\}\right)$$

and

$$\left(\mathbb{R}^p \times \mathbb{R}_+^*, \mathfrak{B}(\mathbb{R}^p \times \mathbb{R}_+^*),\left\{\overline{T}(Q_{\beta,\sigma^2}): \beta \in \mathbb{R}^p, \sigma^2 \in \mathbb{R}_+^*\right\}\right),$$

and therefore also to the experiments

$$\mathsf{X}\left((\bigotimes_1^p \nu_{0,1}) \otimes \chi_m^2\right)$$

and

$$\mathsf{X}\left((\bigotimes_{i=1}^p \nu_{0,1/c_i^2}) \otimes \chi_\ell^2\right)$$

respectively. This proves the assertion 1.

2. (ii) ⇒ (iii). First of all we introduce the measures

$$P: = \left(\bigotimes_1^p \nu_{0,1}\right) \otimes \chi_m^2 \quad \text{and} \quad Q: = \left(\bigotimes_{i=1}^p \nu_{0,1/c_i^2}\right) \otimes \chi_\ell^2$$

as well as

$$P_{\beta,\sigma^2}: = P * \varepsilon_{(\beta,\sigma^2)} \qquad \text{and} \qquad Q_{\beta,\sigma^2}: = Q * \varepsilon_{(\beta,\sigma^2)}$$

(for $\beta: = (\beta_1,\ldots,\beta_p) \in \mathbb{R}^p$, $\sigma^2 \in \mathbb{R}_+^*$) on $\mathbb{R}^p \times \mathbb{R}_+^*$. By assumption we have $\mathsf{X}(P) > \mathsf{X}(Q)$. Then Theorem 24.4 yields the existence of a measure $\rho \in \mathscr{M}^1(\mathbb{R}^p \times \mathbb{R}_+^*, \mathfrak{B}(\mathbb{R}^p \times \mathbb{R}_+^*))$ satisfying

$$\rho * P = Q.$$

Let us abbreviate $\Omega: = (\mathbb{R}^p \times \mathbb{R}_+^*) \times (\mathbb{R}^p \times \mathbb{R}_+^*)$, $\mathfrak{A}: = \{\emptyset, \mathbb{R}^p\} \otimes \mathfrak{B}(\mathbb{R}_+^* \times \mathbb{R}^p \times \mathbb{R}_+^*)$ and $\mu_{\beta,\sigma^2}: = P_{\beta,\sigma^2} \otimes \rho$ for $\beta \in \mathbb{R}^p$, $\sigma^2 \in \mathbb{R}_+^*$. Clearly $\mu: = \mu_{0,1} = P \otimes \rho$, where 0 denotes the vector $(0,\ldots,0) \in \mathbb{R}^p$. Finally we introduce the random variables

$$X_i: = pr_i: = pr_i^\Omega \quad (i = 1,\ldots,p),$$

$$X: = pr_{p+1},$$

$$Z_i: = pr_{p+1+i} \quad (i = 1,\ldots,p), \text{ and}$$

$$Z: = pr_{2p+2}$$

on $(\Omega,\mathfrak{A})$ and observe that

$$Q(B) = (((Z_1,\ldots,Z_p,Z) \oplus (X_1,\ldots,X_p,X))(\mu))(B)$$

holds for all $B \in \mathfrak{B}(\mathbb{R}^p \times \mathbb{R}_+^*)$. This relationship implies that the random variables $X_1 + Z_1\sqrt{X},\ldots,X_p + Z_p\sqrt{X}, ZX$ are independent with respect to μ, and so we have

$$\left(\bigotimes_{i=1}^{p} \nu_{0,1/c_i^2}\right) \otimes \chi_\ell^2 = \left(\bigotimes_{i=1}^{p} (X_i + Z_i\sqrt{X})(\mu)\right) \otimes (ZX(\mu)),$$

whence $\nu_{0,1/c_i^2} = \nu_{0,1} * (Z_i\sqrt{X})(\mu)$ for $i = 1,\dots,p$ and $\chi_\ell^2 = ZX(\mu)$. By Cramér's characterization theorem we conclude $|c_i| \leq 1$ for all $i = 1,\dots,p$ and $(Z_i\sqrt{X})(\mu) = \nu_{0,(1-c_i^2)/c_i^2}$ for all $i = 1,\dots,p$ with $|c_i| < 1$.

Let $\mathscr{P} := \{\mu_{0,\sigma^2} : \sigma^2 \in \mathbb{R}_+^*\}$ and put $g(\mu_{0,\sigma^2}) := \sigma^2$ for all $\mu_{0,\sigma^2} \in \mathscr{P}$ as well as

$$S := ZX + \sum_{\substack{i=1\\ |c_i|<1}}^{p} \frac{c_i^2}{1 - c_i^2} (Z_i\sqrt{X})^2$$

and

$$q := \ell + \operatorname{card}(\{i = 1,\dots,p : |c_i| < 1\}).$$

It follows that

$$\left(\frac{1}{\sigma^2} S\right)\left(\mu_{0,\sigma^2}\right) = \chi_q^2$$

and

$$\left(\frac{1}{\sigma^2} X\right)\left(\mu_{0,\sigma^2}\right) = \chi_m^2 \quad (\text{for } \sigma^2 \in \mathbb{R}_+^*).$$

Thus $\frac{1}{q} S$ and $\frac{1}{m} X$ are unbiased estimators for the estimation problem $(\Omega, \mathfrak{A}, \mathscr{P}, g : \mathscr{P} \to \mathbb{R}_+)$. Moreover, $\frac{1}{m} X$ is an MVU estimator for g with respect to the loss function V defined by

$$V\left(\mu_{0,\sigma^2}, x\right) := (x - \sigma^2)^2$$

for all $\mu_{0,\sigma^2} \in \mathscr{P}$, $x \in \mathbb{R}$, as one concludes from Corollary 17.4 by considering estimators that are independent of the first p variables. Therefore we get for all $\sigma^2 \in \mathbb{R}_+^*$

$$\int V\left(\mu_{0,\sigma^2}, \frac{1}{m}X(\omega)\right)\mu_{0,\sigma^2}(d\omega) \leq \int V\left(\mu_{0,\sigma^2}, \frac{1}{q}S(\omega)\right)\mu_{0,\sigma^2}(d\omega),$$

whence

$$\int \left(\frac{1}{m} X - \sigma^2\right)^2 d\mu_{0,\sigma^2} \leq \int \left(\frac{1}{q} S - \sigma^2\right)^2 d\mu_{0,\sigma^2}$$

and thus

$$\frac{2\sigma^4}{m} = \frac{\sigma^4}{m^2}\int\left(\frac{1}{\sigma^2}X - m\right)^2 d\mu_{0,\sigma^2} \leq \frac{\sigma^4}{q^2}\int\left(\frac{1}{\sigma^2}S - q\right)^2 d\mu_{0,\sigma^2} = \frac{2\sigma^4}{q},$$

from which $q \leq m$ follows. This together with the above implication of Cramér's theorem yields the assertion.

3. (iii) ⇒ (i). As in 1. we consider the measures P_{β,σ^2} and Q_{β,σ^2} for all $\beta := (\beta_1,\ldots,\beta_p) \in \mathbb{R}^p$ and $\sigma^2 \in \mathbb{R}_+^*$. For every $i = 1,\ldots,p+m$ let $X_i := pr_i := pr_i^\Omega$ the projection from $(\Omega,\mathfrak{A}) := (\mathbb{R}^{p+m},\mathfrak{B}(\mathbb{R}^{p+m}))$ onto the i-th component, and let $k := \operatorname{card}(\{i = 1,\ldots,p: |c_i| < 1\})$. By hypothesis (iii) we have $m \geq \ell + k$, whence the following random variables on $(\Omega,\mathfrak{A})$ are well-defined:

$$Y_i := \begin{cases} X_i & \text{for all } i = 1,\ldots,p \text{ and } |c_i| = 1, \\ X_i + \dfrac{\sqrt{1-c_i^2}}{c_i} X_{p+j(i)} & \text{for all } i = 1,\ldots,p \text{ with } |c_i| < 1, \\ & \text{where } j \text{ is a bijection from } \{i = 1,\ldots,p: |c_i| < 1\} \\ & \text{onto } \{p+1,\ldots,p+k\}, \\ X_{k+i} & \text{for all } i = p+1,\ldots,p+\ell \text{ without further restriction,} \end{cases}$$

and

$$Y := (Y_1,\ldots,Y_{p+m}).$$

Clearly $Y(P_{\beta,\sigma^2}) = Q_{\beta,\sigma^2}$ for all $\beta \in \mathbb{R}^p$, $\sigma^2 \in \mathbb{R}_+^*$, whence there exists a kernel $M \in \operatorname{Stoch}((\mathbb{R}^{p+m},\mathfrak{B}^{p+m}),(\mathbb{R}^{p+\ell},\mathfrak{B}^{p+\ell}))$, defined by

$$M(x,B) := 1_B \circ Y(x) \quad \text{for all } x \in \mathbb{R}^{p+m},\ B \in \mathfrak{B}^{p+\ell},$$

with the property

$$M(P_{\beta,\sigma^2}) = Q_{\beta,\sigma^2} \quad \text{for all } \beta \in \mathbb{R}^p,\ \sigma^2 \in \mathbb{R}_+^*.$$

Theorem 21.5 then implies $\mathrm{X}(p+m;1,p) > \mathrm{X}(p+\ell;c,p)$ which is the desired assertion. □

CHAPTER IX

Comparison of Finite Experiments

§26. COMPARISON BY k-DECISION PROBLEMS

In Section 19 we dealt with general decision problems of the form $\underline{\bar{D}} = (I, D, \mathscr{V})$, where $I: = (\Omega_I, \mathfrak{A}_I)$ and $D: = (\Omega_D, \mathfrak{A}_D)$ denoted measurable spaces and $\mathscr{V}$ a set of separately measurable functions on $\Omega_I \times \Omega_D$. From now on, and for the remainder of the chapter, we shall specialize the general framework in two steps: First we shall restrict our attention to decision problems of the form $\underline{\bar{D}}_k(I): = (I, D_k, \mathscr{V})$ with $D_k: = \{1, \ldots, k\}$ $(k \geq 1)$ as the decision space and the set $\mathscr{V}$ of all *bounded* , separately measurable functions on $\Omega_I \times \Omega_D$ as the set of loss functions. Later we shall consider decision problems of the form $\underline{\bar{D}}_k(I_n)$ with $I_n: = \{1, \ldots, n\}$ $(n \geq 1)$.

The following notation enlarges the terminology of Section 19: For two experiments $X, Y \in \mathscr{E}(\underline{\bar{D}}_k(I))$ and any tolerance function ε corresponding to $\underline{\bar{D}}_k(I)$ we put

$$X >_{\varepsilon}^{k} Y: \Longleftrightarrow X >_{\varepsilon}^{\underline{\bar{D}}_k(I)} Y.$$

Decision problems of the form $\underline{\bar{D}}_k(I)$ will be called *k-decision problems*. The comparison relation $X >_{\varepsilon}^{k} Y$ will then be read as X *is more informative than* Y *at level* ε *by k-decision problems*.

Moreover, we introduce the useful symbols

$$\rho_k(X,Y): = \rho_{\underline{\bar{D}}_k}(X,Y),$$

$$\Delta_k(X,Y): = \Delta_{\underline{\bar{D}}_k}(X,Y) \quad \text{for} \quad X,Y \in \mathscr{E}(\underline{\bar{D}}_k(I)),$$

and

$$\rho(X,Y) := \sup_{k\geq 1} \rho_k(X,Y),$$

$$\Delta(X,Y) := \sup_{k\geq 1} \Delta_k(X,Y) \quad \text{for} \quad X,Y \in \mathscr{E}(I),$$

without any particular reference to a decision problem.

Remark 26.1. Clearly $\rho_k(X,Y) \leq 2$ for all $k \geq 1$, and $\rho_1(X,Y) = \Delta_1(X,Y) = 0$.

While the first statement follows from Property 19.11.1, the second one is a direct consequence of the definitions of $\delta_1(X,Y)$ and $\Delta_1(X,Y)$ involving the trivial decision space $D = (\Omega_D, \mathbf{A}_D)$ with $\Omega_D := \{1\}$ and $\mathbf{A}_D := \mathfrak{P}(\Omega_D)$.

Theorem 26.2. For any two experiments $X := (\Omega, \mathbf{A}, (P_i)_{i\in\Omega_I})$ and $Y := (\Omega_1, \mathbf{A}_1, (Q_i)_{i\in\Omega_1})$ in $\mathscr{E}(\underline{D}_k(I))$ and corresponding tolerance function ε we have the implication

$$X >_\varepsilon^{k+1} Y \Rightarrow X >_\varepsilon^k Y.$$

Proof: Let $D_k := \{1,\ldots,k\}$. To given $V \in \mathscr{V}$ and $\delta_Y \in \mathscr{D}(Y,D_k)$ we shall construct $\delta_X \in \mathscr{D}(X,D_k)$ satisfying

$$R^V_{\delta_X}(i) \leq R^V_{\delta_Y}(i) + \varepsilon(i)||V|| \tag{*}$$

for all $i \in \Omega_I$. For every $i \in \Omega_I$ we put

$$V'(i,d) := \begin{cases} V(i,d) & \text{for} \quad d = 1,\ldots,k \\ V(i,k) & \text{for} \quad d = k+1. \end{cases}$$

Clearly δ_Y is determined by its values $\delta_Y(\omega_1,d)$ for $\omega_1 \in \Omega_1$ and $d \in D_k$. Defining $\delta_Y(\omega_1,k+1) := 0$ for all $\omega_1 \in \Omega_1$ we can consider δ_Y as an element of $\mathscr{D}(Y,D_{k+1})$. By assumption there exists a $\bar{\delta}_X \in \mathscr{D}(X,D_{k+1})$ such that

$$R^{V'}_{\bar{\delta}_X}(i) \leq R^{V'}_{\bar{\delta}_Y}(i) + \varepsilon(i)||V'|| \tag{**}$$

for all $i \in \Omega_I$. Now we set

$$\delta_X(\omega,d) := \begin{cases} \bar{\delta}_X(\omega,d) & \text{if} \quad d < k \\ \bar{\delta}_X(\omega,k) + \bar{\delta}_X(\omega,k+1) & \text{if} \quad d = k, \end{cases}$$

whenever $\omega \in \Omega$. Then $\Sigma_{d=1}^k \delta_X(\omega,d) = 1$ for all $\omega \in \Omega$, whence $\delta_X \in \mathscr{D}(X,D_k)$. It remains to show (*).

First of all we note that for all $i \in \Omega_I$ we have

$$R_{\bar{\delta}_X}^{V'}(i) = \int \sum_{d=1}^{k+1} V'(i,d)\bar{\delta}_X(\omega,d)P_i(d\omega)$$

$$= \int \sum_{d=1}^{k} V(i,d)\delta_X(\omega,d)P_i(d\omega)$$

$$= R_{\delta_X}^{V}(i),$$

and

$$R_{\delta_Y}^{V'}(i) = R_{\delta_X}^{V}(i),$$

since $\delta_Y(\omega_1,k+1) = 0$ for all $\omega_1 \in \Omega_1$. Finally, $||V'|| = ||V||$, thus (*) follows from (**). □

<u>Corollary 26.3</u>. The following limit relations hold:

(i) $\lim_{k\to\infty} \rho_k(X,Y) = \rho(X,Y)$.

(ii) $\lim_{k\to\infty} \Delta_k(X,Y) = \Delta(X,Y)$.

<u>Proof</u>: It suffices to show (i). From the theorem we conclude that the sequence $(\rho_k(X,Y))_{k\geq 1}$ is isotone. Moreover $(\rho_k(X,Y))_{k\geq 1}$ is bounded. But then

$$\lim_{k\to\infty} \rho_k(X,Y) = \sup_{k\to\infty} \rho_k(X,Y) = \rho(X,Y). \quad \square$$

Now we proceed to the discussion of comparison of experiments X, $Y \in \mathscr{E}(\underline{\bar{D}}_k(I_n))$, for which the decision problem $\underline{\bar{D}}_k(I_n)$ involves a finite decision set $D_k = \{1,\ldots,k\}$ and a finite parameter set $I_n := \{1,\ldots,n\}$. We shall continue dealing with experiments of the form $X = (\Omega,\mathbf{A},(P_i)_{i\in I_n})$ and $Y := (\Omega_1,\mathbf{A}_1,(Q_i)_{i\in I_n})$ having the same parameter set I_n.

Defining $P := \Sigma_{i\in I_n} P_i$ and $Q := \Sigma_{i\in I_n} Q_i$ we obtain measures $P,Q \in \mathscr{M}_+^b(\Omega,\mathbf{A})$ satisfying $P_i \ll P$, $Q_i \ll Q$ and therefore the existence of Radon-Nikodym densities $f_i := dP_i/dP$, $g_i := dQ_i/dQ$ of P_i, Q_i respectively, for all $i \in I_n$.

Let $f: \Omega \to \mathbb{R}^n$ and $g: \Omega_1 \to \mathbb{R}^n$ be defined by $f := (f_1,\ldots,f_n)$ and $g := (g_1,\ldots,g_n)$ respectively. Modifying the definitions of f and g on a set of measure 0 we get that

$$\sum_{i\in I_n} f_i = \sum_{i\in I_n} g_i = 1.$$

Remark 26.4. For all $\omega \in \Omega$ and $\omega_1 \in \Omega_1$ the vectors $(f_1(\omega),\ldots,f_n(\omega))$ and $(g_1(\omega_1),\ldots,g_n(\omega_1))$ define probability distributions on I_n. They are called the *aposteriori distributions* under ω and ω_1 respectively in the presence of the uniform distribution as apriori distribution.

Now we consider the set $\Psi(\mathbb{R}^n)$ of all sublinear functionals on $\mathbb{R}^n$, and for a given $r \geq 1$, the set

$$\Psi_r(\mathbb{R}^n) := \{\psi \in \Psi(\mathbb{R}^n) : \psi = \bigvee_{i=1}^{r} \ell_i \text{ for } \ell_1,\ldots,\ell_r \in (\mathbb{R}^p)^*\}.$$

Let X be a given experiment dominated by $\mu \in \mathcal{M}_+^\sigma(\Omega,\mathbf{A})$ and let $P_i := \tilde{f}_i \cdot \mu$ with $\tilde{f}_i$ denoting the Radon-Nikodym density of P_i with respect to μ, for all $i \in I_n$.

Definition 26.5. For any $\psi \in \Psi(\mathbb{R}^n)$ we introduce the number

$$\psi(X) := \int \psi(dP_1,\ldots,dP_n) := \int \psi(\tilde{f}_1,\ldots,\tilde{f}_n)d\mu.$$

In order to justify this definition we add the

Remark 26.6. $\psi(X)$ is well-defined, and

$$\psi(X) = \int \psi(f_1,\ldots,f_n)dP,$$

where $f = (f_1,\ldots,f_n)$ and P are as above. In fact, putting $\tilde{f}_i := dP_i/d\mu$ we get for all $i \in I_n$

$$f_i = \frac{dP_i}{d\left(\sum_{i\in I_n} P_i\right)} = \frac{\tilde{f}_i}{\sum_{i\in I_n} \tilde{f}_i} \; [P].$$

Since ψ is positive homogeneous,

$$\int \psi(\tilde{f}_1,\ldots,\tilde{f}_n)d\mu = \int \psi\left(\frac{\tilde{f}_1}{\sum_{i\in I_n}\tilde{f}_i},\ldots,\frac{\tilde{f}_n}{\sum_{i\in I_n}\tilde{f}_i}\right)\left(\sum_{i\in I_n}\tilde{f}_i\right)d\mu$$

$$= \int \psi(f_1,\ldots,f_n)\sum_{i\in I_n} dP_i = \int \psi(f_1,\ldots,f_n)dP.$$

Examples 26.7. Examples of numbers $\psi(X)$ successfully used in mathematical statistics are to be gotten for experiments $X = (\Omega,\mathbf{A},(P_i)_{i\in I_2})$ and sublinear functionals $\psi \in \Psi(\mathbb{R}^2)$ as the integrals

(1) $\gamma(P_1,P_2) := \int \sqrt{dP_1 dP_2}$,

(2) $D^2(P_1,P_2) := \int (\sqrt{dP_1} - \sqrt{dP_2})^2 = 2(1 - \gamma(P_1,P_2))$ and

(3) $\int dP_1 \vee dP_2$.

$\gamma(P_1,P_2)$ is known as the *affinity*, $D(P_1,P_2)$ as the *Hellinger distance* of P_1 and P_2.

Theorem 26.8. Let $\mathsf{X} \in \mathscr{E}(\underline{D}_k, (I_n))$ be as above. By Λ_0 we denote the uniform distribution on $(I_n, \mathfrak{P}(I_n))$.

(i) For every $\psi \in \Psi_k(\mathbb{R}^n)$ there exists a $V \in \mathscr{V}$ such that

$$\psi(\mathsf{X}) = -n \inf_{\delta_{\mathsf{X}} \in \mathscr{D}(\mathsf{X}, D_k)} r(\delta_{\mathsf{X}} | \Lambda_0) \qquad (*)$$

(ii) For every $V \in \mathscr{V}$ there exists a $\psi \in \Psi_k(\mathbb{R}^n)$ satisfying (*) of (i).

Proof: (i) Every $\psi \in \Psi_k(\mathbb{R}^n)$ is of the form

$$\psi(x) = \bigvee_{d=1}^{k} \sum_{i=1}^{n} a_{i,d}\, x_i$$

for all $x = (x_1,\ldots,x_n) \in \mathbb{R}^n$ and coefficients $a_{i,d} \in \mathbb{R}$ $(i \in I_n,\ d \in D_k)$. We define V by $V(i,d) := -a_{i,d}$ for all $i \in I_n$, $d \in D_k$. Then

$$\psi(\mathsf{X}) = \int \psi(f_1,\ldots,f_n)dP = \int \bigvee_{d=1}^{k} \sum_{i=1}^{n} (-V(i,d)) f_i(\omega) P(d\omega).$$

For all $\delta_{\mathsf{X}} \in \mathscr{D}(\mathsf{X}, D_k)$ we have

$$\begin{aligned}\psi(\mathsf{X}) &\geq \int \sum_{d=1}^{k} \sum_{i=1}^{n} (-V(i,d)) f_i(\omega) \delta_{\mathsf{X}}(\omega,d) P(d\omega)\\ &= -\sum_{i=1}^{n} \sum_{d=1}^{k} V(i,d) \int \delta_{\mathsf{X}}(\omega,d) P_i(d\omega)\\ &= -\sum_{i=1}^{n} \sum_{d=1}^{k} V(i,d) \delta_{\mathsf{X}}(P_i)(\{d\})\\ &= -\sum_{i=1}^{n} R^V_{\delta_{\mathsf{X}}}(i) = -nr(\delta_{\mathsf{X}} | \Lambda_0)\end{aligned}$$

or

$$r(\delta_{\mathsf{X}} | \Lambda_0) \geq -\frac{1}{n} \psi(\mathsf{X}),$$

where equality holds if δ_{X} for every $\omega \in \Omega$ assigns mass 1 to those $d \in D_k$ for which the maximum is attained. Thus

$$\inf_{\delta_X \in \mathscr{D}(X,D_k)} r(\delta_X|\Lambda_0) = -\frac{1}{n}\psi(X).$$

(ii) The proof follows directly if we define for a given $V \in \mathscr{V}$ the corresponding function $\psi \in \Psi_k(\mathbb{R}^n)$ by

$$\psi(x) := \bigvee_{d=1}^{k} \sum_{i=1}^{n} (-V(i,d))x_i$$

for all $x = (x_1,\dots,x_n) \in \mathbb{R}^n$. □

The following result contains a natural generalization of Theorem 21.1 to ε-informativity.

Theorem 26.9. Let $X,Y \in \mathscr{E}(\underline{\overline{D}}_k(I_n))$ and let ε be a tolerance function corresponding to $\underline{\overline{D}}_k(I_n)$. The following statements are equivalent:

(i) $X >_\varepsilon^k Y$.

(ii) For every $\delta_Y \in \mathscr{D}(Y,D_k)$ and all $V \in \mathscr{V}$ there exists a $\delta_X \in \mathscr{D}(X,D_k)$ such that

$$\sum_{i=1}^{n} R_{\delta_X}^{V}(i) \le \sum_{i=1}^{n} R_{\delta_Y}^{V}(i) + \sum_{i=1}^{n} \varepsilon(i)||V(i,\cdot)||.$$

(iii) For every $\delta_Y \in \mathscr{D}(Y,D_k)$ there exists a $\delta_X \in \mathscr{D}(X,D_k)$ such that

$$||\delta_X(P_i) - \delta_Y(Q_i)|| \le \varepsilon(i)$$

for all $i \in I_n$.

(iv) $$\psi(X) \ge \psi(Y) - \sum_{i=1}^{n} \varepsilon(i)(\psi(e_i) \vee \psi(-e_i))$$

for all $\psi \in \Psi_k(\mathbb{R}^n)$, where e_i denotes the vector $(0,\dots,1,0,\dots,0) \in \mathbb{R}^n$ with 1 at the i-th place.

Remark 26.10. Statement (iv) of the theorem contains the ε-comparison of the minimal Bayes' risks of the experiments X and Y with respect to the uniform distribution on I_n. This follows from Theorem 26.8.

Proof of the Theorem: 1. The equivalence (i) ⟺ (ii) ⟺ (iii) is shown in analogy to the corresponding equivalence of Theorem 21.1, the non-trivial implication being (ii) ⇒ (iii). For its proof one considers the concave-convex game $\Gamma := (A,B,M)$ with $A := \mathscr{V}$, $B := \mathscr{R}$ and $M := \Phi$ defined by

$$\Phi(V,T) := \sum_{i=1}^{n} \left(\int T(v(i))dP_i - \int V(i,\cdot)d\delta_Y(Q_i) - \varepsilon(i)||V(i,\cdot)|| \right)$$

for all $V \in \mathscr{V}$, $T \in \mathscr{R}$ and fixed $\delta_Y \in \mathscr{D}(Y,D_k)$, and applies Theorem 2.7 in the same manner as in the proof of Theorem 21.1.

2. (ii) $\Leftrightarrow$ (iv). Let $\psi \in \psi_k(\mathbb{R}^n)$. By Theorem 26.8 we can choose $V \in \mathscr{V}$ such that

$$\psi(X) = -n \inf_{\delta_X \in \mathscr{D}(X,D_k)} r(\delta_X|\Lambda_0)$$

holds, where Λ_0 denotes the uniform distribution on $(I_n,\mathfrak{P}(I_n))$. From (ii) we obtain

$$(1) \qquad \inf_{\delta_X \in \mathscr{D}(X,D_k)} \sum_{i=1}^n R^V_{\delta_X}(i) \leq \sum_{i=1}^n R^V_{\delta_Y}(i) + \sum_{i=1}^n \varepsilon(i)||V(i,\cdot)||$$

$$(2) \qquad \inf_{\delta_X \in \mathscr{D}(X,D_k)} \sum_{i=1}^n R^V_{\delta_X}(i) \leq \inf_{\delta_Y \in \mathscr{D}(Y,D_k)} \sum_{i=1}^n R^V_{\delta_Y}(i) + \sum_{i=1}^n \varepsilon(i)||V(i,\cdot)||,$$

and after multiplication by (-1)

$$(3) \qquad \psi(X) \geq \psi(Y) - \sum_{i=1}^n \varepsilon(i)||V(i,\cdot)||.$$

From the proof of Theorem 26.8 we have for every $i \in I_n$

$$\psi(e_i) = \bigvee_{d=1}^k (-V(i,d)),$$

$$\psi(-e_i) = \bigvee_{d=1}^k V(i,d),$$

thus

$$\psi(e_i) \vee \psi(-e_i) = \bigvee_{d=1}^k |V(i,d)| = ||V(i,\cdot)||,$$

and the implication (ii) $\Rightarrow$ (iv) has been proved.

Since the inequalities (1), (2) and (3) happen to be equivalent to (ii), the full equivalence (ii) $\Leftrightarrow$ (iv) has been shown. □

Corollary 26.11. For every $k \geq 1$

(i) $\rho_k(X,Y) = 0$ iff $\psi(X) \geq \psi(Y)$ for all $\psi \in \Psi_k(\mathbb{R}^n)$.

(ii) $\Delta_k(X,Y) = 0$ iff $\psi(X) = \psi(Y)$ for all $\psi \in \Psi_k(\mathbb{R}^n)$.

Proof: It suffices to show (i). But this equivalence is a direct consequence of Theorem 26.9. Indeed, let $\rho_k(X,Y) = 0$ and let $\psi \in \Psi_k(\mathbb{R}^n)$. Then for any $\varepsilon > 0$ we have

$$\psi(X) \geq \psi(Y) - \varepsilon \sum_{i=1}^n (\psi(e_i) \vee \psi(-e_i)).$$

This implies $\psi(X) \geq \psi(Y)$. The converse is obvious. □

Corollary 26.12. For any $k \geq 1$ the following statements are equivalent:

(i) $X >_{\varepsilon}^{k} Y$.

(ii) $\psi(X) \geq \psi(Y) - \frac{1}{2} \sum_{i=1}^{n} \varepsilon(i)(\psi(e_i) + \psi(-e_i))$ for all $\psi \in \Psi_k(\mathbb{R}^n)$.

(iii) $\psi(X) \geq \psi(Y) - \sum_{i=1}^{n} \varepsilon(i)\psi(e_i)$ for all $\psi \in \Psi_k(\mathbb{R}^n)$ satisfying $\psi(-e_i) = \psi(e_i)$ whenever $i \in I_n$.

Proof: 1. (ii) ⇒ (i) follows directly from Theorem 26.9.

2. (i) ⇒ (ii). Let statement (iv) of Theorem 26.9 be satisfied. For every $\psi \in \Psi_k(\mathbb{R}^n)$ we define $\psi' := \psi + \ell$ with

$$\ell(x) := \frac{1}{2} \sum_{i=1}^{n} (\psi(-e_i) - \psi(e_i))x_i$$

for all $x = (x_1,\ldots,x_n) \in \mathbb{R}^n$. Clearly, $\psi' \in \Psi_k(\mathbb{R}^n)$.

But now $\Delta_1(X,Y) = 0$ implies $\ell(X) = \ell(Y)$ by Corollary 26.11. Then another application of Theorem 26.9 yields

$$\psi(X) \geq \psi(Y) - \sum_{i=1}^{n} \varepsilon(i)(\psi'(e_i) \vee \psi'(-e_i)).$$

Noting that

$$\psi'(e_i) = \psi(e_i) + \ell(e_i) = \frac{1}{2}(\psi(-e_i) + \psi(e_i))$$

and

$$\psi'(-e_i) = \frac{1}{2}(\psi(-e_i) + \psi(e_i))$$

for all $i \in I_n$ we conclude (ii).

3. (ii) ⇒ (iii) is easily verified.

4. For the proof of (iii) ⇒ (ii) we note that

$$\psi'(e_i) = \psi'(-e_i)$$

holds for all $i \in I_n$. □

Corollary 26.13. For any $k \geq 1$ we have

$$\Delta_k(X,Y) = \sup_{\psi \in \Gamma_k} |\psi(X) - \psi(Y)|,$$

where

$$\Gamma_k := \left\{ \psi \in \Psi_k(\mathbb{R}^n) : \psi(-e_i) = \psi(e_i) \quad \text{for all} \quad i \in I_n, \ \sum_{i=1}^{n} \psi(e_i) = 1 \right\}.$$

The **proof** is based on Corollary 26.12: One just notices that in statement (iii) of that corollary one may, without loss of generality, consider sublinear functionals ψ satisfying the additional hypothesis $\sum_{i=1}^{n} \psi(e_i) = 1$. □

Remark 26.14. The statements of Corollaries 26.11 and 26.13 remain valid for experiments $X, Y \in \mathscr{E}(I_n)$ if we replace the functions ρ_k and Δ_k by ρ and Δ, and the sets $\Psi_k(\mathbb{R}^n)$ and Γ_k by $\Psi(\mathbb{R}^n)$ and Γ (as a subset of $\Psi(\mathbb{R}^n)$) respectively.

We shall perform the proof of (i) in Corollary 26.11 for ρ and $\Psi(\mathbb{R}^n)$ in place of ρ_k and $\Psi_k(\mathbb{R}^n)$.

If $\rho(X,Y) = 0$, then $\rho_k(X,Y) = 0$ for all $k \geq 1$ by (i) of Corollary 26.3, whence $\psi(X) \geq \psi(Y)$ for every $\psi \in \Psi_1(\mathbb{R}^n) \cup \Psi_2(\mathbb{R}^n) \cup \ldots$ by (i) of Corollary 26.11. Let $\psi \in \Psi(\mathbb{R}^n)$. Then $\psi = \lim_{k \to \infty} \psi_k$ with $\psi_1 \leq \psi_2 \leq \ldots$ and $\psi_k \in \Psi_k(\mathbb{R}^n)$ for all $k \geq 1$. Beppo Levi's theorem implies $\lim_{k \to \infty} \psi_k(X) = \psi(X)$ and $\lim_{k \to \infty} \psi_k(Y) = \psi(Y)$, and from $\psi_k(X) \geq \psi_k(Y)$ for all $k \geq 1$ we deduce $\psi(X) \geq \psi(Y)$.

If, conversely, $\psi(X) \geq \psi(Y)$ for all $\psi \in \Psi(\mathbb{R}^n)$, then in particular $\psi(X) \geq \psi(Y)$ for all $\psi \in \Psi_k(\mathbb{R}^n)$ and all $k \geq 1$, whence $\rho_k(X,Y) = 0$ for all $k \geq 1$, thus $\rho(X,Y) = 0$ by (i) of Corollary 26.3.

Example 26.15. Consider the experiment $X = (\Omega, \mathbf{A}, (P_i)_{i \in I})$ with $\Omega := I_r$, $\mathbf{A} := \mathfrak{P}(I_r)$ and $I := I_n$. We put $p_{ij} := P_i(\{j\})$ for all $i \in I_n$, $j \in I_r$. Then X is determined by the stochastic matrix

$$P^X = (p_{ij}) = \begin{pmatrix} p_{11} & \cdots & p_{1r} \\ \vdots & & \\ p_{n1} & & p_{nr} \end{pmatrix}.$$

Let μ denote the counting measure on $(\Omega, \mathbf{A})$.

1. For each $i \in I_n$ we have $P_i = \tilde{f}_i \cdot \mu$ with $\tilde{f}_i$ defined by $\tilde{f}_i(j) := p_{ij}$ for all $j \in I_r$. Thus

2.
$$\psi(X) = \int \psi(\tilde{f}_1, \ldots, \tilde{f}_n) d\mu = \sum_{j=1}^{r} \psi(\tilde{f}_1(j), \ldots, \tilde{f}_n(j))$$

$$= \sum_{j=1}^{r} \psi(p_{1j}, \ldots, p_{nj}).$$

Now let $Y = (\Omega_1, \mathbb{A}_1, (Q_i)_{i\in I})$ be another experiment with $\Omega_1 := \Omega$ and $\mathbb{A}_1 := \mathbb{A}$, determined by the stochastic matrix $P^Y = (q_{ij})$, and let

$$\psi \in \Gamma := \left\{\psi \in \Psi(\mathbb{R}^n) : \psi(-e_i) = \psi(e_i) \text{ for all } i \in I_n, \sum_{i=1}^n \psi(e_i) = 1\right\}$$

3. By the properties of ψ we get

$$\begin{aligned}|\psi(X) - \psi(Y)| &= \left|\sum_{j=1}^r \psi(p_{1j},\ldots,p_{nj}) - \sum_{j=1}^r \psi(q_{1j},\ldots,q_{nj})\right| \\ &\le \sum_{j=1}^r |\psi(p_{1j},\ldots,p_{nj}) - \psi(q_{1j},\ldots,q_{nj})| \\ &\le \sum_{j=1}^r \sum_{i=1}^n |p_{ij} - q_{ij}|\psi(e_i) \\ &= \sum_{i=1}^n \psi(e_i) \sum_{j=1}^r |p_{ij} - q_{ij}| \\ &\le \max_{1\le i\le n} \left|\sum_{j=1}^r p_{ij} - q_{ij}\right|.\end{aligned}$$

It follows by Corollary 26.13 that

$$\Delta(X,Y) \le \max_{1\le i\le n} \sum_{j=1}^r |p_{ij} - q_{ij}|.$$

<u>Example 26.16</u>. Let X and Y be as in the previous example with $\Omega := \Omega_1 := I_n$. Moreover, let $P^X = \mathrm{Id}$ and $P^Y = (q_{ij})$ with $q_{ij} := \frac{1}{n}$ for all $i,j \in I_n$. That is to say, X is the experiment where the "true value" of i is observed or equivalently the experiment that contains all information about i, and Y is the experiment that contains no information about i. (Compare the general discussion in Chapter X.)

We want to establish an upper bound for $\Delta(X,Y)$. For this we note that for any $i \in I_n$ we have

$$\begin{aligned}\sum_{j=1}^r |p_{ij} - q_{ij}| &= |p_{ii} - q_{ii}| + \sum_{j\neq i} |p_{ij} - q_{ij}| \\ &= 1 - \frac{1}{n} + \sum_{j\neq i} q_{ij} = 2(1 - \frac{1}{n}) = 2 - \frac{2}{n},\end{aligned}$$

which by 3 of the previous example yields

$$\Delta(X,Y) \le 2 - \frac{2}{n}.$$

Now we define $\psi \in \Psi(\mathbb{R}^n)$ by

$$\psi(x) := \frac{2}{n} \bigvee_{i=1}^{n} x_i - \frac{1}{n} \sum_{i=1}^{n} x_i$$

for all $x := (x_1,\ldots,x_n) \in \mathbb{R}^n$. Clearly $\psi(e_i) = \psi(-e_i) = \frac{1}{n}$ for all $i \in I_n$, and $\Sigma_{i=1}^{n} \psi(e_i) = 1$, thus $\psi \in \Gamma$.

By 2 of the preceding example we obtain

$$\psi(X) = \sum_{j=1}^{n} \psi(e_j) = 1,$$

and

$$\psi(Y) = \sum_{j=1}^{n} \psi(\tfrac{1}{n},\ldots,\tfrac{1}{n}) = n\psi(\tfrac{1}{n},\ldots,\tfrac{1}{n}) = \psi(1,\ldots,1) = \frac{2}{n} - 1.$$

It follows that

$$\psi(X) - \psi(Y) = 2 - \frac{2}{n},$$

whence

$$\Delta(X,Y) \geq 2 - \frac{2}{n} .$$

This together with the above upper bound of $\Delta(X,Y)$ yields

$$\Delta(X,Y) = 2 - \frac{2}{n} .$$

Since $\rho(X,Y) = 0$, we also have $\rho(Y,X) = 2 - 2/n$.

It can be shown that for arbitrary experiments Z and W with parameter set I_n one gets

$$\Delta(Z,W) \leq 2 - \frac{2}{n} ,$$

since Z and W can be compared with respect to ρ with the *maximal informative experiment* X and the *minimal informative experiment* Y.

§27. COMPARISON BY TESTING PROBLEMS

The strongest and most applicable results established within the theory of comparison of experiments concern the comparison of testing experiments. The subsequent discussion emphasizes the geometric insight into the theory of *comparison* by 2-decision problems or equivalently *by testing problems*. Some well-known facts from the theory of testing statistical hypotheses will appear in a new and possibly brighter light.

We keep the notation of the preceding section, giving ourselves two experiments $X,Y \in \mathscr{E}(\underline{D}_2(I_n))$ and an arbitrary tolerance function ε.

Theorem 27.1. The following statements are equivalent:

(i) $X >^2_\varepsilon Y$.

(ii) For all $a = (a_1,\dots,a_n) \in \mathbb{R}^n$ we have

$$\|\sum_{i=1}^n a_i P_i\| \geq \|\sum_{i=1}^n a_i Q_i\| - \sum_{i=1}^n \varepsilon(i)|a_i|.$$

Proof: Every $\psi \in \Psi_2(\mathbb{R}^n)$ is of the form $\psi = \ell_1 \vee \ell_2$ with $\ell_1,\ell_2 \in (\mathbb{R}^n)^*$, and $\ell_1 \vee \ell_2 = \frac{1}{2}(\ell_1 + \ell_2 - |\ell_1 - \ell_2|)$. It follows that $\psi = L_1 + |L_2|$ with $L_1,L_2 \in \Psi_1(\mathbb{R}^n)$, whence $|L_2| = L_2 \vee (-L_2) \in \Psi_2(\mathbb{R}^n)$. This implies by Corollary 26.12 that (i) is equivalent to

$$\psi(X) \geq \psi(Y) - \sum_{i=1}^n \varepsilon(i)\psi(e_i)$$

for all $\psi = |L|$ with $L \in \Psi_1(\mathbb{R}^n)$.

But $L \in \Psi_1(\mathbb{R}^n)$ is of the form $L(x) = \sum_{i=1}^n a_i x_i$ for all $x = (x_1,\dots,x_n) \in \mathbb{R}^n$ and some $a := (a_1,\dots,a_n) \in \mathbb{R}^n$. Then for $\psi := |L|$ we get

$$\psi(X) = \int\left|\sum_{i=1}^n a_i f_i\right| dP = \int\left|\frac{d\left(\sum_{i=1}^n a_i P_i\right)}{dP}\right| dP$$
$$= \|\sum_{i=1}^n a_i P_i\|,$$

where we use the notation of §26. Analogously we obtain

$$\psi(Y) = \|\sum_{i=1}^n a_i Q_i\|.$$

Since $\psi(e_i) = |a_i|$ for all $i \in I_n$, the assertion follows. □

In order to present a *geometric interpretation* of the preceding criterion we consider for a given experiment $X \in \mathscr{E}(\underline{D}_2(I_n))$ the set $\mathfrak{M}_X := \mathfrak{M}_{(1)}(\Omega,\mathfrak{A})$ of all *tests* and the subset

$$V_X := \left\{\left(\int t\, dP_1,\dots,\int t\, dP_n\right) : t \in \mathfrak{M}_X\right\}$$

of $[0,1]^n$ consisting of all (available) *power vectors corresponding to* X. For any $x,y \in \mathbb{R}^n$ we put

$$I_{[x,y]} := \{z := (z_1,\dots,z_n) \in \mathbb{R}^n : x_i \leq z_i \leq y_i \text{ for all } i \in I_n\}.$$

Theorem 27.2. The following statements are equivalent:

(i) $X >_{\varepsilon}^{2} Y$.

(ii) $V_X + \frac{1}{2} I_{[-\varepsilon,\varepsilon]} \supset V_Y$.

Proof: We consider the support function H_X of the compact convex set V_Y defined by

$$H_X(a) := \sup_{y \in V_X} \langle a,y \rangle = \sup_{t \in \mathfrak{M}_X} \sum_{i=1}^{n} a_i \int t \, dP_i$$
$$= \sup_{t \in \mathfrak{M}_X} \int t d\left(\sum_{i=1}^{n} a_i P_i \right)$$

for all $a = (a_1,\ldots,a_n) \in \mathbb{R}^n$.

For any measure $\mu \in \mathcal{M}^b(\Omega,\mathfrak{A})$ we have

$$||\mu|| = \sup_{||f|| \leq 1} \int f \, d\mu = 2\left[\sup_{||f|| \leq 1} \int \frac{f+1}{2} d\mu - \frac{1}{2} \mu(\Omega) \right]$$

$$= 2 \sup_{t \in \mathfrak{M}_X} \int t \, d\mu - \mu(\Omega),$$

whence

$$\sup_{t \in \mathfrak{M}_X} \int t \, d\mu = \frac{||\mu|| + \mu(\Omega)}{2} .$$

Applying this equality to $\mu := \Sigma_{i=1}^{n} a_i P_i$ we obtain

$$H_X(a) = \frac{|| \sum_{i=1}^{n} a_i P_i || + \sum_{i=1}^{n} a_i}{2}$$

for all $a \in \mathbb{R}^n$, and a similar representation for H_Y.

Now we note that the support function H of $\frac{1}{2} I_{[-\varepsilon,\varepsilon]}$ is given by

$$H(a) := \frac{1}{2} \sum_{i=1}^{n} |a_i| \varepsilon(i) \quad \text{for all} \quad a \in \mathbb{R}^n.$$

Theorem 27.1 implies that $X >_{\varepsilon}^{2} Y$ is equivalent to

$$H_X(a) + H(a) \geq H_Y(a)$$

for all $a \in \mathbb{R}^n$, and this turns out, by the properties of support functions, to be equivalent to

$$V_X + \frac{1}{2} I_{[-\varepsilon,\varepsilon]} \supset V_Y. \qquad \square$$

Corollary 27.3. The following statements are equivalent:

(i) $X >^2_\varepsilon Y$.

(ii) For every testing problem $H_0: i \in I_0$ versus $H_1: i \in I \setminus I_0$ and for every power vector β_Y corresponding to Y there exists a power vector β_X corresponding to X such that

$$\begin{cases} \beta_X(i) \leq \beta_Y(i) + \frac{1}{2}\varepsilon(i) & \text{if } i \in I_0 \\ \beta_X(i) \geq \beta_Y(i) - \frac{1}{2}\varepsilon(i) & \text{if } i \in I \setminus I_0. \end{cases}$$

Proof: Given the power vector β_Y corresponding to Y there exists a $t \in \mathfrak{M}_Y$ satisfying

$$y_t := (\beta_Y(1),\ldots,\beta_Y(n)) = \left(\int t\, dQ_1,\ldots,\int t\, dQ_n\right) \in V_Y.$$

The theorem implies $y_t \in V_X + \frac{1}{2} I_{[-\varepsilon,\varepsilon]}$, thus there exists a $t' \in \mathfrak{M}_X$ satisfying

$$\left(\int t\, dQ_1,\ldots,\int t\, dQ_n\right) = \left(\int t'\, dP_1,\ldots,\int t'\, dP_n\right) + \frac{1}{2}(a_1\varepsilon(1),\ldots,a_n\varepsilon(n))$$

for some $a_1,\ldots,a_n \in [-1,1]$. Defining

$$\beta_X(i) := \int t'\, dP_i \quad \text{for all } i \in I_n$$

we obtain

$$|\beta_X(i) - \beta_Y(i)| = \frac{1}{2}|a_i|\varepsilon(i) \leq \frac{1}{2}\varepsilon(i)$$

for all $i \in I_n$, which implies the assertion. □

Remark 27.4. For every $X \in \mathscr{E}(\underline{D}_2(I_n))$ the set V_X is a symmetric, compact, convex subset of $[0,1]^n$ containing the origin. If $n = 2$ one can show that any subset of $\mathbb{R}^n$ having these geometric properties is in fact of the form aV_X for some X and $a \in \mathbb{R}$. For $n > 2$, however, this statement is generally false.

Let $\mathfrak{K}(\mathbb{R}^n)$ denote the system of all compact subsets of $\mathbb{R}^n$. For every $x,y \in \mathbb{R}^n$ and sets $K, K_1, K_2 \in \mathfrak{K}(\mathbb{R}^n)$ we put

$$d(x,y) := ||x-y||_\infty,$$

$$d(x,K) := \inf_{y \in K} d(x,y),$$

$$d(K_i,K_j) := \sup_{x_i \in K_i} d(x_i,K_j) \quad \text{for } i \neq j \text{ in } \{1,2\},$$

and

$$d_H(K_1,K_2) := d(K_1,K_2) \vee d(K_2,K_1).$$

The mapping $(K_1,K_2) \to d_H(K_1,K_2)$ is called the *Hausdorff metric* on $\mathbb{K}(\mathbb{R}^n)$.

Corollary 27.5. We have

$$\Delta_2(X,Y) = 2d_H(V_X,V_Y).$$

Proof: 1. For every $x \in V_X$ there exists by the theorem a $y \in V_Y$ such that

$$||x - y||_\infty \le \frac{1}{2}\Delta_2(X,Y).$$

Therefore we obtain

$$d(x,V_Y) \le \frac{1}{2}\Delta_2(X,Y) \quad \text{for all} \quad x \in V_X$$

and consequently

$$d_H(V_X,V_Y) \le \frac{1}{2}\Delta_2(X,Y).$$

2. By definition of the Hausdorff distance there is for every $x \in V_X$ a $y \in V_Y$ such that

$$||x - y||_\infty = d(x,V_Y) \le d_H(V_X,V_Y).$$

With the notation $e: = (1,\ldots,1) \in \mathbb{R}^n$ we conclude

$$V_X \subset V_Y + [-d_H(V_X,V_Y)e,\ d_H(V_X,V_Y)e],$$

thus the theorem yields

$$\rho_2(X,Y) \le 2d_H(V_X,V_Y).$$

Analogously we obtain

$$\rho_2(Y,X) \le 2d_H(V_X,V_Y),$$

and the assertion follows. □

Finally we want to specialize the situation to *dichotomies* which by definition are experiments $X \in \mathscr{E}(\underline{D}_2(I_2))$.

Theorem 27.6. For dichotomies X and Y the following statements are equivalent:

(i) $X >^k_\varepsilon Y$ for all $k \ge 2$.

(ii) $X >^2_\varepsilon Y$.

Proof: Obviously only the implication (ii) ⇒ (i) has to be shown. Let $X >^2_\varepsilon Y$ and let $\psi \in \Psi_k(\mathbb{R}^2)$ be of the form $\psi = \bigvee_{i=1}^{k} \ell_i$ with linear functionals ℓ_i on $\mathbb{R}^2$ defined by $\ell_i(x_1,x_2) := a_i x_1 + b_i x_2$ for all $x_1,x_2 \in \mathbb{R}$ and real numbers a_i,b_i $(i = 1,\ldots,k)$. Without loss of generality we assume the existence of $r \geq 1$ such that

$$\psi(1,x_2) = \bigvee_{i=1}^{r} \ell_i(1,x_2)$$

whenever $x_2 \in \mathbb{R}_+^*$, where the representation is minimal in the sense that for each $i \leq r$ there exists an $x_2 > 0$ satisfying $\ell_i(1,x_2) > \ell_j(1,x_2)$ for all $j \neq i$. From the above representation we see that we may assume that $b_1 < b_2 < \ldots < b_r$ and hence $a_1 > a_2 > \ldots > a_r$.

Moreover, for any $x_2 \in \mathbb{R}_+^*$ we get

$$\begin{aligned}\psi(1,x_2) = \ell_1(1,x_2) &+ [\ell_2(1,x_2) - \ell_1(1,x_2)]^+ + \ldots \\ &+ [\ell_r(1,x_2) - \ell_{r-1}(1,x_2)]^+,\end{aligned}$$

whence for all $x_1,x_2 \in \mathbb{R}_+^*$

$$\begin{aligned}\psi(x_1,x_2) = \ell_1(x_1,x_2) &+ [\ell_2(x_1,x_2) - \ell_1(x_1,x_2)]^+ + \ldots \\ &+ [\ell_r(x_1,x_2) - \ell_{r-1}(x_1,x_2)]^+.\end{aligned}$$

Let $\tilde{\psi}$ be defined by the right side of this equality *for all* $x_1,x_2 \in \mathbb{R}$. Then $\tilde{\psi}$ is a sum of elements of $\Psi_2(\mathbb{R}^2)$, and

(1) $\psi(x_1,x_2) = \tilde{\psi}(x_1,x_2)$ whenever $x_1,x_2 \in \mathbb{R}_+$,

(2) $\psi(-e_i) \geq \tilde{\psi}(-e_i)$ for all $i = 1,2$.

We now apply Corollary 26.12 to the assumption and get

$$\tilde{\psi}(X) \geq \tilde{\psi}(Y) - \varepsilon(1)\,\frac{\tilde{\psi}(e_1) + \tilde{\psi}(-e_1)}{2} - \varepsilon(2)\,\frac{\tilde{\psi}(e_2) + \tilde{\psi}(-e_2)}{2}.$$

By (1) we deduce

$$\psi(X) = \int \psi(f_1,f_2)dP = \int \tilde{\psi}(f_1,f_2)dP = \tilde{\psi}(X),$$

thus by (2)

$$\begin{aligned}\psi(X) &\geq \psi(Y) - \varepsilon(1)\,\frac{\psi(e_1) + \tilde{\psi}(-e_1)}{2} - \varepsilon(2)\,\frac{\psi(e_2) + \tilde{\psi}(-e_2)}{2} \\ &\geq \psi(Y) - \varepsilon(1)\,\frac{\psi(e_1) + \psi(-e_1)}{2} - \varepsilon(2)\,\frac{\psi(e_2) + \psi(-e_2)}{2},\end{aligned}$$

and the assertion follows, again by Corollary 26.12. □

Remark 27.7. We have seen that for dichotomies X and Y the implication

$$\Delta_2(X,Y) = 0 \Rightarrow \Delta(X,Y) = 0$$

holds. It can be shown that this implication remains valid for experiments X and Y with arbitrary finite parameter set.

Example 27.8. Let $X = (\Omega, \mathfrak{A}, (P_i)_{i \in I})$ with $\Omega := I := I_3$ be determined by the matrix

$$P^X := \begin{pmatrix} p_1 & q_1 & r_1 \\ p_2 & q_2 & r_2 \\ p_3 & q_3 & r_3 \end{pmatrix}.$$

Any test $t \in \mathfrak{M}_{(1)}(\Omega, \mathfrak{A})$ will be given as a triple $(t_1, t_2, t_3) \in [0,1]^3$. First of all one notes that

(1) $V_X = \{(p_1t_1 + q_1t_2 + r_1t_3,\ p_2t_1 + q_2t_2 + r_2t_3,$
$p_3t_1 + q_3t_2 + r_3t_3): t_i \in [0,1] \text{ for } i = 1,2,3\} \subset \mathbb{R}^3.$

It is easily seen that

(2) V_X is the convex hull of those 8 points which correspond to non-randomized tests t. Moreover,

(3) V_X is the parallel-epiped spanned by the column vectors of P^X.

Let Y be another experiment with $\Omega := I := I_3$ determined by the matrix

$$P^Y := \begin{pmatrix} \alpha_1 & \beta_1 & \gamma_1 \\ \alpha_2 & \beta_2 & \gamma_2 \\ \alpha_3 & \beta_3 & \gamma_3 \end{pmatrix}.$$

Then

(4) $X >_0^2 Y \Leftrightarrow$ The column vectors of P^Y lie in V_X.

§28. STANDARD EXPERIMENTS

In order to simplify the study of the various comparison relations between experiments with the same finite parameter set we shall introduce a standardization procedure which to every experiment X corresponding to a decision problem $\underline{\overline{D}} = (I_n, D, \mathscr{V})$ associates an experiment with the standard simplex of $\mathbb{R}^n$ together with its Borel σ-algebra as its sample space. It will turn out that this standardization procedure preserves all the information contained in X and moreover, that the associated standardization X^s of X can be completely characterized by one single measure on the standard simplex which will be called the standard measure associated with $\mathbf{X}^s$.

We start by introducing the *standard simplex* of $\mathbb{R}^n$ as the set $K_n := \mathscr{M}^1(I_n)$ which can be rewritten as

$$\left\{x := (x_1,\ldots,x_n) \in \mathbb{R}^n : x_i \geq 0 \text{ for all } i \in \Omega_{I_n}, \sum_{i=1}^n x_i = 1\right\}.$$

The obvious notation $(K_n, \mathfrak{B}(K_n))$ will be applied whenever we deal with the measurable space of the standard simplex as the sample space of a standard experiment in the sense of the following

<u>Definition 28.1</u>. Let $\underline{D} := (I_n, D, \mathscr{V})$. An experiment $\mathbf{X} := (X_{\mathbf{X}}, N_{\mathbf{X}}) \in \mathscr{E}(\underline{D})$ is called a *standard experiment corresponding to* $\underline{D}$ if $X_{\mathbf{X}} = (K_n, \mathfrak{B}(K_n))$ and if there exists a measure $S \in \mathscr{M}_+(K_n)$ such that

$$N_{\mathbf{X}}(i,\cdot) = X_i \cdot S$$

for all $i \in \Omega_{I_n}$, where X_i denotes the i-th projection of $\mathbb{R}^n$ for every $i \in \Omega_{I_n}$.

By $\mathscr{S}(\underline{D})$ or $\mathscr{S}(I_n)$ we shall abbreviate the totality of all standard experiments corresponding to $\underline{D} := \underline{D}(I_n)$.

<u>Definition 28.2</u>. A measure $S \in \mathscr{M}_+(K_n)$ is said to be a *standard measure* (on the standard simplex K_n) if

$$\int_{K_n} X_i \, dS = 1$$

for all $i \in \Omega_{I_n}$.

The set of all standard measures will be abbreviated by $\mathscr{S}(K_n)$.

<u>Theorem 28.3</u>. (i) Given a standard experiment $\mathbf{X} := (X_{\mathbf{X}}, N_{\mathbf{X}}) \in \mathscr{S}(I_n)$ with defining measure $S \in \mathscr{M}_+(K_n)$ we have that S is a standard measure in $\mathscr{S}(K_n)$.

(ii) To every standard measure $S \in \mathscr{S}(K_n)$ there corresponds a unique standard experiment $\mathbf{X} := (X_{\mathbf{X}}, N_{\mathbf{X}}) \in \mathscr{S}(I_n)$ satisfying $\Sigma_{i=1}^n N_{\mathbf{X}}(i,\cdot) = S$.

<u>Proof:</u> (i). By assumption we have that $N_{\mathbf{X}}(i,\cdot) = X_i \cdot S$, whence

$$\int_{K_n} X_i \, dS = X_i \cdot S(K_n) = N_{\mathbf{X}}(i, K_n) = 1$$

for all $i \in \Omega_{I_n}$, and this is the assertion.

(ii) Let now $S \in \mathscr{S}(K_n)$. Then we define the kernel $N_{\mathbf{X}}$ from I_n to K_n by $N_{\mathbf{X}}(i,\cdot) := X_i \cdot S$ for all $i \in \Omega_{I_n}$. Clearly $N_{\mathbf{X}} \in \mathrm{Stoch}(I_n, K_n)$, and

$$\sum_{i=1}^{n} N_{\mathbf{X}}(i,\cdot) = \sum_{i=1}^{n} X_i \cdot S = \left(\sum_{i=1}^{n} X_i\right) \cdot S = S.$$

The uniqueness of $\mathbf{X} = (\mathbf{X}_{\mathbf{X}}, N_{\mathbf{X}})$ with this property is clear. □

Notation 28.3. The above theorem provides us with a one-to-one correspondence between $\mathscr{S}(I_n)$ and $\mathscr{S}(K_n)$: To every $\mathbf{X} \in \mathscr{S}(I_n)$ there corresponds a measure $S_{\mathbf{X}} \in \mathscr{S}(K_n)$, and for every $S \in \mathscr{S}(K_n)$ the corresponding experiment $\mathbf{X}_S$ belongs to $\mathscr{S}(I_n)$. $S_{\mathbf{X}}$ will be called the *standard measure corresponding to* $\mathbf{X}$, and $\mathbf{X}_S$ will be called the *standard experiment corresponding to* S.

For the standard experiment $\mathbf{X}_S$ corresponding to the standard measure S we derive without difficulty that

$$\psi(\mathbf{X}_S) = \int \psi dS = S(\psi)$$

for all $\psi \in \Psi_k(\mathbb{R}^n)$ for all $k \geq 1$ or $\psi \in \Psi(\mathbb{R}^n)$ respectively.

Theorem 28.4. For any two standard measures $S, T \in \mathscr{S}(K_n)$ the following statements are equivalent:

(i) $\Delta(\mathbf{X}_S, \mathbf{X}_T) = 0$.

(ii) $S = T$.

Proof: It remains to show the implication (i) ⇒ (ii).

1. For any measure $S \in \mathscr{M}_+^b(K_n)$ we introduce its Laplace function L_S defined on K_n by

$$L_S(\xi) := \int_{K_n} X_1^{\xi_1} \cdot \ldots \cdot X_n^{\xi_n} dS$$

for all $\xi := (\xi_1, \ldots, \xi_n) \in K_n$.

If $S, T \in \mathscr{M}_+^b(K_n)$, then $L_S = L_T$ implies $S = T$. In fact, without loss of generality we restrict the discussion to $\mathring{K}_n$ in place of K_n, and define the function $u: \mathring{K}_n \to \mathbb{R}^n$ by $u(\xi) := (\log \xi_1, \ldots, \log \xi_n)$ for all $\xi := (\xi_1, \ldots, \xi_n) \in \mathring{K}_n$. Then for all $\eta := (\eta_1, \ldots, \eta_n) \in K_n$ we get

$$L_S(\eta) = \int_{K_n} e^{\sum_{i=1}^{n} \eta_i \log \xi_i} S(d\xi)$$

$$= \int_{K_n} e^{\langle \eta, \xi \rangle} u(S)(d\xi) = u(S)^\wedge(i\eta),$$

where $u(S)^\wedge$ denotes the Fourier transform of the measure $u(S)$. By assumption we have $u(S)^\wedge(i\eta) = u(T)^\wedge(i\eta)$ for all $\eta \in K_n$, whence $u(S)^\wedge = u(T)^\wedge$ as follows from the identity theorem for holomorphic functions. The injectivity of the Fourier transform yields $u(S) = u(T)$, and since u is invertible, we obtain $S = T$.

2. We are now prepared to prove the desired implication. Suppose that $S \neq T$. By 1 we have $L_S \neq L_T$ and thus there exists a $\xi := (\xi_1,\ldots,\xi_n) \in K_n$ satisfying

$$\int_{K_n} X_1^{\xi_1}\cdot\ldots\cdot X_n^{\xi_n} dS \neq \int_{K_n} X_1^{\xi_1}\cdot\ldots\cdot X_n^{\xi_n} dT. \tag{*}$$

On the other hand the function $-X_1^{\xi_1}\cdot\ldots\cdot X_n^{\xi_n}$ on K_n is convex and therefore of the form

$$-X_1^{\xi_1}\cdot\ldots\cdot X_n^{\xi_n} = \lim_{k\to\infty} \bigvee_{i=1}^{k} \ell_i$$

with affine linear functionals ℓ_i on K_n $(i \geq 1)$. From the hypothesis $\Delta(X_S,X_T) = 0$ we infer by Remark 26.14 and Corollary 26.11 that for all $k \geq 1$

$$\int_{K_n} \bigvee_{i=1}^{k} \ell_i dS = \int_{K_n} \bigvee_{i=1}^{k} \ell_i dT,$$

whence

$$\begin{aligned} -\int_{K_n} X_1^{\xi_1}\cdot\ldots\cdot X_n^{\xi_n} dS &= \lim_{k\to\infty} \int_{K_n} \bigvee_{i=1}^{k} \ell_i dS \\ &= \lim_{k\to\infty} \int_{K_n} \bigvee_{i=1}^{k} \ell_i dT \\ &= -\int_{K_n} X_1^{\xi_1}\cdot\ldots\cdot X_n^{\xi_n} dT, \end{aligned}$$

contradicting (*). □

Remark 28.5. The preceding result shows that Δ behaves like a metric on $\mathcal{S}(I_n)$.

We note that for every $a \in \mathbb{R}_+$ the set $\mathcal{M}^a(K_n) := \{\mu \in \mathcal{M}_+^b(K_n): ||\mu|| = a\}$ is convex, compact and metrizable with respect to the vague topology in $\mathcal{M}_+(K_n)$. $\mathcal{S}(K_n)$ is a convex subset of $\mathcal{M}_{(n)}(K_n)$, and since it is the intersection of $\mathcal{M}^n(K_n)$ with finitely many closed hyperplanes, it is also compact and metrizable. Let $\mathcal{S}_e(K_n)$ denote the set of extreme points of $\mathcal{S}(K_n)$. Applying Choquet's theorem to $\mathcal{S}(K_n)$ we obtain for every $S \in \mathcal{S}(K_n)$ the existence of a *representing measure* $\mu^S \in \mathcal{M}^1(\mathcal{S}_e(K_n))$ such that

$$\ell(S) = \int_{\mathcal{S}_e(K_n)} \ell(\nu)\mu^S(d\nu)$$

for all affine linear functionals ℓ on $\mathcal{S}(K_n)$.

Let now $M_n = (\Omega_{M_n}, \mathfrak{A}_{M_n})$ denote the measurable space of $\mathcal{S}_e(K_n)$. Then the representing measure μ^S can be considered as an element of $\mathcal{M}^1(M_n)$. The following theorem follows from Choquet's theory once one applies the terminology of §19.

<u>Theorem 28.6</u>. Let $X = (X_X, N_X)$ be a standard experiment $\in \mathcal{S}(I_n)$ of the form $X = X_S$ for some standard measure $S \in \mathcal{S}(K_n)$. We define $N_n \in \mathrm{Stoch}(M_n \otimes I_n, X_X)$ by

$$N_n((m,i),A) := \int_A X_i \, dm$$

for all $m \in \Omega_{M_n}$, $i \in \Omega_{I_n}$ and $A \in \mathfrak{A}_X$. Then (N_n, μ^S) is an M_n-decomposition of X.

Our next aim will be a geometric characterization of the set $\mathcal{S}_e(K_n)$.

<u>Theorem 28.7</u>. For any measure $S \in \mathcal{S}(K_n)$ the following statements are equivalent:

(i) $S \in \mathcal{S}_e(K_n)$.

(ii) S is supported by the vertices of a simplex.

<u>Proof</u>: 1. (ii) ⇒ (i). Let $S \in \mathcal{S}(K_n)$ be of the form $S = \Sigma_{i=1}^k \alpha_i \varepsilon_{x_i}$ with $\alpha_1,\ldots,\alpha_k \in \mathbb{R}$ and points $x_1,\ldots,x_k \in K_n$ which are affinely independent in the sense that $\Sigma_{i=1}^k \xi_i x_i = 0$ for $\xi_1,\ldots,\xi_k \in \mathbb{R}$ with $\Sigma_{i=1}^k \xi_i = 0$ implies $\xi_i = 0$ for all $i = 1,\ldots,k$. Suppose that $S = \rho T + (1-\rho)U$ with $\rho \in]0,1[$ is a convex combination of S in $\mathcal{S}(K_n)$. Then $T \ll S$ of $T = \Sigma_{i=1}^k \alpha_i' \varepsilon_{x_i}$ with $\alpha_1',\ldots,\alpha_k' \in \mathbb{R}$. Since by $S,T \in \mathcal{S}(K_n)$, $(1,\ldots,1) = \Sigma_{i=1}^k \alpha_i x_i = \Sigma_{i=1}^k \alpha_i' x_i$, we obtain $\alpha_i = \alpha_i'$ for all $i = 1,\ldots,k$, thus $S = T$.

Similarly we deduce $S = U$, so we have $S \in \mathcal{S}_e(K_n)$.

2. (i) ⇒ (ii). Let $S \in \mathcal{S}_e(K_n)$, and let $x_1,\ldots,x_k$ be elements of K_n such that for $i = 1,\ldots,k$, $S(W) > 0$ for all neighborhoods W of x_i. Let $K_n = \bigcup_{i=1}^k W_i$ be a partition of K_n consisting of neighborhoods W_i of x_i $(i = 1,\ldots,k)$ and put

$$\tau_i := S(W_i), \quad f_i := \frac{1}{\tau_i} 1_{W_i}, \quad \text{and} \quad v_i := \int_{K_n} x f_i(x) S(dx)$$

(for $i = 1,\dots,k$). Then $\int_{K_n} f_i dS = 1$, $v_i \in K_n$ for all $i = 1,\dots,k$, and

$$(1,\dots,1) = \int_{K_n} xS(dx) = \sum_{i=1}^{k} \tau_i \int_{W_i} \frac{1}{\tau_i} xS(dx) = \sum_{i=1}^{k} \tau_i v_i.$$

Let $\sigma_1,\dots,\sigma_k \in \mathbb{R}_+$ such that $\Sigma_{i=1}^k \sigma_i = 1$ and $(1,\dots,1) = \Sigma_{i=1}^k \sigma_i v_i$, and let $T := (\Sigma_{i=1}^k \sigma_i f_i)\cdot S \in \mathscr{S}(K_n)$. Then $T = S$ and hence $\sigma_i = \tau_i$ for all $i = 1,\dots,k$.

In fact, assuming $T \neq S$ we obtain a convex combination $S = \rho T + (1-\rho)U$ where $\rho \in]0,1[$ such that $\rho\sigma_i \leq \tau_i$ for all $i = 1,\dots,k$, and $U := \frac{1}{1-\rho}(S - \rho T)$.

From the previous assertion we conclude that $v_1,\dots,v_k$ are affinely independent, whence that $k \leq n$. Thus the support of S is a finite set $\{x_1,\dots,x_k\}$ and $S = \Sigma_{\ell=1}^k \alpha_\ell \varepsilon_{x_\ell}$ with $\alpha_1,\dots,\alpha_k > 0$. In the above construction we get

$$\tau_i = S(W_i) = \alpha_i$$

and

$$v_i = \int_{K_n} x\, f_i(x) \sum_{\ell=1}^{k} \tau_\ell \varepsilon_{x_\ell}(dx) = x_i \cdot \frac{1}{\tau_i} \cdot \tau_i = x_i.$$

Since we have shown that $v_1,\dots,v_k$ are affinely independent, the x_i $(i = 1,\dots,k)$ are affinely independent, too. □

In the sequel we shall apply the geometric structure of $\mathscr{S}(K_n)$ to the comparison by testing problems.

<u>Theorem 28.8</u>. Let $S,T \in \mathscr{S}(K_n)$ such that $X_S >_0^2 X_T$. Then $\operatorname{supp} T \subset \operatorname{conv}(\operatorname{supp} S)$.

<u>Proof</u>: Let $C := \operatorname{conv}(\operatorname{supp} S)$ and choose sequences $(\ell_m)_{m\geq 1}$ in $(\mathbb{R}^n)^*$ and $(a_m)_{m\geq 1}$ in $\mathbb{R}$ such that

$$C = \bigcap_{m\geq 1} [\ell_m \leq a_m].$$

This can be achieved by choosing the ℓ_m as separating hyperplanes between C and the points with rational coordinates.

From $X_S >_0^2 X_T$ we deduce by Corollary 26.12 that

$$0 = \int_{K_n} (\ell_m - a_m)^+ dS \geq \int_{K_n} (\ell_m - a_m)^+ dT \geq 0$$

for all $m \geq 1$, which implies

$$T(CC) = T\Big(\bigcup_{n\geq 1} [\ell_m > a_m]\Big) = 0.$$

Thus $\operatorname{supp} T \subset C$. □

For any measure $S \in \mathscr{S}(K_2)$ we consider the *power function of* S as a mapping $\beta_S: [0,1] \to [0,1]$ defined by

$$\beta_S(\alpha) := \sup\Big\{\int tX_2 dS: t \in \mathfrak{M}_{(1)}(K_2, \mathfrak{B}(K_2)): \int tX_1 dS \leq \alpha\Big\}$$

for all $\alpha \in [0,1]$. We note that β_S enjoys the properties:

(1) β_S is isotone.

(2) β_S is concave.

(3) $\beta_S(0+) = \beta_S(0)$.

(4) $\beta_S(1) = 1$.

For technical purposes we extend the definition of β_S to the whole of $\mathbb{R}$ by putting

$$\beta_S(x) := \begin{cases} \beta_S(0) & \text{if } x < 0 \\ 1 & \text{if } x > 1. \end{cases}$$

In order to compare power functions of standard measures we introduce the *Lévy metric* d_L on the set $\mathscr{F}(\mathbb{R})$ of all bounded isotone real-valued functions on $\mathbb{R}$ by

$$d_L(F,G) := \inf\{\varepsilon \geq 0: F(x-\varepsilon) - \varepsilon \leq G(x) \leq F(x+\varepsilon) + \varepsilon \quad \text{for all } x \in \mathbb{R}\}$$

for all $F,G \in \mathscr{F}(\mathbb{R})$.

Theorem 28.9. For $S,T \in \mathscr{S}(K_2)$ and any tolerance function ε corresponding to $\underline{\overline{D}}_2$ the following statements are equivalent:

(i) $X_S \;\geq_\varepsilon^2\; X_T$.

(ii) $\beta_S(\alpha + \frac{\varepsilon(1)}{2}) + \frac{\varepsilon(2)}{2} \geq \beta_T(\alpha)$ for all $\alpha \in [0,1]$.

In particular,

$$d_L(\beta_S, \beta_T) = \frac{1}{2}\Delta_2(X_S, X_T).$$

Proof: From Theorem 27.2 we get that (i) is equivalent to the inclusion

$$V_{X_S} + \frac{1}{2} I_{[-\varepsilon,\varepsilon]} \supset V_{X_T}. \tag{*}$$

This is seen to be equivalent to (ii).

We only show the nontrivial implication (ii) ⇒ (*). Given

$$\left(\int tX_1 dT, \int tX_2 dT\right) \in V_{X_T}$$

the condition (ii) is applied twice in order to obtain tests t' and t'' in $\mathfrak{M}_{(1)}(K_2, \mathfrak{B}(K_2))$ satisfying the relations

$$\int t'X_1 dS = 1 \wedge (\alpha + \frac{\varepsilon(1)}{2}),$$

$$\int t'X_2 dS \geq \int tX_2 dT - \frac{\varepsilon(2)}{2}$$

and

$$\int t''X_1 dS = 0 \vee (\alpha - \frac{\varepsilon(1)}{2}),$$

$$\int t''X_2 dS \leq \int tX_2 dT + \frac{\varepsilon(2)}{2} .$$

By forming a convex combination of t' and t'' we get a test $t_0 \in \mathfrak{M}_{(1)}(K_2, \mathfrak{B}(K_2))$ satisfying

$$\left|\int tX_1 dT - \int t_0 X_1 dS\right| \leq \frac{\varepsilon(1)}{2}$$

and

$$\left|\int tX_2 dT - \int t_0 X_2 dS\right| \leq \frac{\varepsilon(2)}{2} ,$$

which implies the assertion.

The statement concerning the Lévy metric is now a direct consequence. □

§29. GENERAL THEORY OF STANDARD MEASURES

From the results of the Sections 21, 26 and 28 we obtain the following list of equivalent statements about two experiments $X, Y \in \mathscr{S}(\underline{\overline{D}}_k(I_n))$ with standard measures $S_X, S_Y \in \mathscr{S}(K_n)$ respectively:

(1) $X >_0^k Y$.

(2) $S_X(\psi) \geq S_Y(\psi)$ for all $\psi \in \Psi_k(\mathbb{R}^n)$.

(3) $X >_0^B Y$.

The equivalence (2) ⟺ (3) can be studied in a more general framework which admits further equivalences.

Let E be a locally convex Hausdorff space with dual E' and K a metrizable, compact, convex subset of E such that there exists a $u_0 \in E'$ with $\text{Res}_K u_0 \equiv 1$. By $S(K)$ we denote the set of all continuous concave functions on K. Then $A(K) = S(K) \cap (-S(K))$ is the vector space of all continuous affine functions on K.

Theorem 29.1. (P. Cartier, J. M. G. Fell, P. A. Meyer). For any two measures $\mu,\nu \in \mathscr{M}_+(K)$ the following statements are equivalent:

(i) $\mu(p) \leq \nu(p)$ for all $p \in -S(K)$.

(ii) There exists a measure $\rho \in \mathscr{M}_+(K \times K)$ with $pr_2(\rho) = \nu$, satisfying the equality

$$\int_{K\times K} f(x)u(y)\rho(d(x,y)) = \int_K f(x)u(x)\mu(dx) \qquad (*)$$

valid for all $f \in \mathscr{C}(K)$, $u \in E'$.

(iii) There exists a kernel $T \in \mathrm{Stoch}(K,\mathfrak{B}(K))$ with the property $b(T(\varepsilon_x)) = x$ for all $x \in K$, satisfying $T\mu = \nu$.

(iv) Given $n \geq 1$ and measures $\mu_1,\ldots,\mu_n \in \mathscr{M}_+(K)$ with $\Sigma_{i=1}^n \mu_i = \mu$, there exist measures $\nu_1,\ldots,\nu_n \in \mathscr{M}_+(K)$ satisfying $\Sigma_{i=1}^n \nu_i = \nu$ and $r(\nu_i) = r(\mu_i)$ for all $i = 1,\ldots,n$.

Remark 29.2. Statement (i) of the theorem contains the fact that μ is smaller than ν in the *ordering* introduced *by Bishop and deLeeuw*. In symbols this can be rephrased as $\mu < \nu$.

Markov kernels T on $(K,\mathfrak{B}(K))$ with the property that the barycentre $b(T(\varepsilon_x))$ of $T(\varepsilon_x)$ equals x for all $x \in K$, or equivalently that $T(\ell) = \ell$ holds for all $\ell \in A(K)$, are called *dilations on* K.

Finally we give statement (iv) a more condensed form by introducing for any $\mu \in \mathscr{M}_+(K)$ the set

$$S(\mu) := \left\{(z_1,\ldots,z_n)\text{: There exist } \mu_1,\ldots,\mu_n \in \mathscr{M}_+(K) \text{ such that } \sum_{i=1}^n \mu_i = \mu \text{ and } r(\mu_i) = z_i \text{ for all } i = 1,\ldots,n\right\}.$$

Then (iv) reads as $S(\mu) \subset S(\nu)$.

Proof of the Theorem: 1. (i) $\Rightarrow$ (ii). At first we will establish an equivalent formulation of statement (ii).

Let $H' := \{h' \in \mathscr{C}(K \times K)\text{: } h' = g \circ pr_2,\ g \in \mathscr{C}(K)\}$. Then H' is a linear subspace of $\mathscr{C}(K \times K)$, and the mapping $j'\colon H' \to \mathbb{R}$ defined by $j'(h') = j'(g \circ pr_2) := \nu(g)$ for $h' \in \mathscr{C}(K \times K)$ is a linear functional on H'. If now $\rho \in \mathscr{M}_+(K \times K)$, then $\nu = pr_2(\rho)$ iff $\rho(h') = j'(h')$ for all $h' \in H'$.

Let

$$H'' := \left\{ h'' \in \mathscr{C}(K \times K) : h''(x,y) := \sum_{i=1}^{m} f_i(x)u_i(y) \quad \text{with} \right.$$
$$\left. f_i \in \mathscr{C}(K),\ u_i \in E' \quad \text{for all} \quad i = 1,\ldots,m \right\}.$$

Again, H'' is a linear subspace of $\mathscr{C}(K \times K)$, and the mapping $j'': H'' \to \mathbb{R}$ defined by

$$j''(h'') := \int_K h''(x,x)\mu(dx)$$

for all $h'' \in H''$ is a linear functional on H''. As above we note that a measure $\rho \in \mathscr{M}_+(K \times K)$ satisfies condition (*) of (ii) iff $\rho(h'') = j''(h'')$ for all $h'' \in H''$.

Thus we have seen that statement (ii) of the theorem is equivalent to the existence of a measure $\rho \in \mathscr{M}_+(K \times K)$ satisfying $\mathrm{Res}_{H'}\rho = j'$ and $\mathrm{Res}_{H''}\rho = j''$.

Next we show that for $h' \in H'$ and $h'' \in H''$ such that $h' \geq h''$ we get $j'(h') \geq j''(h'')$.

In fact, let $q(y) := \sup_{x \in K} h''(x,y)$ for every $h'' \in H''$. Since every $h'' \in H''$ is bounded, also q is bounded. Moreover, for any fixed $x \in K$ the function $y \to h''(x,y)$ is lower semicontinuous, whence q is itself lower semicontinuous, since it is the supremum of lower semicontinuous functions. Let

$$P := \left\{ p : p(y) = \sup_{1 \leq i \leq n} h''(x_i, y),\ x_1,\ldots,x_n \in K \right\}.$$

Then P is an ascending family, since with two functions of P their supremum belongs to P. Every $p \in P$ is continuous and convex. By construction of q and P we have $q = \sup P$, and applying (i) we get

$$\mu(q) = \mu(\sup P) = \sup_{p \in P} \mu(p) \leq \sup_{p \in P} \nu(p) = \nu(\sup P) = \nu(q).$$

Moreover, for every $x \in K$,

$$h''(x,x) \leq \sup_{y \in K} h''(y,x) = q(x) \leq \sup_{y \in K} h'(y,x) = g(x),$$

where $h' = g \circ pr_2$, whence

$$j''(h'') = \int_K h''(x,x)\mu(dx) \leq \int_K q\, d\mu = \mu(q)$$
$$\leq \nu(q) \leq \nu(g) = j'(h').$$

Let us note that for $h' = h'' \in H' \cap H''$ the previous discussion implies $j'(h') = j''(h'')$.

Now we introduce the linear subspace $H := \{h' - h'': h' \in H', h'' \in H''\}$ of $\mathscr{C}(K \times K)$ and consider the mapping $j: H \to \mathbb{R}$ defined by

$$j(h' - h'') := j'(h') - j''(h'')$$

for all $h' - h'' \in H$. One verifies that j is in fact well-defined, and that j is a positive linear functional on H. We shall show that there exists a positive linear functional ρ on $\mathscr{C}(K \times K)$ satisfying $\mathrm{Res}_H \rho = j$. From this (ii) follows, since $\mathrm{Res}_H \rho = j$, $\mathrm{Res}_{H'} j = j'$ and $\mathrm{Res}_{H''} j = j''$ imply $\mathrm{Res}_{H'} \rho = j'$ and $\mathrm{Res}_{H''} \rho = j''$.

To show the remaining assertion we consider the mapping $R: \mathscr{C}(K{\times}K) \to \mathbb{R}$ defined by

$$R(f) := \sup_{z \in K \times K} f(z) < \infty.$$

R is a sublinear functional on $\mathscr{C}(K \times K)$. For any $h \in H$ we have $j(h) \leq j(R(h)\cdot 1) = R(h)\cdot j(1)$ with $j(1) \geq 0$. Thus $f \to j(1)\cdot R(f)$ is also a sublinear functional on $\mathscr{C}(K \times K)$, and j is bounded by $j(1)\cdot R$. An application of the Hahn-Banach theorem yields the existence of a linear functional ρ on $\mathscr{C}(K \times K)$ which extends j such that $\rho(f) \leq j(1)\cdot R(f)$ for all $f \in \mathscr{C}(K \times K)$. For every $f \in \mathscr{C}_+(K \times K)$ we have

$$-\rho(f) = \rho(-f) \leq j(1)\cdot R(-f) = j(1) \sup_{z \in K \times K} (-f(z)) \leq 0,$$

whence $\rho(f) \geq 0$. This completes the proof of the implication (i) $\Rightarrow$ (ii).

Before we enter the proof of the implication (ii) $\Rightarrow$ (iii) we have to insert three Lemmas.

<u>Lemma 29.3</u>. For every measure $\rho \in {}_+(K \times K)$ with $pr_1(\rho) = \mu$ there exists a kernel $T_1 \in \mathrm{Stoch}(K, \mathfrak{B}(K))$ such that

$$\rho(A \times B) = \int_A T_1(x,B)\mu(dx)$$

for all $A, B \in \mathfrak{B}(K)$.

<u>Proof</u>: Consider the conditional measure $\rho_{pr_2|\mathfrak{B}_1}$ of the projection $pr_2: K \times K \to K$ under the condition of the sub-σ-algebra $\mathfrak{B}_1 := \mathfrak{B}(K) \otimes \{\emptyset, K\}$. We put

$$T_0(x,B) := \rho_{pr_2|\mathfrak{B}_1}((x,y),B)$$

for all $x \in K$, $B \in \mathfrak{B}(K)$ and some fixed $y \in K$. Then, by the very definition of the conditional measure we have for all $A, B \in \mathfrak{B}(K)$,

$$\int_A T_0(x,B)\mu(dx) = \int_K 1_A(x)T_0(x,B)\rho(d(x,y))$$
$$= \int_{A\times K} T_0(x,B)\rho(d(x,y))$$
$$= \rho((A \times K) \cap pr_2^{-1}(B))$$
$$= \rho(A \times B).$$

For $B: = K$ we get

$$\int_A T_0(x,K)\mu(dx) = \rho(A \times K) = \mu(A),$$

whence $T_0(x,K) = 1$ μ -a.e. Setting

$$T_1(x,\cdot): = \begin{cases} T_0(x,\cdot) & \text{if } T_0(x,K) = 1 \\ \varepsilon_x & \text{if } T_0(x,K) \neq 1 \end{cases}$$

we arrive at the desired assertion. □

Lemma 29.4. There exists a sequence $(u_n)_{n\geq 1}$ in E' which separates the points of K.

Proof: Since E is a locally convex space and K is convex, there exists a basis of K consisting of convex sets. Moreover, K is assumed to be metrizable and compact, whence K has a countable basis of its topology. Both facts imply that K has a countable basis $\{B_n: n \geq 1\}$ of convex sets. Suppose $\overline{B}_i \cap \overline{B}_j = \emptyset$ for some $i,j \geq 1$. Since $\overline{B}_i$ and $\overline{B}_j$ are compact, convex subsets of K, there exist a $u_{ij} \in E'$ and an $a \in \mathbb{R}$ such that $\overline{B}_i \subset [u_{ij} < a]$ and $\overline{B}_j \subset [u_{ij} > a]$. The linear functionals u_{ij} for $i,j \geq 1$ with $\overline{B}_i \cap \overline{B}_j = \emptyset$ so constructed clearly separate the points of K. □

We proceed with the proof of the theorem. 2. (ii) ⇒ (iii). Let (ii) be satisfied with $u: = u_0$. Then $\mu = pr_1(\rho)$. Choosing the kernel $T_1 \in \text{Stoch}(K,\mathfrak{B}(K))$ as in Lemma 29.3 we get for all $f,g \in$ (K) the equality

$$\int f(x)(T_1g)(x)\mu(dx) = \int f(x)g(y)\rho(d(x,y)).$$

Next we choose $g: = u_n$ for $n \geq 1$, where the sequence $(u_n)_{n\geq 1}$ is as in Lemma 29.4. Thus for every $n \geq 1$ we obtain

$$\int f(x)u_n(b(T_1(x,\cdot)))\mu(dx) = \int f(x)u_n(y)\rho(d(x,y))$$
$$= \int f(x)u_n(x)\mu(dx),$$

still for all $f \in \mathscr{C}(K)$.

For $N_n := \{x \in K: u_n(b(T_1(x,\cdot))) \neq u_n(x)\}$ one has $\mu(N_n) = 0$ for all $n \geq 1$. Then $\mu(\bigcup_{n\geq 1} N_n) = 0$. Defining

$$T(x,\cdot) := \begin{cases} T_1(x,\cdot) & \text{if } x \notin N := \bigcup_{n\geq 1} N_n \\ \varepsilon_x & \text{if } x \in N \end{cases}$$

we obtain a dilation T on K satisfying for every $g \in \mathscr{C}(K)$ the equalities

$$(T\mu)(g) = \int (Tg)(x)\mu(dx) = \int g(y)\rho(d(x,y)) = \nu(g),$$

i.e., $T\mu = \nu$.

3. (iii) $\Rightarrow$ (iv). Let T be a dilation on K with $T\mu = \nu$. Considering $\mu = \Sigma_{i=1}^n \mu_i \in \mathscr{M}_+(K)$ we get $\nu = \Sigma_{i=1}^n T\mu_i$. But for every $i = 1,\ldots,n$ the measure $\nu_i := T\mu_i$ satisfies $r(\nu_i) = r(\mu_i)$. This proves the assertion.

For the proof of the remaining implication we need two more lemmas.

Let $\mathscr{M}$ denote the set of all elementary measures in $\mathscr{M}_+(K)$ of the form $\Sigma_{i=1}^r c_i\varepsilon_{x_i}$ with $x_1,\ldots,x_r \in K$ and $c_1,\ldots,c_r \in \mathbb{R}_+$.

<u>Lemma 29.5</u>. Any given measure $\mu \in \mathscr{M}_+(K)$ is the smallest measure with respect to the order $<$ of Bishop and deLeeuw that is larger than all measures $m \in \mathscr{M}$ satisfying $m < \mu$.

<u>Proof</u>: Since K is metrizable, there exists a metric d on K compatible with the topology of K. For every $n \geq 1$ there exist disjoint sets $A_1,\ldots,A_r \in \mathfrak{B}(K)$ with $\bigcup_{i=1}^r A_i = K$ and convex sets $B_1,\ldots,B_r$ in K with $B_i \supset A_i$ and $d(B_i) \leq 1/n$ for all $i = 1,\ldots,r$. For every $i = 1,\ldots,r$ we define a measure $\mu_i := 1_{A_i}\cdot\mu$, for which $r(\mu_i) = c_i x_i$ with $c_i = ||\mu_i||$ and $x_i \in B_i$ holds. For every $i = 1,\ldots,r$ we have that $m_{\mu_i} := c_i\varepsilon_{x_i} < \mu_i$, thus

$$m_n := \sum_{i=1}^r m_{\mu_i} < \sum_{i=1}^r \mu_i = \mu.$$

In order to show that the sequence $(m_n)_{n\geq 1}$ converges weakly towards μ, we pick an $f \in \mathscr{C}(K)$ and an $\varepsilon > 0$. There exists an $n_0 \geq 1$ such that $|x - y| \leq 1/n_0$ implies $|f(x) - f(y)| \leq \varepsilon$. But this yields for all $n \geq n_0$,

$$\left|\int f\,d\mu - \int f\,dm_n\right| \leq \sum_{i=1}^{r} \left|\int f\,d\mu_i - ||\mu_i||f(x_i)\right|$$
$$\leq \sum_{i=1}^{r} \int |f(x) - f(x_i)|\mu_i(dx)$$
$$\leq \varepsilon \sum_{i=1}^{r} ||\mu_i|| = \varepsilon||\mu||.$$

Now let $m < \nu \in \mathscr{M}_+(K)$ for all $m \in \mathscr{M}$ with $m < \mu$. Then for all $f \in -S(K)$ we obtain $m_n(f) \leq \nu(f)$ whenever $n \geq 1$, thus $\mu(f) \leq \nu(f)$ or $\mu < \nu$, which was to be shown. □

Let $C: = \mathbb{R}_+K$ and

$$\mathfrak{S}: = \{\{z_1,\ldots,z_n\}: (z_1,\ldots,z_n) \in C^n,\ n \geq 1\}.$$

The mapping $\{c_1x_1,\ldots,c_nx_n\} \to \sum_{i=1}^{n} c_i\varepsilon_{x_i}$ establishes a one-to-one correspondence between $\mathfrak{S}$ and the set $\mathscr{M}$ appearing in Lemma 29.5. This correspondence will also be regarded as a one-to-one correspondence $\delta \to m_\delta$ between C^n and $\mathscr{M}$.

Lemma 29.6. For $\nu \in \mathscr{M}_+(K)$ the following statements are equivalent:

(i) $m_\delta < \nu$.

(ii) $\delta \in S(\nu)$.

Proof: 1. (i) ⇒ (ii). Let $m_\delta < \nu$. Then $S(m_\delta) \subset S(\nu)$. Since $\delta \in S(m_\delta)$, this implies $\delta \in S(\nu)$.

2. (ii) ⇒ (i). Let $\delta = (\delta_1,\ldots,\delta_n) \in S(\nu)$. There exist measures $\nu_1,\ldots,\nu_n \in \mathscr{M}_+(K)$ such that $\Sigma_{i=1}^n \nu_i = \nu$ and $r(\nu_i) = \delta_i = c_ix_i$ for all $i = 1,\ldots,n$. But $c_i\varepsilon_{x_i} < \nu_i$ for all $i = 1,\ldots,n$ implies $m_\delta < \sum_{i=1}^{n} \nu_i = \nu$. □

We finish the proof of the theorem by showing the remaining implication

4. (iv) ⇒ (i). Let $S(\mu) \subset S(\nu)$. Then $m_\delta < \mu$ implies $\delta \in S(\mu)$, whence by assumption $\delta \in S(\nu)$ and therefore $m_\delta < \nu$. Here, Lemma 29.6 has been applied twice. But now we infer from Lemma 29.5 that $\mu < \nu$ which is statement (i) of the theorem. □

§30. SUFFICIENCY AND COMPLETENESS

So far we have studied standard experiments and standard measures for their own sake, without any close connection to statistics. We are now prepared to introduce for an arbitrary experiment with finite parameter set its standardization and relate this to the information metrics and to sufficient and complete σ-algebras.

Let $\underline{D}: = \underline{D}_k(I_n) = (I_n, D_k, \mathscr{V})$. Let us suppose that we are given an experiment $X \in \mathscr{E}(I_n)$ of the form $(\Omega, \mathbf{A}, \{P_1, \ldots, P_n\})$. As in the previous sections we put $P: = \Sigma_{i=1}^n P_i$ and obtain the existence of functions $f_1, \ldots, f_n \in \mathfrak{M}_{(1)}(\Omega, \mathbf{A})$ with $\Sigma_{i=1}^n f_i = 1$. In other words we get a measurable mapping $f_X: = (f_1, \ldots, f_n)$ from $(\Omega, \mathbf{A})$ into $(K_n, \mathfrak{B}(K_n))$. Using this mapping we introduce the measures

$$P_{iX}: = f_X(P_i) \quad (i \in I_n),$$

$$P_X: = f_X(P)$$

in $\mathscr{M}^1(K_n)$ and claim the following

<u>Theorem 30.1</u>. For all $i \in I_n$ we have $P_{iX} = X_i \cdot P_X$. In particular, the experiment $(K_n, \mathfrak{B}(K_n), \{P_{1X}, \ldots, P_{nX}\})$ is a standard experiment in $\mathscr{S}(I_n)$.

<u>Proof</u>: Let $i \in I_n$ and $B \in \mathfrak{B}(K_n)$. Then the chain of equalities

$$\begin{aligned}(X_i \cdot P_X)(B) &= \int_B X_i dP_X = \int 1_B X_i df_X(P) \\ &= \int 1_B \circ f_X \; X_i \circ f_X \; dP \\ &= \int 1_{f_X^{-1}(B)} f_i dP = P_i(f_X^{-1}(B)) \\ &= f_X(P_i)(B) = P_{iX}(B)\end{aligned}$$

yields the assertion. □

<u>Remark 30.2</u>. From the theorem we deduce that for every $i \in I_n$

(1) $\int X_i dP_X = 1,$

and

(2) $f_i = X_i \circ f_X.$

Thus Theorem 8.7 implies that

(3) f_X is sufficient for X.

Definition 30.3. Given $X = (\Omega,\mathfrak{A},\{P_1,\dots,P_n\}) \in \mathscr{E}(I_n)$ we call $X^s := (K_n,\mathfrak{B}(K_n),\{P_{1X},\dots,P_{nX}\}) \in \mathscr{S}(I_n)$ the *standard experiment* and $P_X := f_X(P) \in \mathscr{M}^1(K_n)$ the *standard measure corresponding to* X.

The standardization procedure described in this definition enjoys useful

Properties 30.4.

30.4.1. $(X^s)^s = X^s$.

30.4.2. $\Delta_{\bar{\underline{D}}}(X^s,X) = 0$,

and

30.4.3. $\Delta(X^s,X) = 0$.

It suffices to show the second property. By Corollary 26.11 the asserted equality is equivalent to the validity of

$$\psi(X) = \psi(X^s)$$

for all $\psi \in \Psi_k(\mathbb{R}^n)$. But this follows from

$$\psi(X) = \int_\Omega \psi \circ f_X dP = \int_\Omega \psi \, df_X(P) = \psi(X^s)$$

for all $\psi \in \Psi_k(\mathbb{R}^n)$.

Theorem 30.5. Let $X := (\Omega,\mathfrak{A},\{P_1,\dots,P_n\})$ and $Y := (\Omega_1,\mathfrak{A}_1,\{Q_1,\dots,Q_n\})$ be two experiments $\in \mathscr{E}(I_n)$ with corresponding standard experiments $X^s, Y^s \in \mathscr{S}(I_n)$ and standard measures $P_X, P_Y \in \mathscr{S}(K_n)$ respectively. Then

(i) $\Delta_{\bar{\underline{D}}}(X,Y) = \Delta_{\bar{\underline{D}}}(X^s,Y^s)$,

and

(ii) $\Delta(X,Y) = 0$ iff $X^s = Y^s$.

Proof: 1. (i) follows directly from Property 30.4.2 by a simple application of the triangle inequality for $\Delta_{\bar{\underline{D}}}$:

$$\Delta_{\bar{\underline{D}}}(X,Y) \le \Delta_{\bar{\underline{D}}}(X,X^s) + \Delta_{\bar{\underline{D}}}(X^s,Y^s) + \Delta_{\bar{\underline{D}}}(Y^s,Y)$$
$$= \Delta_{\bar{\underline{D}}}(X^s,Y^s),$$

and

$$\Delta_{\bar{\underline{D}}}(X^s,Y^s) \le \Delta_{\bar{\underline{D}}}(X^s,X) + \Delta_{\bar{\underline{D}}}(X,Y) + \Delta_{\bar{\underline{D}}}(Y,Y^s)$$
$$= \Delta_{\bar{\underline{D}}}(X,Y).$$

2. In order to see (ii) we note that $X^s = Y^s$ implies $\Delta(X^s,Y^s) = 0$ and thus by (i), $\Delta(X,Y) = 0$. Let conversely $\Delta(X,Y) = 0$ which, again by

(i), implies $\Delta(X^s,Y^s) = 0$. Since X^s,Y^s are standard experiments with standard measures P_X,P_Y respectively, we infer from Theorem 28.4 that $P_X = P_Y$. This, however, yields $X^s = Y^s$. □

Theorem 30.6. For two experiments $X,Y \in \mathscr{E}(I_n)$ as in the preceding theorem the following formula holds

$$(X \otimes Y)^s = (X^s \otimes Y^s)^s.$$

Proof: We keep the terminology as above. For all $j \in I_n$ we have with $g_j := dQ_j / d\left(\sum_{i=1}^n Q_j\right)$ the following chain of equalities:

$$\frac{d(P_j \otimes Q_j)}{d\left(\sum_{i=1}^n P_i \otimes Q_i\right)} = \frac{d\left[\left(f_j \cdot \sum_{i=1}^n P_i\right) \otimes \left(g_j \cdot \sum_{i=1}^n Q_i\right)\right]}{d\left(\sum_{i=1}^n (f_i \cdot \sum_{j=1}^n P_j) \otimes (g_i \cdot \sum_{j=1}^n Q_j)\right)}$$

$$= \frac{d(f_j \otimes g_j) \cdot \left[\left(\sum_{i=1}^n P_i\right) \otimes \left(\sum_{i=1}^n Q_i\right)\right]}{d\left(\sum_{i=1}^n (f_i \otimes g_i) \cdot (\sum_{j=1}^n P_j \otimes \sum_{j=1}^n Q_j)\right)}$$

$$= \frac{d(f_j \otimes g_j) \cdot \left(\sum_{i=1}^n P_i \otimes \sum_{i=1}^n Q_i\right)}{d\left(\sum_{i=1}^n f_i \otimes g_i\right) \cdot \left(\sum_{j=1}^n P_j \otimes \sum_{j=1}^n Q_j\right)}$$

$$= \frac{f_j \otimes g_j}{\sum_{i=1}^n f_i \otimes g_i} \left[\sum_{i=1}^n P_i \otimes Q_i\right].$$

Here, the denominator vanishes at most on a set of $(\sum_{i=1}^n P_i \otimes Q_i)$-measure zero. For every $j \in I_n$ the end term of the chain of equalities is ≥ 0, and the terms sum up to 1 $(\Sigma_{i=1}^n P_i \otimes Q_i)$ - a.e. Thus the standard measure of $X \otimes Y$ is

$$\left(\frac{f_1 \otimes g_1}{\sum_{i=1}^n f_i \otimes g_i}, \ldots, \frac{f_n \otimes g_n}{\sum_{i=1}^n f_i \otimes g_i}\right)\left(\sum_{i=1}^n P_i \otimes Q_i\right),$$

and the standard measure of $X^s \otimes Y^s$ is therefore

$$\left(\frac{X_1 \otimes X_1}{\sum_{i=1}^n X_i \otimes X_i}, \ldots, \frac{X_n \otimes X_n}{\sum_{i=1}^n X_i \otimes X_i}\right)\left(\sum_{i=1}^n (X_i \cdot P_X \otimes X_i \cdot P_Y)\right).$$

But by definition of the standard measure we have

$$\sum_{i=1}^{n} X_i \cdot P_X \otimes X_i \cdot P_Y = \sum_{i=1}^{n} f(P_i) \otimes g(Q_i)$$
$$= \sum_{i=1}^{n} (f \times g)(P_i \otimes Q_i)$$
$$= (f \times g)\Big(\sum_{i=1}^{n} P_i \otimes Q_i\Big),$$

where $f \times g$ denotes the vector function (f_X, f_Y).

Moreover,

$$\left(\frac{X_1 \otimes X_1}{\sum_{i=1}^{n} X_i \otimes X_i}, \ldots, \frac{X_n \otimes X_n}{\sum_{i=1}^{n} X_i \otimes X_i}\right) \circ (f \times g)$$

$$= \left(\frac{f_1 \otimes g_1}{\sum_{i=1}^{n} f_i \otimes g_i}, \ldots, \frac{f_n \otimes g_n}{\sum_{i=1}^{n} f_i \otimes g_i}\right).$$

The desired equality follows. □

In order to relate the standardization of experiments to statistics we first pose the question under what conditions the σ-algebra $\mathfrak{A}(f_X)$ is complete. By Theorem 6.3(i) this is equivalent to asking for conditions ensuring that the σ-algebra $\mathfrak{A}_{K_n}$ is complete for the corresponding standard experiment X^s.

Theorem 30.7. Let X be a standard experiment of the form $(K_n, \mathfrak{B}(K_n), \{X_1 \cdot P, \ldots, X_n \cdot P\})$ for some standard measure $P \in \mathscr{S}(K_n)$. The following statements are equivalent:

(i) $\mathfrak{B}(K_n)$ is complete for X.

(ii) $\mathfrak{B}(K_n)$ is boundedly complete for X.

(iii) $P \in \mathscr{S}_e(K_n)$.

Proof: Since the implication (i) ⇒ (ii) is trivial, it remains to show (ii) ⇒ (iii) ⇒ (i).

1. (ii) ⇒ (iii). Let $Q, R \in \mathscr{S}(K_n)$ be given such that $P = \frac{1}{2}(Q+R)$. Then $Q \ll P$ and dQ/dP is bounded. Moreover, for every $i \in I_n$ we obtain

$$\int \Big(\frac{dQ}{dP} - 1_{K_n}\Big) d(X_i \cdot P) = \int \frac{dQ}{dP} X_i \, dP - \int X_i \, dP$$
$$= \int X_i \, dQ - 1 = 0.$$

Since $\mathfrak{B}(K_n)$ is assumed to be boundedly complete for X, the above implies $\frac{dQ}{dP} = 1[X_i \cdot P]$ for all $i \in I_n$, whence $Q = P$. Similarly one deduces $R = P$ and thus that P is an extreme point of $\mathscr{S}(K_n)$.

2. (iii) ⇒ (i). Let $P \in \mathscr{S}_e(K_n)$. By Theorem 28.7 this means that $P = \Sigma_{i=1}^k \alpha_i \varepsilon_{x_i}$ for $\alpha_1,\ldots,\alpha_k \in \mathbb{R}_+^*$ and affinely independent points $x_1,\ldots,x_k$ of K_n.

Let $f \in \mathfrak{M}(K_n, \mathfrak{B}(K_n))$ be such that

$$\int f \, d(X_j \cdot P) = 0 \quad \text{for all } j \in I_n.$$

Then

$$\sum_{i=1}^{k} \alpha_i f(x_i) x_{ij} = \int f \, X_j dP = \int f \, d(X_j \cdot P) = 0$$

for all $j \in I_n$, and from $1 = \Sigma_{j=1}^n x_{ij}$ for all $i = 1,\ldots,k$ we deduce $\Sigma_{i=1}^k \alpha_i f(x_i) = 0$.

But since the $x_1,\ldots,x_k$ are assumed to be affinely independent, this implies $\alpha_i f(x_i) = 0$ for all $i \in I_n$, and by the choice of $\alpha_1,\ldots,\alpha_k$ to be $\neq 0$ we get $f(x_i) = 0$ for all $i \in I_n$, whence $f = 0[X_i \cdot P]$ for all $i \in I_n$. This shows the completeness of $\mathfrak{A}_{K_n}$. □

The following result is a translation of Theorem 30.7 into the framework of arbitrary finite experiments.

<u>Theorem 30.8</u>. Let $X := (\Omega, \mathfrak{A}, \{P_1,\ldots,P_n\})$ be an experiment with corresponding standard experiment X^s and standard measure $P_X \in \mathscr{S}(K_n)$.

(i) If $\mathfrak{A}$ is boundedly complete, then $P_X \in \mathscr{S}_e(K_n)$.

(ii) If $P_X \in \mathscr{S}_e(K_n)$, then $\mathfrak{A}(f_X)$ is complete.

<u>Proof</u>: (i). If $\mathfrak{A}$ is boundedly complete, then in particular $\mathfrak{A}(f_X)$ is boundedly complete for X. Thus Theorem 6.3(i) implies that $\mathfrak{A}_{K_n}$ is boundedly complete for X^s, whence $P_X \in \mathscr{S}_e(K_n)$ by Theorem 30.7.

(ii) Let $P_X \in \mathscr{S}_e(K_n)$. Then $\mathfrak{A}_{K_n}$ is complete for X^s by Theorem 30.7. Again we apply Theorem 6.3(i) and obtain that $\mathfrak{A}(f_X)$ is complete for X. □

<u>Theorem 30.9</u>. Let $X = (\Omega, \mathfrak{A}, \{P_1,\ldots,P_n\})$ be an experiment with corresponding standard measure P_X.

(i) If $\mathfrak{A}$ is complete, then $\mathfrak{A} = \mathfrak{A}(f_X)[\{P_1,\ldots,P_n\}]$.

(ii) P_X is extreme iff $\mathfrak{A}$ contains a complete and sufficient sub-σ-algebra.

Proof: (i) Since $\mathbf{A}$ is clearly sufficient for X, the completeness of $\mathbf{A}$ implies its minimal sufficiency by Theorem 6.15. From Remark 30.2 we infer that $\mathbf{A}(f_X)$ is sufficient for X, whence $\mathbf{A} \subset \mathbf{A}(f_X)[\{P_1,\ldots,P_n\}]$, and thus $\mathbf{A} = \mathbf{A}(f_X)[\{P_1,\ldots,P_n\}]$.

(ii) If P_X is extreme, then by Theorem 30.7 together with Theorem 6.3(i), $\mathbf{A}(f_X)$ is complete, and $\mathbf{A}(f_X)$ is the desired complete and sufficient sub-σ-algebra of $\mathbf{A}$. Let conversely $\mathbf{A}$ contain such a sub-σ-algebra $\mathbf{S}$. Since by the proof of Theorem 8.8, $\mathbf{A}(f_X)$ is minimal sufficient, it coincides with $\mathbf{S}$. The standard measure corresponding to the experiment $(\Omega,\mathbf{A}(f_X),\{\mathrm{Res}_{\mathbf{A}(f_X)}P_1,\ldots,\mathrm{Res}_{\mathbf{A}(f_X)}P_n\})$ is P_X. Application of Theorem 30.8(i) yields that P_X is extreme. □

Theorem 30.10. Let X and Y be two experiments in $\mathscr{E}(I_n)$ with standard measures P_X and P_Y respectively, and let $P_X \in \mathscr{S}_e(K_n)$. The following statements are equivalent:

(i) $X >_0^k Y$.

(ii) $X >_0^2 Y$.

(iii) $\mathrm{supp}(P_Y) \subset \mathrm{conv}(\mathrm{supp}(P_X))$.

Proof: From Theorem 28.8 we infer the validity of the implication (ii) ⇒ (iii). It remains to show the implication (iii) ⇒ (i). Let P_X be extreme in $\mathscr{S}(K_n)$, i.e., $P_X = \sum_{i=1}^{\ell} \alpha_i \varepsilon_{x_i}$ for $\alpha_1,\ldots,\alpha_\ell \in \mathbb{R}_+^*$ and affinely independent points $x_1,\ldots,x_\ell \in K_n$. We consider the experiment

$$Z: = (\{x_1,\ldots,x_\ell\},\ \mathfrak{P}(\{x_1,\ldots,x_\ell\}),\ \{P_1,\ldots,P_n\}),$$

where

$$P_j: = \sum_{i=1}^{\ell} \alpha_i X_j \cdot \varepsilon_{x_i} \quad \text{for } j \in I_n.$$

Then $Z^s = X^s$. From Property 30.4.2 we know that it suffices to show that $Z >_0^k Y^s$. To establish this it suffices by Theorem 19.17 to construct a stochastic kernel N from $(\{x_1,\ldots,x_\ell\},\mathfrak{P}(\{x_1,\ldots,x_\ell\}))$ to $(K_n,\mathbf{B}(K_n))$ satisfying $N(P_j) = X_j \cdot P_Y$ for all $j \in I_n$.

In order to perform the construction we consider the set $C: = \mathrm{conv}(\{x_1,\ldots,x_\ell\})$ and note that there exist affine functions $f_1,\ldots,f_\ell\colon C \to \mathbb{R}$ such that every $y \in C$ admits a barycentric representation of the form

$$y = \sum_{i=1}^{\ell} f_i(y)x_i.$$

First we define a mapping $N: \{x_1,\ldots,x_\ell\} \times \mathfrak{B}(K_n) \to \overline{\mathbb{R}}$ by

$$N(x_i,B) := \frac{1}{\alpha_i}\int_B f_i \, dP_Y$$

for all $i \in I_\ell$, $B \in \mathfrak{B}(K_n)$. Then for every $B \in \mathfrak{B}(K_n)$ we have $N(\cdot,B) \in \mathfrak{M}_+^b(\{x_1,\ldots,x_\ell\},\mathfrak{P}(\{x_1,\ldots,x_\ell\}))$, and for every $i \in I_\ell$, $N(x_i,\cdot) \in \mathcal{M}_+^b(K_n,\mathfrak{B}(K_n))$. Moreover, $N(x_i,K_n) = 1$ for all $i \in I_\ell$, since

$$\begin{aligned}(1,\ldots,1) &= \int_C yP_Y(dy) = \int_C\left(\sum_{i=1}^{\ell} f_i(y)x_i\right)P_Y(dy)\\ &= \sum_{i=1}^{\ell}\left(\frac{1}{\alpha_i}\int f_i dP_Y\right)\alpha_i x_i,\end{aligned}$$

and $x_1,\ldots,x_\ell$ are assumed to be affinely independent. Consequently N is a stochastic kernel from $(\{x_1,\ldots,x_\ell\},\mathfrak{P}(\{x_1,\ldots,x_\ell\}))$ to $(K_n,\mathfrak{B}(K_n))$.

Finally we obtain for all $j \in I_n$ and $B \in \mathfrak{B}(K_n)$

$$\begin{aligned}N(P_j)(B) &= \sum_{i=1}^{\ell} P_j(\{x_i\})N(x_i,B)\\ &= \sum_{i=1}^{\ell} \alpha_i X_j(x_i)\frac{1}{\alpha_i}\int_B f_i dP_Y\\ &= \int_B\left(\sum_{i=1}^{\ell} X_j(x_i)f_i\right)dP_Y\\ &= \int_B X_j dP_Y = (X_j\cdot P_Y)(B),\end{aligned}$$

and this completes the proof of the theorem. □

In the remaining part of this section we want to deal with the set $\mathscr{E}(I_2)$ of all dichotomies, which can be viewed as a lattice with respect to informativity.

We prepare the proof of this result by a number of facts which are of some independent interest.

Let X be a dichotomy of the form $(\Omega,\mathfrak{A},\{P_1,P_2\})$ and let β_X be the *power function* of X defined by $\beta_X := \beta_{P_X}$, where P_X is the standard measure of X.

<u>Properties 30.11.</u>

<u>30.11.1.</u> For every $\alpha \in [0,1]$ we have

$$\beta_X(\alpha) = \sup\left\{\int t\,dP_2 : t \in \mathfrak{M}_{(1)}(\Omega,\mathfrak{A}), \int t\,dP_1 \leq \alpha\right\}.$$

30.11.2. Let $\tilde{X} := ([0,1],\mathfrak{B}([0,1]),\{\lambda_{[0,1]},P\})$, where $P \in \mathscr{M}^1([0,1])$ is the measure corresponding to the restricted distribution function β_X. Then $\Delta_2(X,\tilde{X}) = 0$.

In fact, by the definition of P we have $P([0,\alpha]) = \beta_X(\alpha)$ and by Property 30.11.1, $\beta_{\tilde{X}}(\alpha) = P([0,\alpha])$ for all $\alpha \in [0,1]$. Thus $\beta_X = \beta_{\tilde{X}}$, and by Theorem 28.9 together with Property 30.4.2 the assertion follows.

30.11.3. To every function $\beta: [0,1] \to [0,1]$ satisfying the conditions (1) to (4) below there corresponds an experiment $X \in \mathscr{E}(I_2)$ satisfying $\beta_X = \beta$.

(1) β is isotone.

(2) β is concave.

(3) $\beta(0+) = \beta(0)$.

(4) $\beta(1) = 1$.

In fact, let P be the measure in $\mathscr{M}^1([0,1])$ corresponding to the distribution function β. Then the experiment $X := ([0,1],\mathfrak{B}([0,1]),\{\lambda_{[0,1]},P\})$ has the desired property.

Given two experiments $X,Y \in \mathscr{E}(I_2)$ we introduce the experiment $Z := X \wedge Y$ by $\beta_Z := \beta_X \wedge \beta_Y$. Similarly, the experiment $U := X \vee Y$ is defined by the function $H_U := H_X \vee H_Y$, where H_X is given by

$$H_X(a) := \sup\left\{a_1 \int t\,dP_1 + a_2 \int t\,dP_2 : t \in \mathfrak{M}_{(1)}(\Omega,\mathfrak{A})\right\}$$

for all $a = (a_1,a_2) \in \mathbb{R}^2$.

Theorem 30.12. The collection $\mathscr{E}(I_2)$ of all dichotomies is a lattice with respect to the comparison by testing problems.

Proof: We shall restrict ourselves to showing that for $X,Y \in \mathscr{E}(I_2)$ the experiment $Z := X \wedge Y$ belongs to $\mathscr{E}(I_2)$. For this we consider $\beta := \beta_X \wedge \beta_Y$. β is a mapping from $[0,1]$ into itself satisfying the conditions (1) to (4) of Property 30.11.3. This very property yields the existence of an experiment $Z \in \mathscr{E}(I_2)$ such that $\beta_Z = \beta$. Since $\beta_Z \leq \beta_X$ and $\beta_Z \leq \beta_Y$, we get from Theorem 28.9 together with Property 30.4.2 that $Z <_0^2 X$ and $Z <_0^2 Y$. Let $Z' \in \mathscr{E}(I_2)$ be such that $Z' <_0^2 X$ and $Z' <_0^2 Y$ holds. Then $\beta_{Z'} \leq \beta_Z$, whence again by Theorem 28.9, $Z' <_0^2 Z$. Thus Z exists as an element of $\mathscr{E}(I_2)$. □

Finally we present a *résumé* of the most useful equivalences between comparison relations in $\mathscr{E}(I_2)$.

Theorem 30.13. Let X and Y be dichotomies with corresponding standardizations X^s and Y^s, standard measures P_X and P_Y and power functions β_X and β_Y. The following statements are equivalent:

(i) $X >_0^k Y$ for all $k \geq 2$

(ii) $X >_0^2 Y$.

(iii) $X^s >_0^2 Y^s$.

(iv) $P_X(\psi) \geq P_Y(\psi)$ for all $\psi \in \Psi_2(K_2)$.

(v) $P_X > P_Y$.

(vi) There exists a dilation on K_2 such that $TP_X = P_Y$.

(vii) $X >_0^B Y$.

(viii) $\beta_X \geq \beta_Y$.

The proof relies on Theorem 21.5 (for (i) $\Leftrightarrow$ (ii) $\Leftrightarrow$ (vii)), Property 30.4.2 (for (ii) $\Leftrightarrow$ (iii)), Corollary 26.11 (for (iii) $\Leftrightarrow$ (iv)), Theorem 29.1 (for (iv) $\Leftrightarrow$ (v) $\Leftrightarrow$ (vi)), and Theorem 28.9 (for (iii) $\Leftrightarrow$ (viii)). □

CHAPTER X

Comparison with Extremely Informative Experiments

§31. BAYESIAN DEFICIENCY

The topic of this section refers to the comparison of experiments with respect to apriori measures which have been introduced in Section 3. There we formulated the Bayesian principle as one of the basic ideas of modern statistics. Although we did not put much emphasis on the Bayesian approach throughout the exposition we intend at least to touch upon the general scope in handling a few interesting types of examples: We shall study deviations from total information and from total ignorance as measures of information. In other words we shall compute the deficiencies of experiments relative to totally informative and totally uninformative ones respectively. For the corresponding computations apriori distributions are of great value.

Before we go into the definition of Bayesian comparison and deficiency we quote the by now natural *generalization of LeCam's stochastic kernel criterion* and two consequences. The proofs of the subsequent results can be performed along the lines of the proofs in §21 leading to the basic form of the criterion. While in Theorem 21.5 the comparison relation has been characterized for vanishing tolerance functions, we shall now consider arbitrary ones.

As in §21 we assume given a parameter space $(I,\mathfrak{J})$, a decision space $(D,\mathfrak{D})$ and the set $\mathscr{V}$ of all bounded measurable loss functions on $I \times D$. In the comparison relation $\overset{\underline{D}}{>}$ we shall now drop the symbol $\underline{D}$ denoting the corresponding decision problem $\underline{D} = (I,D,\mathscr{V})$.

Theorem 31.1. Let $X = (\Omega,\mathfrak{A},(P_i)_{i\in I})$ and $Y: = (\Omega_1,\mathfrak{A}_1,(Q_i)_{i\in I})$ be two experiments in $\mathscr{E}(I)$ such that $(\Omega_1,\mathfrak{A}_1)$ is a standard Borel space and X is μ-dominated by a measure $\mu \in \mathscr{M}_+^\sigma(\Omega,\mathfrak{A})$, and let $\varepsilon: I \to \mathbb{R}_+$ be a tolerance function.

The following statements are equivalent:

(i) $X >_\varepsilon Y$.

(ii) There exists a kernel $N \in \mathrm{Stoch}((\Omega,\mathfrak{A}),(\Omega_1,\mathfrak{A}_1))$ satisfying $||N(P_i) - Q_i|| \leq \varepsilon(i)$ for all $i \in I$.

<u>Corollary 31.2</u>. Under the hypothesis of the theorem there exists a kernel $N \in \mathrm{Stoch}((\Omega,\mathfrak{A}),(\Omega_1,\mathfrak{A}_1))$ such that

$$\rho(X,Y) = \sup_{i\in I} ||N(P_i) - Q_i||.$$

<u>Corollary 31.3</u>. Let $\mathfrak{F}$ denote the system of all finite (non-empty) subsets of I. For the experiments X and Y of the theorem we have

$$\rho(X,Y) = \sup_{I'\in\mathfrak{F}} \rho(X_{I'},Y_{I'}),$$

where most suggestively $X_{I'}$ denotes the *subexperiment* $(\Omega,\mathfrak{A},(P_i)_{i\in I'})$ of X.

Instead of going into the <u>proofs</u> of these results established for the first time by LeCam we shall discuss the following

<u>Example 31.4</u>. We consider the set-up of Example 26.15 in which experiments $X = (\Omega,\mathfrak{A},(P_i)_{i\in I})$ and $Y: = (\Omega_1,\mathfrak{A}_1,(Q_i)_{i\in I})$ were given with $I: = I_n$, and $\Omega: = I_r$, $\mathfrak{A}: = \mathfrak{P}(I_r)$ and $\Omega_1: = I_s$, $\mathfrak{A}_1: = \mathfrak{P}(I_s)$ $(n,r,s \geq 1)$ respectively. X and Y are determined by stochastic matrices $P^X: = (p_{ij})$ and $P^Y: = (q_{ij})$ respectively. A Markov kernel N from $(\Omega,\mathfrak{A})$ to $(\Omega_1,\mathfrak{A}_1)$ is given as a stochastic matrix $(n_{k\ell}) \in \mathbb{M}(r \times s, \mathbb{R})$ through $N(k,\{\ell\}): = n_{k\ell}$ for all $k \in I_r$, $\ell \in I_s$. For any $i \in I_n$ we obtain

$$N(P_i) = \left(\sum_{k=1}^r p_{ik}n_{k1},\ldots,\sum_{k=1}^r p_{ik}n_{ks}\right).$$

Defining for any matrix $A = (a_{k\ell}) \in \mathbb{M}(r \times s, \mathbb{R})$ its norm $||A||$ by $\max_{k\in I_r} \sum_{\ell\in I_s} |a_{k\ell}|$, it follows from the results quoted above that

$$\rho(X,Y) = \inf_{N\in\$(r\times s,\,\mathbb{R})} ||P^X N - P^Y||$$

holds, where $\$(r \times s, \mathbb{R})$ denotes the set of all stochastic matrices in $\mathbb{M}(r \times s, \mathbb{R})$. It follows that

$$\rho(X,Y) \leq ||P^X N - P^Y||$$

for all $N \in \$(r \times s, \mathbb{R})$.

In the special case that $s = r$ and $N: = \mathrm{Id} \in \mathbb{M}(r \times r, \mathbb{R})$ we obtain

$$\Delta(X,Y) \leq ||P^X - P^Y|| = \max_{i \in I_n} \sum_{j \in I_r} |p_{ij} - q_{ij}|,$$

a result that we established in Example 26.15 with a different method.

Clearly $X > Y$ if there exists a stochastic matrix $N \in \$(r \times s, \mathbb{R})$ satisfying $P^X N = P^Y$.

For further applications of this framework see Examples 21.6 and 21.7.

Example 31.5. We consider the Markov chain $X: = (X_n)_{n \in \mathbb{Z}_+}$ with state space $I: = \{1,2\}$ and transition function

$$N: = \begin{pmatrix} 1-\alpha & \alpha \\ \beta & 1-\beta \end{pmatrix} \in \$(2 \times 2, \mathbb{R}),$$

where $\alpha, \beta \in [0,1]$, $\alpha + \beta \neq 0,1,2$.

Let the initial "state" X_0 be the unknown parameter $i \in I$, and let X_n be the experiment obtained by observing X_n. Then X_n is of the form $(\Omega, \mathbf{A}, (P_i)_{i \in I})$ with $\Omega: = I$, $\mathbf{A}: = \mathfrak{P}(\Omega)$, and the stochastic matrix determining X_n is just N^n.

1. One sees very easily that

$$N^n = \frac{1}{\alpha + \beta}\begin{pmatrix} \beta & \alpha \\ \beta & \alpha \end{pmatrix} + \frac{1}{\alpha + \beta}\begin{pmatrix} \alpha & -\alpha \\ -\beta & \beta \end{pmatrix}(1 - \alpha - \beta)^n.$$

2. Let X_∞ be the experiment determined by the stochastic matrix

$$P^{X_\infty}: = \lim_{n \to \infty} N^n = \frac{1}{\alpha + \beta}\begin{pmatrix} \beta & \alpha \\ \beta & \alpha \end{pmatrix}.$$

Then by the result of Example 31.4 we obtain

$$\Delta(X_n, X_\infty) \leq \max\left\{\frac{2\alpha}{\alpha + \beta}, \frac{2\beta}{\alpha + \beta}\right\} |1 - \alpha - \beta|^n$$

$$= \frac{2(\alpha \vee \beta)}{\alpha + \beta} |1 - \alpha - \beta|^n.$$

3. From the representation of the deficiency given in Corollary 31.2 we infer that

$$\rho(X_\infty, X_n) = \inf_{M \in \$(2\times 2, \mathbb{R})} ||P^{X_\infty} M - N^n||,$$

while $\rho(X_n, X_\infty) = \inf_{M \in \$(2\times 2, \mathbb{R})} ||N^n M - P^{X_\infty}|| = 0$ for all $n \geq 1$.

We shall compute $\rho(X_\infty, X_n)$ in terms of the entries of N (and $n \geq 1$).

Let

$$M: = \begin{pmatrix} 1-a & a \\ b & 1-b \end{pmatrix} \in \$(2 \times 2, \mathbb{R}),$$

and put $\sigma: = 1 - \alpha - \beta$. Then

$$P^{X_\infty} M - N^n = \frac{1}{\alpha + \beta} \begin{pmatrix} -\alpha & \alpha \\ \beta & -\beta \end{pmatrix} \sigma^n$$
$$+ \frac{1}{\alpha + \beta} \begin{pmatrix} -a\beta + b\alpha & a\beta - b\alpha \\ -a\beta + b\alpha & a\beta - b\alpha \end{pmatrix},$$

whence

$$||P^{X_\infty} M - N^n|| = \frac{2}{\alpha + \beta} \{|-\alpha\sigma^n - a\beta + b\alpha| \vee |-\beta\sigma^n + a\beta - b\alpha|\}.$$

Minimizing this expression as a function of $\gamma: = a\beta - b\alpha$ yields $\gamma = \frac{\beta - \alpha}{2} \sigma^n$ as the point where the minimum is attained.

On the other side there exist numbers $a,b \in [0,1]$ such that $a\beta - b\alpha = \frac{\beta - \alpha}{2} \sigma^n$. Substituting the actual kernel M into the above formula for the norm we obtain

$$||P^{X_\infty} M - N^n|| = |\sigma|^n$$

and thus

$$\rho(X_\infty, X_n) = |\sigma|^n = |1 - \alpha - \beta|^n$$

for every $n \geq 1$.

Now we turn to the Bayesian comparison. Let $X = (\Omega, \mathfrak{A}, (P_i)_{i \in I})$ be an experiment in $\mathscr{E}(I)$, $(D, \mathfrak{D})$ a decision space and V a single loss function corresponding to X and $(D, \mathfrak{D})$.

<u>Definition 31.6.</u> For any apriori measure $\Lambda \in \mathscr{M}^1(I, \mathfrak{I})$ we introduce the *minimal Bayes risk with respect to* Λ as the number

$$r^V(X|\Lambda): = \inf_{\delta \in \mathscr{D}(X)} r^V(\delta|\Lambda),$$

where as in §3,

$$r^V(\delta|\Lambda): = \int R_\delta^V(i)\Lambda(di),$$

whenever the integral exists.

Special Case 31.7. Let (I,D,V) be a *decision triple* with at most countable parameter set I, $D:=I$ and $V: I \times D \to \mathbb{R}_+$ defined for all $(i,j) \in I \times D$ by

$$V(i,j) := \begin{cases} 1 & \text{if } i \neq j \\ 0 & \text{if } i = j \end{cases}.$$

Then for every $\Lambda \in \mathscr{M}^1(I,\mathfrak{I})$,

$$\begin{aligned} r^V(X|\Lambda) &= \inf_{N \in \mathrm{Stoch}((\Omega,\mathfrak{A}),(I,\mathfrak{I}))} \sum_{i\in I} \Lambda(\{i\}) \int 1_{C\{i\}}(j) N(x,dj) P_i(dx) \\ &= \inf_N \sum_{i\in I} \Lambda(\{i\}) \int N(x, C\{i\}) P_i(dx) \\ &= \inf_N \sum_{i\in I} \Lambda(\{i\}) \int (1 - N(x,\{i\})) P_i(dx) \\ &= 1 - \sup_{N \in \mathrm{Stoch}((\Omega,\mathfrak{A}),(I,\mathfrak{I}))} \sum_{i\in I} \Lambda(\{i\}) \int N(x,\{i\}) P_i(dx). \end{aligned}$$

Theorem 31.8. Under the assumptions of 31.7 we have for every $\Lambda \in \mathscr{M}^1(I,\mathfrak{I})$,

$$r^V(X|\Lambda) = 1 - \left\| \bigvee_{i\in I} \Lambda(\{i\}) P_i \right\|.$$

If, in particular $I = I_2$, then

$$r^V(X|\Lambda) = \left\| \bigwedge_{i\in I} \Lambda(\{i\}) P_i \right\|.$$

Proof: 1. Given X and $\Lambda \in \mathscr{M}^1(I,\mathfrak{I})$ we have to show that

$$\sup_N \sum_{i\in I} \Lambda(\{i\}) \int N(x,\{i\}) P_i(dx) = \left\| \bigvee_{i\in I} \Lambda(\{i\}) P_i \right\|.$$

In fact, for any $N \in \mathrm{Stoch}((\Omega,\mathfrak{A}),(I,\mathfrak{I}))$,

$$\begin{aligned} \sum_{i\in I} \Lambda(\{i\}) \int N(x,\{i\}) P_i(dx) &\leq \int \sum_{j\in I} N(x,\{j\}) \Big(\bigvee_{i\in I} \Lambda(\{i\}) P_i \Big)(dx) \\ &= \int \Big(\bigvee_{i\in I} \Lambda(\{i\}) P_i \Big)(dx) = \left\| \bigvee_{i\in I} \Lambda(\{i\}) P_i \right\|. \end{aligned}$$

The inverse inequality is obvious.

2. For the additional statement of the theorem let $\alpha \in [0,1]$. There exists an $A \in \mathfrak{A}$ such that

$$\begin{aligned} \|\alpha P_1 \wedge (1-\alpha) P_2\| &= \alpha P_1(CA) + (1-\alpha) P_2(A) \\ &= 1 - (\alpha P_1(A) + (1-\alpha) P_2(CA)) \\ &= 1 - \|\alpha P_1 \vee (1-\alpha) P_2\|. \end{aligned}$$

This proves the assertion. □

Let $X = (\Omega,\mathfrak{A},(P_i)_{i\in I})$ and $Y: = (\Omega_1,\mathfrak{A}_1,(Q_i)_{i\in I})$ be experiments in $\mathscr{E}(I)$ (with an arbitrary I), and let Λ be an apriori measure in the set $\mathscr{M}_c^1(I,\mathfrak{J})$ of all probability measures on $(I,\mathfrak{J})$ with countable support.

Definition 31.9. For every $k \geq 1$ we introduce the *Λ-weighted k-deficiency of X relative to Y as*

$$\rho_k(X,Y|\Lambda): = \inf\left\{\sum_{i\in I} \Lambda(\{i\})\varepsilon(i): X >_\varepsilon^k Y\right\}$$

and the corresponding distance

$$\Delta_k(X,Y|\Lambda): = \delta_k(X,Y|\Lambda) \vee \delta_k(Y,X|\Lambda).$$

Analogously we define the *Λ-weighted deficiency* $\rho(X,Y|\Lambda)$ *of* X *relative to* Y and the corresponding distance $\Delta(X,Y|\Lambda)$.

Properties 31.10. Under the hypothesis of Theorem 31.1 we get

31.10.1.

$$\rho(X,Y|\Lambda) = \inf_{N\in \mathrm{Stoch}((\Omega,\mathfrak{A}),(\Omega_1,\mathfrak{A}_1))} \sum_{i\in I} \Lambda(\{i\})||N(P_i) - Q_i||.$$

Let $k \geq 1$.

31.10.2.

$$\rho_k(X,Y|\Lambda) \leq \rho_k(X,Y)$$

for all $\Lambda \in \mathscr{M}_c^1(I,\mathfrak{J})$.

31.10.3.

$$\rho_k(X,Y) = \sup_{\Lambda\in\mathscr{M}_f^1(I,\mathfrak{J})} \rho_k(X,Y|\Lambda).$$

In fact, without loss of generality we may assume that $|I| < \infty$. Let E denote the convex set of all nonnegative functions ε on I such that $X >_\varepsilon^k Y$. Then

$$\sup_{\Lambda\in\mathscr{M}_f^1(I,\mathfrak{J})} \rho_k(X,Y|\Lambda) = \sup_\Lambda \inf_{\varepsilon\in E} \sum_{i\in I} \Lambda(\{i\})\varepsilon(i)$$

$$= \inf_{\varepsilon\in E} \sup_\Lambda \sum_{i\in I} \Lambda(\{i\})\varepsilon(i) = \inf_{\varepsilon\in E} \bigvee_{i\in I} \varepsilon(i)$$

$$= \rho_k(X,Y).$$

31.10.4.

$$\Delta_k(X,Y) = \sup_{\Lambda \in \mathscr{M}_f^1(I,\mathfrak{I})} \Delta_k(X,Y|\Lambda)$$

Finally,

31.10.5. $\rho_k(.,.|\Lambda)$ and $\Delta_k(.,.|\Lambda)$ for $k \geq 1$ enjoy the properties of ρ_k and Δ_k respectively, as they are listed in §19.

Under the assumption that also Y is dominated we have

31.10.6. that the assertions of 31.10.2 to 31.10.4 remain valid if one replaces ρ_k and Δ_k by ρ and Δ respectively.

Discussion 31.11. Let $X,Y \in \mathscr{E}(I)$ and $\Lambda \in \mathscr{M}_c^1(I,\mathfrak{I})$ be given as abov such that the hypothesis of Theorem 31.1 is satisfied.

31.11.1. Let $\rho(X,Y) = \rho(X,Y|\Lambda)$. Then by Theorem 31.1 there exists a kernel $N \in \text{Stoch}((\Omega,\mathfrak{A}),(\Omega_1,\mathfrak{A}_1))$ such that

$$\rho(X,Y) = \sup_{i \in I} ||N(P_i) - Q_i||.$$

This implies

$$||N(P_j) - Q_j|| = \rho(X,Y)$$

for all $j \in I$ such that $\Lambda(\{j\}) > 0$.

31.11.2. Suppose on the other hand that

$$\sum_{i \in I} \Lambda(\{i\})||N(P_i) - Q_i|| = \rho(X,Y|\Lambda)$$

and

$$||N(P_j) - Q_j|| = \sup_{i \in I} ||N(P_i) - Q_i||$$

whenever $\Lambda(\{j\}) > 0$. Then

$$\rho(X,Y) = \rho(X,Y|\Lambda) = \sup_{i \in I} ||N(P_i) - Q_i||.$$

§32. TOTALLY INFORMATIVE EXPERIMENTS

Occasionally one wants to consider the experiment of directly observing the underlying parameter i from a set I. This experiment is more informative than any other experiment with the same parameter set I. Any given experiment can be regarded to contain much or little information according to whether it is close to the above extremely informative one or far away from it. In measuring this distance we arrive at the deficiency of an experiment with respect to the totally informative one as a measure of the content of information in the given experiment.

Definition 32.1. An experiment $X = (\Omega,\mathbf{A},(P_i)_{i\in I})$ is said to be *totally informative* if for every pair (i_1,i_2) of $I \times I$ with $i_1 \neq i_2$ the measures P_{i_1} and P_{i_2} are mutually singular.

Since we are interested in estimating the "unknown" parameter $i \in I$ when the loss is 0 or 1 as the estimator hits or fails, we shall restrict ourselves to considering the decision triple (I,D,V) of 31.7 and *the* totally informative experiment $X_a := (I,\mathfrak{I},(\varepsilon_i)_{i\in I})$, where $\mathfrak{I}$ denotes a σ-algebra in I containing the one-point sets $\{i\}$ for $i \in I$.

By definition we have $\rho_k(X_a,X) = 0$ for every $k \geq 2$ and $\rho(X_a,X) = 0$ whenever $X \in \mathscr{E}(I)$, whence $\Delta_k(X,X_a) = \rho_k(X,X_a)$ and $\Delta(X,X_a) = \rho(X,X_a)$.

Without loss of generality we may assume that I is at most countable, as follows from the

Theorem 32.2. Let $(\Omega,\mathbf{A})$ and $(I,\mathfrak{I})$ be two standard Borel spaces, X μ-dominated by a measure $\mu \in \mathscr{M}_+^\sigma(\Omega,\mathbf{A})$, and let $X_a := (I,\mathfrak{I},(\varepsilon_i)_{i\in I})$ be the totally informative experiment with an uncountable parameter set I.

Then $\Delta(X,X_a) = 2$.

Proof: Let $N \in \mathrm{Stoch}((\Omega,\mathbf{A}),(I,\mathfrak{I}))$ be a kernel provided by Theorem 31.1, which satisfies the inequalities

$$||N(P_i) - \varepsilon_i|| \leq \Delta(X,X_a)$$

valid for all $i \in I$. From

$$||N(P_i) - \varepsilon_i|| = 1 - N(P_i)(\{i\}) + N(P_i)(I \setminus \{i\})$$

we infer that

$$2 - \Delta(X,X_a) \leq 2N(P_i)(\{i\})$$

for all $i \in I$. Since X is assumed to be μ-dominated by $\mu \in \mathscr{M}_+^\sigma(\Omega,\mathbf{A})$, we obtain $(N(P_i))_{i\in I} \ll N(\mu)$. But I is uncountable by hypothesis. Therefore there exists a $j \in I$ such that $N(\mu)(\{j\}) = 0$, whence $N(P_i)(\{j\}) = 0$ for all $i \in I$. This implies $2 - \Delta(X,X_a) \leq 0$, i.e., $\Delta(X,X_a) \geq 2$.

In general we have $\Delta(X,X_a) \leq 2$, thus altogether $\Delta(X,X_a) = 2$. □

From now on until the end of the chapter we shall preserve the following *general assumptions*.

(1) The basic spaces $(\Omega,\mathbf{A})$ and $(I,\mathfrak{I})$ of the experiments X and X_a respectively are standard Borel spaces.

(2) X is μ-dominated by a measure $\mu \in \mathscr{M}_+^\sigma(\Omega,\mathbf{A})$.

(3) I is at most countable.

Theorem 32.3. For every $\Lambda \in \mathscr{M}^1(I,\mathfrak{I})$ one has

$$\Delta(X,X_a|\Lambda) = 2r(X|\Lambda),$$

where $r(X|\Lambda) := r^V(X|\Lambda)$ for the loss function corresponding to the above decision triple (I,D,V).

Proof: First of all we recall that

$$r(X|\Lambda) = \inf_{N\in \mathrm{Stoch}((\Omega,\mathfrak{A}),(I,\mathfrak{I}))} \left(1 - \sum_{i\in I} \Lambda(\{i\}) \int N(x,\{i\})P_i(dx)\right).$$

But $X >_\varepsilon X_a$ is equivalent to the existence of a kernel $N \in \mathrm{Stoch}((\Omega,\mathfrak{A}),(I,\mathfrak{I}))$ such that $||N(P_i) - \varepsilon_i|| \leq \varepsilon(i)$ for all $i \in I$, or equivalently such that

$$2\left(1 - \int N(x,\{i\})P_i(dx)\right) \leq \varepsilon(i)$$

for all $i \in I$.

Now let $X >_\varepsilon X_a$. Then there exists a kernel $N \in \mathrm{Stoch}((\Omega,\mathfrak{A}),(I,\mathfrak{I}))$ satisfying

$$2\left(1 - \sum_{i\in I} \Lambda(\{i\})\int N(x,\{i\})P_i(dx)\right) \leq \sum_{i\in I} \Lambda(\{i\})\varepsilon(i),$$

which implies $2r(X|\Lambda) \leq \Delta(X,X_a|\Lambda)$.

For the inverse inequality we take an $a \in \mathbb{R}_+$ and a kernel $M \in \mathrm{Stoch}((\Omega,\mathfrak{A}),(I,\mathfrak{I}))$ satisfying

$$2 - 2\sum_{i\in I} \Lambda(\{i\})\int M(x,\{i\})P_i(dx) < 2r(X|\Lambda) + a.$$

From the above equivalence we infer that $X >_{\bar\varepsilon} X_a$ for $\bar\varepsilon: I \to \mathbb{R}_+$ defined by

$$\bar\varepsilon(i) := 2 - 2\int M(x,\{i\})P_i(dx)$$

for all $i \in I$. Therefore

$$\begin{aligned}\Delta(X,X_a|\Lambda) &\leq \sum_{i\in I} \Lambda(\{i\})\bar\varepsilon(i)\\ &= 2 - 2\sum_{i\in I} \Lambda(\{i\}) \int M(x,\{i\})P_i(dx)\\ &< 2r(X|\Lambda) + a,\end{aligned}$$

which completes the proof of the asserted equality. □

Corollary 32.4. One has

$$\Delta(X,X_a) = 2 \sup_{\Lambda\in\mathscr{M}_f^1(I,\mathfrak{I})} r(X|\Lambda).$$

Proof: This is a direct implication of the theorem if one applies Property 31.10.5. □

Remark 32.5. The number $\frac{1}{2}\Delta(X,X_a)$ is in fact the *minimax risk corresponding to the estimation problem given by the decision triple* (I,D,V).

In fact, from $X >_\varepsilon X_a$ with

$$\varepsilon := 2 \sup_{i\in I}\left(1 - \int N(x,\{i\})P_i(dx)\right)$$

for any kernel $N \in \mathrm{Stoch}((\Omega,\mathfrak{A}),(I,\mathfrak{I}))$ we conclude that

$$2 \sup_{i\in I}\left(1 - \int N(x,\{i\})P_i(dx)\right) \geq \Delta(X,X_a)$$

holds.

Moreover, by Theorem 31.1 there exists a kernel $N \in \mathrm{Stoch}((\Omega,\mathfrak{A}),(I,\mathfrak{I}))$ satisfying

$$\Delta(X,X_a) = \sup_{i\in I} ||N(P_i) - \varepsilon_i||$$

and consequently

$$2 \sup_{i\in I}\left(1 - \int N(x,\{i\})P_i(dx)\right) = \Delta(X,X_a).$$

Corollary 32.6. If $I := I_n$ for $n \geq 1$ and Λ_0 the uniform distribution on I, then

$$\Delta(X,X_a) \leq 2n\, r(X|\Lambda_0).$$

Proof: By Remark 32.5 we obtain

$$\begin{aligned}\Delta(X,X_a) &= 2 \inf_{N\in\mathrm{Stoch}((\Omega,\mathfrak{A}),(I,\mathfrak{I}))} \sup_{i\in I}\left(1 - \int N(x,\{i\})P_i(dx)\right)\\ &= 2n \inf_N \frac{1}{n}\sup_{i\in I}\left(1 - \int N(x,\{i\})P_i(dx)\right)\\ &\leq 2n \inf_N \sum_{i\in I}\Lambda_0(\{i\})\left(1 - \int N(x,\{i\})P_i(dx)\right)\\ &= 2n \inf_N\left(1 - \sum_{i\in I}\Lambda_0(\{i\})\int N(x,\{i\})P_i(dx)\right)\\ &= 2n\, r(X|\Lambda_0).\end{aligned}$$

□

Theorem 32.7. Let $\Lambda \in \mathscr{M}^1(I,\mathfrak{I})$ and $I' \subset I$ such that $\Lambda(I') > 0$. Then

(i) $r(X|\Lambda) \geq \Lambda(I')r(X_{I'}|\Lambda_{I'})$

and

(ii) $\Delta(X,X_a|\Lambda) \geq \Lambda(I')\Delta(X,X_a|\Lambda_{I'})$,

where $\Lambda_{I'} := \frac{1}{\Lambda(I')} \operatorname{Res}_{I'}\Lambda$.

Proof: (i) follows from

$$\Lambda(I')r(X_{I'}|\Lambda_{I'})$$

$$= \Lambda(I')\Big(1 - \sup_{N\in \operatorname{Stoch}((\Omega,\mathfrak{A}),(I,\mathfrak{I}))} \sum_{i\in I'} \frac{\Lambda(\{i\})}{\Lambda(I')} \int N(x,\{i\})P_i(dx)\Big)$$

$$\leq \Lambda(I') - \sup_N \sum_{i\in I} \Lambda(\{i\}) \int N(x,\{i\})P_i(dx) + \sum_{i\in I\setminus I'} \Lambda(\{i\})$$

$$= 1 - \sup_N \sum_{i\in I} \Lambda(\{i\}) \int N(x,\{i\})P_i(dx)$$

$$= r(X|\Lambda).$$

(ii) is a direct consequence of (i) with the aid of Theorem 32.3. □

Theorem 32.8. Let $\Lambda \in \mathscr{M}^1(I,\mathfrak{I})$ be not a Dirac measure. Then

$$2r(X|\Lambda) \leq \sum_{\substack{i,j\in I\\ i\neq j}} \Lambda(\{i,j\})r(X_{\{i,j\}}|\Lambda_1)$$

with

$$\Lambda_1 := \frac{\Lambda(\{i\})}{\Lambda(\{i,j\})}\varepsilon_i + \frac{\Lambda(\{j\})}{\Lambda(\{i,j\})}\varepsilon_j.$$

Proof: 1. One easily verifies that for any bounded sequence $(a_n)_{n\geq 1}$ in $\mathbb{R}_+$ the inequality

$$2\Big(\sum_{n\geq 1} a_n - \bigvee_{n\geq 1} a_n\Big) \leq \sum_{\substack{m,n\in \mathbb{N}\\ m\neq n}} a_m \wedge a_n$$

holds.

2. Applying this inequality we obtain

$$2r(X,\Lambda) = 2\Big(1 - \Big\|\bigvee_{i\in I} \Lambda(\{i\})P_i\Big\|\Big)$$

$$= 2\Big\|\sum_{i\in I} \Lambda(\{i\})P_i - \bigvee_{i\in I} \Lambda(\{i\})P_i\Big\|$$

$$\leq \sum_{\substack{i,j\in I\\ i\neq j}} \|\Lambda(\{i\})P_i \wedge \Lambda(\{j\})P_j\|$$

$$= \sum_{\substack{i,j\in I\\ i\neq j}} \Lambda(\{i,j\}) r(X_{\{i,j\}}|\Lambda_1)$$

with

$$\Lambda_1 := \frac{\Lambda(\{i\})}{\Lambda(\{i,j\})} \varepsilon_i + \frac{\Lambda(\{j\})}{\Lambda(\{i,j\})} \varepsilon_j$$

and an arbitrary measure $\Lambda_1 \in \mathscr{M}^1(\{i,j\})$ if $\Lambda(\{i,j\}) = 0$. □

Corollary 32.9. We have

$$\Delta(X,X_a) \leq \sum_{\substack{i,j\in I\\ i\neq j}} \Delta(X_{\{i,j\}},X_a).$$

Proof: We apply Corollary 32.4 and get

$$\Delta(X,X_a) = 2 \sup_{\Lambda\in\mathscr{M}_f^1(I,\mathfrak{I})} r(X|\Lambda)$$

$$\leq \sum_{\substack{i,j\in I\\ i\neq j}} 2 \sup_{\alpha\in[0,1]} r(X_{\{i,j\}}|\alpha\varepsilon_i + (1-\alpha)\varepsilon_j)$$

$$= \sum_{\substack{i,j\in I\\ i\neq j}} \Delta(X_{\{i,j\}},X_a). \quad \square$$

Corollary 32.10. If $I := I_n$ for $n \geq 1$, then

$$\Delta(X,X_a) \leq (n-1) \sup_{\substack{i,j\in I\\ i\neq j}} \Delta(X_{\{i,j\}},X_a).$$

The proof follows from the theorem together with Corollary 32.4: Let $M := \sup_{\substack{i,j\in I\\ i\neq j}} \Delta(X_{\{i,j\}},X_a)$. Then

$$2r(X|\Lambda) \leq \sum_{\substack{i,j\in I\\ i\neq j}} \Lambda(\{i,j\}) r(X_{\{i,j\}}|\Lambda_1)$$

$$\leq \sum_{\substack{i,j\in I\\ i\neq j}} \Lambda(\{i,j\}) \frac{1}{2} \Delta(X_{\{i,j\}},X_a)$$

$$\leq \sum_{\substack{i,j\in I\\ i\neq j}} \Lambda(\{i,j\}) \frac{M}{2}$$

$$= 2(n-1) \frac{M}{2} = (n-1)M,$$

thus

$$\Delta(X,X_a) \le 2r(X|\Lambda) \le (n-1) \sup_{\substack{i,j\in I\\ i\ne j}} \Delta(X_{\{i,j\}},X_a). \qquad \square$$

§33. TOTALLY UNINFORMATIVE EXPERIMENTS

Any experiment is more informative than an experiment whose chance mechanism does not depend on which of the underlying parameters is the "true" one. The latter experiment appears to be totally uninformative in the sense that we might consider a given experiment as containing much or little information according to whether it is far away from the totally uninformative one or close to it, respectively. As a measure of the distance from the totally uninformative experiment we shall choose the deficiency.

Definition 33.1. An experiment $X: = (\Omega,\mathbf{A},(P_i)_{i\in I})$ is said to be *totally uninformative* if the measures $P_i: = P \in \mathscr{M}^1(\Omega,\mathbf{A})$ are independent of $i \in I$.

If there is no necessity to emphasize the defining measure, we shall abbreviate the totally uninformative experiment by X_i.

By definition we have $\rho_k(X,X_i) = 0$ for any given $k \ge 2$ and $\rho(X,X_i) = 0$ whenever $X \in \mathscr{E}(I)$, whence $\Delta_k(X,X_i) = \rho_k(X_i,X)$ and $\Delta(X,X_i) = \rho(X_i,X)$.

In this section we shall consider the decision triple (I,D,V) with $D: = \mathscr{M}^1(\Omega,\mathbf{A})$, $\mathbf{D}: =$ a sub-σ-algebra of $\mathscr{M}^1(\Omega,\mathbf{A})$ which contains the set $\{P\}$ and all subsets of $\{P_i: i \in I\}$, and $V: I \times D \to \mathbb{R}_+$ defined by

$$V(i,P): = ||P - P_i||$$

for all $(i,P) \in I \times D$.

The *minimax risk corresponding to* the *estimation problem given by the decision triple* (I,D,V) can be computed as the number

$$\inf_{\delta\in\mathrm{Stoch}((\Omega,\mathbf{A}),(D,\mathbf{D}))} \sup_{i\in I} r(i,\delta)$$

$$= \inf_{\delta} \sup_{i} \iint ||P - P_i||\delta(x,dP)P_i(dx)$$

$$= \inf_{P\in\mathscr{M}^1(\Omega,\mathbf{A})} \sup_{i} \int ||P - P_i||P_i(dx)$$

$$= \inf_{P} \sup_{i} ||P - P_i||.$$

Theorem 33.2. For any tolerance function $\varepsilon: I \to \mathbb{R}_+$ the following statements are equivalent:

(i) $X_i >_\varepsilon X$.

(ii) There exists a measure $P \in \mathscr{M}^1(\Omega,\mathfrak{A})$ such that $||P - P_i|| \leq \varepsilon(i)$ for all $i \in I$.

Proof: Let $X_i: = (\Omega_1,\mathfrak{A}_1,(Q_i)_{i\in I})$ with $Q_i: = Q \in \mathscr{M}^1(\Omega_1,\mathfrak{A}_1)$ for a all $i \in I$. An application of Theorem 31.1 yields that $X_i >_\varepsilon X$ iff there exists a kernel $N \in \mathrm{Stoch}((\Omega_1,\mathfrak{A}_1),(\Omega,\mathfrak{A}))$ satisfying $||N(Q) - P_i|| \leq \varepsilon(i)$ for all $i \in I$. Putting $P: = N(Q) \in \mathscr{M}^1(\Omega,\mathfrak{A})$ we arrive at the assertion. □

Corollary 33.3. We have

$$\Delta(X,X_i) = \inf_{P\in\mathscr{M}^1(\Omega,\mathfrak{A})} \sup_{i\in I} ||P - P_i||.$$

The proof is obvious. □

In order to compute the distance $\Delta(X,X_i)$ in terms of the "diameter" of X we have to prove a

Lemma 33.4. Let $X: = (\Omega,\mathfrak{A},(P_i)_{i\in I})$ and $Y: = (\Omega_1,\mathfrak{A}_1,(Q_i)_{i\in I})$ be two experiments in $\mathscr{E}(I)$ with $I: = I_n$ for $n \geq 1$. For any sub-σ-algebra $\mathfrak{S}$ of $\mathfrak{A}_1$ we shall consider the induced experiment $Y_{\mathfrak{S}}: = (\Omega_1,\mathfrak{S},(Q_{i\mathfrak{S}})_{i\in I})$ with $Q_{i\mathfrak{S}}: = \mathrm{Res}_{\mathfrak{S}}Q_i$ for all $i \in I$. For any tolerance function $\varepsilon: I \to \mathbb{R}_+$ and every $k \geq 1$ the following statements are equivalent:

(i) $X >_\varepsilon^k Y$.

(ii) For any sub-σ-algebra $\mathfrak{S}$ of $\mathfrak{A}_1$ containing at most 2^k sets we have

$$X >_\varepsilon^k Y_{\mathfrak{S}}.$$

Proof: It suffices to show the implication (ii) ⇒ (i). We keep the terminology of §26. Let $\psi \in \Psi_k(\mathbb{R}^n)$ be of the form $\psi: = \bigvee_{j=1}^k \ell_j$ with $\ell_1,\ldots,\ell_k \in \Psi_1(\mathbb{R}^n)$, and let $\{B_1,\ldots,B_k\}$ be an $\mathfrak{A}_1$-measurable partition of Ω_1 such that

$$\psi \circ g = \sum_{j=1}^k (\ell_j \circ g)1_{B_j},$$

where

$$g: = \left(\frac{dQ_1}{d\left(\sum_{i=1}^{n} Q_i\right)}, \ldots, \frac{dQ_n}{d\left(\sum_{i=1}^{n} Q_i\right)}\right).$$

For the σ-algebra $\tilde{\mathfrak{S}}: = \mathbf{A}(\{B_1,\ldots,B_k\})$ we consider the experiment $\tilde{Y}: = Y_{\tilde{\mathfrak{S}}} \in (I_n)$ and the $\mathbb{R}^n$-valued function

$$\tilde{g}: = \left(\frac{dQ_{1\tilde{\mathfrak{S}}}}{d\left(\sum_{i=1}^{n} Q_{i\tilde{\mathfrak{S}}}\right)}, \ldots, \frac{dQ_{n\tilde{\mathfrak{S}}}}{d\left(\sum_{i=1}^{n} Q_{i\tilde{\mathfrak{S}}}\right)}\right)$$

on Ω_1. Since $\tilde{g}$ is $\tilde{\mathfrak{S}}$-measurable, we obtain from the assumption with the aid of Corollary 26.12

$$\psi(X) \geq \psi(\tilde{Y}) - \frac{1}{2}\sum_{i=1}^{n} \varepsilon(i)(\psi(e_i) + \psi(-e_i)).$$

But

$$\psi(\tilde{Y}) = \int \psi \circ \tilde{g}d\left(\sum_{i=1}^{n} Q_{i\tilde{\mathfrak{S}}}\right) = \sum_{j=1}^{k} \int_{B_j} \psi \circ gd\left(\sum_{i=1}^{n} Q_{i\mathfrak{S}}\right)$$

$$= \sum_{j=1}^{k} \int_{B_j} \psi\left(\frac{Q_1(B_j)}{\sum_{i=1}^{n} Q_i(B_j)}, \ldots, \frac{Q_n(B_j)}{\sum_{i=1}^{n} Q_i(B_j)}\right) d\left(\sum_{i=1}^{n} Q_i\right)$$

$$= \sum_{j=1}^{k} \psi(Q_1(B_j),\ldots,Q_n(B_j))$$

$$\geq \sum_{j=1}^{k} \ell_j(Q_1(B_j),\ldots,Q_n(B_j))$$

$$= \sum_{j=1}^{k} \ell_j\left(\int_{B_j} gd\left(\sum_{i=1}^{n} Q_i\right)\right)$$

$$= \sum_{j=1}^{k} \int_{B_j} \ell_j \circ gd\left(\sum_{i=1}^{n} Q_i\right) = \int \psi \circ gd\left(\sum_{i=1}^{n} Q_i\right)$$

$$= \psi(Y),$$

whence

$$\psi(X) \geq \psi(Y) - \frac{1}{2}\sum_{i=1}^{n} \varepsilon(i)(\psi(e_i) + \psi(-e_i)),$$

which by Corollary 26.12 yields the assertion. □

<u>Theorem 33.5</u>. For the experiments X and X_i we have

$$2\Delta_2(X,X_i) = \sup_{\substack{i,j\in I \\ i\neq j}} ||P_i - P_j||.$$

Proof: From Corollary 31.3 we infer that it suffices to establish the asserted formula for parameter sets $I := I_n$ for $n \geq 1$. We shall prove the following equivalences, from which the formula follows.

(i) $\Delta_2(X, X_i) \leq \varepsilon$

(ii) $\inf_{\alpha \in [0,1]} \sup_{i \in I} |P_i(A) - \alpha| \leq \frac{\varepsilon}{2}$ for all $A \in \mathfrak{A}$.

(iii) $|P_i(A) - P_j(A)| \leq \varepsilon$ for all $A \in \mathfrak{A}$ and all $i,j \in I$.

(iv) $\frac{1}{2} \sup_{i,j \in I} ||P_i - P_j|| \leq \varepsilon$.

Here, ε is any number in $\mathbb{R}_+^*$.

1. (i) $\Leftrightarrow$ (ii). From Lemma 33.4 we obtain that $X_i >_\varepsilon^2 X$ iff $X_i >_\varepsilon \tilde{X}$ for all experiments $\tilde{X} := (\Omega, \tilde{\mathfrak{A}}, (P_{i\tilde{\mathfrak{A}}})_{i \in I})$ with a sub-σ-algebra $\tilde{\mathfrak{A}}$ of $\mathfrak{A}$ such that $|\tilde{\mathfrak{A}}| \leq 2^2 = 4$. As a consequence we get

$$\Delta(\tilde{X}, X_i) \leq \Delta_2(X, X_i)$$

for all such $\tilde{X}$. Now we consider the σ-algebra $\tilde{\mathfrak{A}}_1 := \{\emptyset, \Omega, A_1, CA_1\}$ with $A_1 \in \mathfrak{A}$ such that $A_1 \neq \emptyset, \Omega$, and the corresponding experiment $\tilde{X}_1 := (\Omega, \tilde{\mathfrak{A}}_1, (P_{i\tilde{\mathfrak{A}}_1})_{i \in I})$. Then

$$\Delta(\tilde{X}_1, X_i) = \inf_{P \in \mathscr{M}^1(\Omega, \tilde{\mathfrak{A}}_1)} \sup_{i \in I} ||P_{i\tilde{\mathfrak{A}}_1} - P||$$

$$= 2 \inf_{\alpha \in [0,1]} \sup_{i \in I} |P_i(A) - \alpha|,$$

whence $\Delta(\tilde{X}_1, X_i) \leq \varepsilon$ iff (ii) is satisfied.

2. (ii) $\Leftrightarrow$ (iii). The implication (ii) $\Rightarrow$ (iii) follows directly from the inequality

$$|P_i(A) - P_j(A)| \leq \inf_{\alpha \in [0,1]} (|P_i(A) - \alpha| + |P_j(A) - \alpha|),$$

valid for all $A \in \mathfrak{A}$ and $i,j \in I$.

For the inverse implication (iii) $\Rightarrow$ (ii) we assume given an $A \in \mathfrak{A}$ such that $|P_i(A) - P_j(A)| \leq \varepsilon$ for all $i,j \in I$. Then choosing

$$\alpha := \frac{\bigvee_{i \in I} P_i(A) + \bigwedge_{i \in I} P_i(A)}{2} \in [0,1]$$

we get

$$P_i(A) \in [\alpha - \frac{\varepsilon}{2}, \alpha + \frac{\varepsilon}{2}]$$

for all $i \in I$, which implies the assertion.

3. (iii) $\Leftrightarrow$ (iv) is clear since for two measures $P, Q \in \mathscr{M}^1(\Omega, \mathfrak{A})$ the Hahn decomposition implies the existence of a set $B \in \mathfrak{A}$ satisfying

$$||P - Q|| = 2(P(B) - Q(B)). \qquad \square$$

In the following we specialize the situation by looking at dichotomies.

Theorem 33.6. For any dichotomy $X = (\Omega, \mathfrak{A}, \{P_1, P_2\}) \in \mathscr{E}(I_2)$ and every $\alpha \in [0,1]$ we have

$$\Delta(X, X_i | \alpha\varepsilon_1 + (1-\alpha)\varepsilon_2) = (\alpha \wedge (1-\alpha))||P_1 - P_2||.$$

Proof: Given $\alpha \in [0,1]$ we have by Definition 31.9 and Theorem 33.2

$$\Delta(X, X_i | \alpha\varepsilon_1 + (1-\alpha)\varepsilon_2) = \inf_{P \in \mathscr{M}^1(\Omega, \mathfrak{A})} (\alpha||P-P_1|| + (1-\alpha)||P-P_2||).$$

But the inequalities

$$\begin{aligned}(\alpha \wedge (1-\alpha))||P_1 - P_2|| &\leq (\alpha \wedge (1-\alpha)) \inf_{P \in \mathscr{M}^1(\Omega, \mathfrak{A})} (||P-P_1|| + ||P-P_2||) \\ &\leq \inf_P (\alpha||P - P_1|| + (1-\alpha)||P - P_2||) \\ &\leq (\alpha \wedge (1-\alpha))||P_1 - P_2||\end{aligned}$$

imply the assertion. $\square$

Corollary 33.7. We have

$$\begin{aligned}\Delta(X, X_i) &= \Delta(X, X_i | \tfrac{1}{2}\varepsilon_1 + \tfrac{1}{2}\varepsilon_2) \\ &= 1 - 2r(X | \tfrac{1}{2}\varepsilon_1 + \tfrac{1}{2}\varepsilon_2) \\ &= \tfrac{1}{2}||P_1 - P_2||.\end{aligned}$$

Proof: The formula $\Delta(X, X_i) = \frac{1}{2}||P_1 - P_2||$ follows from Theorem 33.5. It remains to show that

$$\tfrac{1}{2}||P_1 - P_2|| = 1 - 2r(X | \tfrac{1}{2}\varepsilon_1 + \tfrac{1}{2}\varepsilon_2).$$

For that we compute the equalities

$$\begin{aligned}\tfrac{1}{2}||P_1 - P_2|| &= 1 - ||P_1 \wedge P_2|| \\ &= 1 - 2||\tfrac{1}{2}P_1 \wedge \tfrac{1}{2}P_2|| \\ &= 1 - 2r(X | \tfrac{1}{2}\varepsilon_1 + \tfrac{1}{2}\varepsilon_2),\end{aligned}$$

where in the latter one we applied Theorem 31.8. □

A combination of Theorem 33.5 and Corollary 33.7 yields the

Remark 33.8. For *any* experiment $X \in \mathscr{E}(I)$ we have

$$\Delta_2(X, X_i) = \sup_{i,j \in I} \Delta(X_{\{i,j\}}, X_i).$$

§34. INEQUALITIES BETWEEN DEFICIENCIES

This section will be devoted to establishing inequalities between the deficiencies $\Delta(X, X_a)$ and $\Delta(X, X_i)$ for experiments X of the form $(\Omega, \mathfrak{A}, (P_i)_{i \in I})$ with $I := I_n$ for some fixed $n \geq 1$.

At first we note that under the assumptions of the previous sections we have the formula

$$\Delta(X_i, X_a) = 2 - \frac{2}{n},$$

which by Corollary 33.3 follows from the short computation

$$\Delta(X_i, X_a) = \inf_{P \in \mathscr{M}^1(I,\mathfrak{I})} \sup_{i \in I} ||\varepsilon_i - P||$$

$$= \inf_P \sup_i 2(1 - P(\{i\}))$$

$$= 2\,(1 - \frac{1}{n}).$$

A special case of this formula appears in Example 26.16.

We come to the first basic result of this section.

Theorem 34.1.

$$\frac{2}{n^2} \sup_{\substack{i,j \in I \\ i \neq j}} \Delta(X_{\{i,j\}}, X_i) \leq \Delta(X_i, X_a) - \Delta(X, X_a)$$

Proof: Let $\Lambda \in \mathscr{M}^1_f(I,\mathfrak{I})$ and $k, \ell \in I$ with $k \neq \ell$.

At first we look for a lower bound of

$$||\bigvee_{i \in I} \Lambda(\{i\}) P_i|| - \frac{1}{n^2} ||P_k \vee P_\ell|| \tag{1}$$

under the condition $||P_k \vee P_\ell|| = b + 1$ for some $b \in [0,1]$.

From this very condition we infer the existence of a set $A \in \mathfrak{A}$ such that $P_\ell(A) + 1 - P_k(A) = b + 1$ and therefore $P_\ell(A) = P_k(A) + b$ holds. Let $\tilde{\mathfrak{A}} := \{\emptyset, \Omega, A, CA\}$ and $\tilde{P}_i := \mathrm{Res}_{\tilde{\mathfrak{A}}} P_i$ for all $i \in I$. Since $||\tilde{P}_k \vee \tilde{P}_\ell|| = ||P_k \vee P_\ell|| = b + 1$ and

$$\|\bigvee_{i\in I} \Lambda(\{i\})P_i\| \geq \|\bigvee_{i\in I} \Lambda(\{i\})\tilde{P}_i\|$$

hold, we may assume without loss of generality that $\mathbf{A} = \tilde{\mathbf{A}}$. Defining $p_i := P_i(A)$ for all $i \in I$ we note that (1) is equivalent to

$$\left(\bigvee_{i\in I} \Lambda(\{i\})(1 - p_i) + \bigvee_{i\in I} \Lambda(\{i\})p_i\right) - \frac{1}{n^2}(b + 1), \tag{2}$$

where

$$(1 - p_k) \vee (1 - p_\ell) + p_k \vee p_\ell = 1 + b = 1 - p_k + p_\ell.$$

The minimization of the expression (2) will be achieved in several steps.

<u>1st Step</u>. Let $U := \{k,\ell\}$ and $W := I \setminus U$. We abbreviate

$$a := \bigvee_{i\in I} \Lambda(\{i\})(1 - p_i) + \bigvee_{i\in I} \Lambda(\{i\})p_i.$$

Then

$$a = \left(\bigvee_{i\in U} \Lambda(\{i\})(1 - p_i)\right) \vee \left(\bigvee_{i\in W} \Lambda(\{i\})(1 - p_i)\right)$$
$$+ \left(\bigvee_{i\in U} \Lambda(\{i\})p_i\right) \vee \left(\bigvee_{i\in W} \Lambda(\{i\})p_i\right)$$
$$\geq \left(\bigvee_{i\in W} \Lambda(\{i\})(1 - p_i) + \bigvee_{i\in W} \Lambda(\{i\})p_i\right)$$
$$\vee \left(\bigvee_{i\in U} \Lambda(\{i\})(1 - p_i) + \bigvee_{i\in U} \Lambda(\{i\})p_i\right).$$

Since

$$\bigvee_{i\in W} \Lambda(\{i\})(1 - p_i) + \bigvee_{i\in W} \Lambda(\{i\})p_i = \bigvee_{i\in W} \Lambda(\{i\}),$$

we deduce

$$a \geq \left(\bigvee_{i\in W} \Lambda(\{i\})\right) \vee \left(\bigvee_{i\in U} \Lambda(\{i\})(1 - p_i) + \bigvee_{i\in U} \Lambda(\{i\})p_i\right),$$

whence, applying

$$\bigvee_{i\in W} \Lambda(\{i\}) \geq \frac{1 - \Lambda(\{k,\ell\})}{n - 2},$$

$$a \geq \frac{1}{n-2}(1 - \Lambda(\{k,\ell\})) \vee \left(\bigvee_{i\in U} \Lambda(\{i\})(1 - p_i) + \bigvee_{i\in U} \Lambda(\{i\})p_i\right).$$

But there exists a measure $M \in \mathscr{M}^1(\{k,\ell\})$ satisfying

$$\bigvee_{i\in U} \Lambda(\{i\})(1 - p_i) + \bigvee_{i\in U} \Lambda(\{i\})p_i$$

$$= \Lambda(\{k,\ell\})\Big(\bigvee_{i\in U} M(\{i\})(1 - p_i) + \bigvee_{i\in U} M(\{i\})p_i\Big).$$

This implies

$$a \geq \frac{1}{n-2}\,(1 - \Lambda(\{k,\ell\}) \vee \Lambda(\{k,\ell\})\cdot\Big(\bigvee_{i\in U} M(\{i\})(1 - p_i) + \bigvee_{i\in U} M(\{i\})p_i\Big).$$

2nd Step. Consider the dichotomy X_U given by

$$P^{X_U} := \begin{pmatrix} 1 - p_k & p_k \\ 1 - p_\ell & p_\ell \end{pmatrix}$$

where $p_\ell = p_k + b$. We want to compute

$$\sup_{p_k\in[0,1]} \Delta(X_U, X_a).$$

From Theorem 28.9 we know that $X_U >_\varepsilon X_a$ iff

$$\beta_{X_U}\left(\frac{\varepsilon(1)}{2}\right) + \frac{\varepsilon(2)}{2} \geq 1.$$

Thus

$$\Delta(X_U,X_a) = \inf\{\varepsilon \geq 0: \beta_{X_U}\left(\frac{\varepsilon}{2}\right) + \frac{\varepsilon}{2} \geq 1\},$$

or

$$\Delta(X_U,X_a) = \varepsilon \text{ with } \beta_{X_U}\left(\frac{\varepsilon}{2}\right) + \frac{\varepsilon}{2} = 1.$$

The desired result follows from the geometry presented in the subsequent sketch: The graphs of the functions $\alpha \to \beta_{X_U}(\alpha)$ and $\alpha \to 1 - \alpha$ intersect in the point $(\frac{1}{2}\,\Delta(X_U,X_a),\ 1 - \frac{1}{2}\,\Delta(X_U,X_a))$. From the sketch it becomes evident that $\Delta(X_U,X_a)$ attains its maximum for the experiment $\overline{X}_U$ given by the matrix

$$P^{\overline{X}_U} := \begin{pmatrix} b & 1-b \\ 0 & 1 \end{pmatrix}.$$

Therefore the intersection of the functions $\alpha \to \beta_{\overline{X}_U}(\alpha)$ and $\alpha \to 1 - \alpha$ is the point $(\frac{1}{2} \sup_{p_k\in[0,1]} \Delta(X_U,X_a),\ 1 - \frac{1}{2} \sup_{p_k\in[0,1]} \Delta(X_U,X_a))$ which coincides with the point $(\frac{1-b}{2-b}, \frac{1}{2-b})$. We obtain

$$\sup_{p_k\in[0,1]} \Delta(X_U,X_a) = 2 \cdot \frac{1-b}{2-b}\,.$$

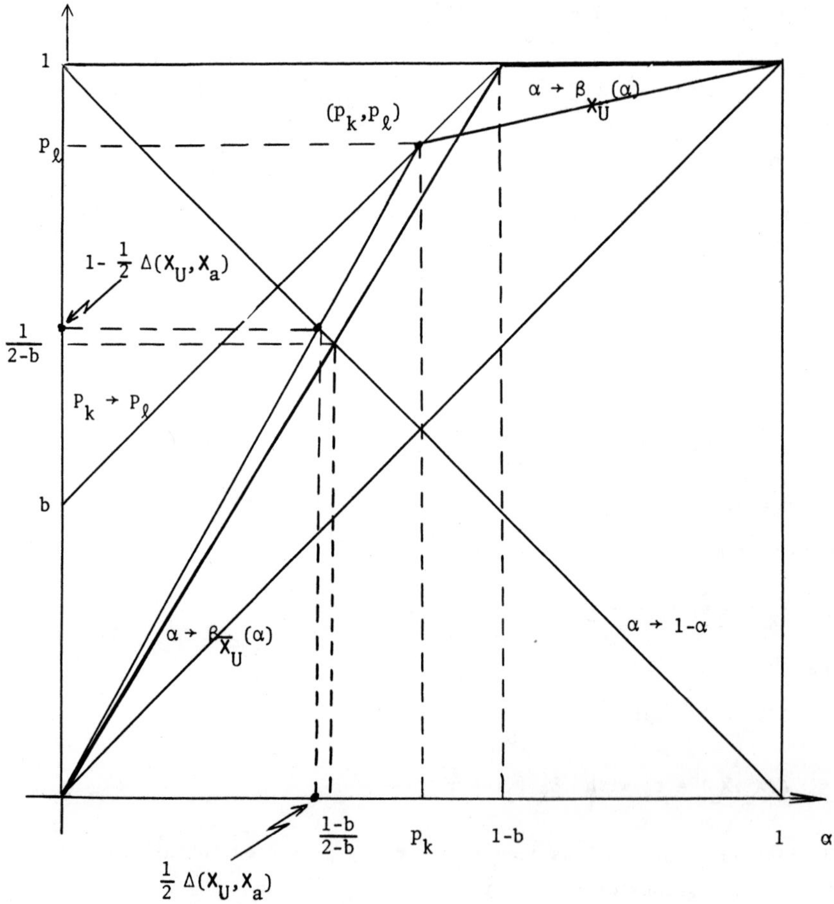

<u>3rd Step</u>. Combining the results of Corollary 32.4 and Theorem 31.8 we have

$$\begin{aligned}\Delta(\mathcal{X}_U, \mathcal{X}_a) &= 2\Big(1 - \inf_{M \in \mathcal{M}_f^1(I,\mathfrak{I})} \Big\|\bigvee_{i \in U} M(\{i\})P_i\Big\|\Big) \\ &= 2\Big(1 - \inf_M \big(\bigvee_{i \in U} M(\{i\})(1 - p_i)\big) + \bigvee_{i \in U} M(\{i\})p_i\Big),\end{aligned}$$

and so we get by the result of the 2nd step

$$1 - \inf_{p_k \in [0,1]} \inf_{M} \left(\bigvee_{i \in U} M(\{i\})(1 - p_i) + \bigvee_{i \in U} M(\{i\}) p_i \right)$$
$$= \frac{1}{2} \sup_{p_k} \Delta(X_U, X_a) = \frac{1-b}{2-b} .$$

Altogether we have achieved that

$$a \geq \frac{1}{n-2} (1 - \Lambda(\{k,\ell\})) \vee \Lambda(\{k,\ell\}) \frac{1}{2-b} ,$$

but the right side of this inequality becomes minimal if

$$\frac{1}{n-2}(1 - \Lambda(\{k,\ell\})) = \frac{1}{2-b} \Lambda(\{k,\ell\})$$

or equivalently if

$$\frac{1}{2-b} \Lambda(\{k,\ell\}) = \frac{1}{n-b}$$

holds. This yields $a \geq 1/(n-b)$.

4th Step. Applying the previous step we get from (2) the desired estimate

$$\left|\left| \bigvee_{i \in I} \Lambda(\{i\}) P_i \right|\right| - \frac{1}{n^2} \left|\left| P_k \vee P_\ell \right|\right| \geq \frac{1}{n-b} - \frac{b+1}{n^2} \geq \frac{1}{n} - \frac{1}{n^2} .$$

The remainder of the proof goes as follows. By Corollary 33.7 we have

$$\Delta(X_U, X_i) = \frac{1}{2} \left|\left| P_k - P_\ell \right|\right| = \left|\left| P_k \vee P_\ell \right|\right| - 1.$$

This implies

$$\left|\left| \bigvee_{i \in I} \Lambda(\{i\}) P_i \right|\right| - \frac{1}{n^2} (\Delta(X_U, X_i) + 1) \geq \frac{1}{n} - \frac{1}{n^2}$$

or equivalently

$$\frac{1}{n^2} \Delta(X_U, X_a) + 1 - \left|\left| \bigvee_{i \in I} \Lambda(\{i\}) P_i \right|\right| \leq 1 - \frac{1}{n} .$$

Since $\Lambda \in \mathscr{M}^1(I, \mathfrak{I})$ was chosen arbitrarily, we also have

$$\frac{1}{n^2} \Delta(X_U, X_i) + \sup_{\Lambda \in \mathscr{M}_f^1(I, \mathfrak{I})} \left(1 - \left|\left| \bigvee_{i \in I} \Lambda(\{i\}) P_i \right|\right| \right) \leq 1 - \frac{1}{n}$$

and together with Corollary 32.4,

$$\frac{2}{n^2} \Delta(X_U, X_i) + \Delta(X, X_a) \leq 2 - \frac{2}{n} = \Delta(X_i, X_a),$$

where the last equality has been established at the beginning of the section. From

$$\frac{2}{n^2} \sup_{p_k \in [0,1]} \Delta(X_U, X_i) + \Delta(X, X_a) \leq \Delta(X_i, X_a)$$

the statement of the theorem follows. □

Corollary 34.2.

$$\frac{1}{n(n-1)} \Delta(X, X_i) \leq \Delta(X_i, X_a) - \Delta(X, X_a) \leq \Delta(X, X_i).$$

Proof: The inequality on the right is the triangle inequality for the information distance Δ. We now apply Corollary 33.3, Theorem 33.5, Remark 33.8 and the theorem in the following chain of inequalities in order to get the desired result:

$$\frac{1}{n(n-1)} \Delta(X, X_i) \leq \frac{1}{n(n-1)} (1 - \frac{1}{n}) \sup_{\substack{i,j \in I \\ i \neq j}} ||P_i - P_j||$$

$$= \frac{1}{n^2} 2\Delta_2(X, X_i) = \frac{2}{n^2} \sup_{\substack{i,j \in I \\ i \neq j}} \Delta(X_{\{i,j\}}, X_i)$$

$$\leq \Delta(X_i, X_a) - \Delta(X, X_a). \quad \square$$

The following result shows that in supplementing the inequality of Theorem 34.1 one can establish an upper bound for $\Delta(X, X_i)$ in terms of $\Delta(Y, X_i)$ for all two-element subexperiments Y of X.

Theorem 34.3.

$$\Delta(X, X_i) \leq \Delta(X_i, X_a) - \frac{2}{n(n-1)} \left(1 - \inf_{\substack{i,j \in I \\ i \neq j}} \Delta(X_{\{i,j\}}, X_i)\right).$$

Proof: 1. From Corollary 33.7 we infer that

$$1 - \inf_{\substack{i,j \in I \\ i \neq j}} \Delta(X_{\{i,j\}}, X_i) = 1 - \frac{1}{2} \inf_{\substack{i,j \in I \\ i \neq j}} ||P_i - P_j||$$

$$= \sup_{\substack{i,j \in I \\ i \neq j}} ||P_i \wedge P_j||,$$

so it suffices to establish the inequality

$$\frac{2}{n(n-1)} ||P_i \wedge P_j|| + \Delta(X, X_i) \leq \Delta(X_i, X_a)$$

for all $i,j \in I$ and $i \neq j$.

For $k,\ell \in I$ with $k \neq \ell$ we consider the problem of maximizing the expression $\Delta(X,X_i)$ under the condition that $||P_k \wedge P_\ell|| = c$ for some $c \in [0,1]$. To this end it suffices to assume that X is a standard experiment with standard measure $S \in \mathscr{M}^1(K_n)$ (see Property 30.4.3).

2. We keep the notation of 1 and note that

$$\int (y_k \wedge y_\ell) S(d(y_1,\ldots,y_n)) = c.$$

In fact, there exists an $A \in \mathfrak{A}$ such that $||P_k \wedge P_\ell|| = P_k(A) + (1 - P_\ell(A))$. Without loss of generality it will be assumed that $P_k(A) \leq P_\ell(A)$. Moreover, we have

$$P_j(B) = \int_B X_j dS$$

for $j \in \{k,\ell\}$, thus

$$P_k(A) + P_\ell(CA) = \left(\int_E X_k dS + \int_F X_\ell dS\right) = \int (y_k \wedge y_\ell) S(d(y_1,\ldots,y_n)),$$

where $E := \{y \in K_n : y_k \leq y_\ell\}$ and $F := K_n \setminus E$.

Now we define the sets $K := \{x \in K_n : x_k \geq x_\ell\}$ and $L := \{x \in K_n : x_k \leq x_\ell\}$, and introduce the mapping $D: K_n \times \mathfrak{B}(K_n) \to \mathbb{R}$ by

$$D(x,\cdot) := (x_k - x_\ell)\varepsilon_{e_k} + 2x_\ell \varepsilon_{\frac{1}{2}(e_k+e_\ell)} + \sum_{i \in I \setminus \{k,\ell\}} x_i \varepsilon_{e_i}$$

whenever $x \in K$, and similarly by

$$D(x,\cdot) := (x_\ell - x_k)\varepsilon_{e_\ell} + 2x_k \varepsilon_{\frac{1}{2}(e_k+e_\ell)} + \sum_{i \in I \setminus \{k,\ell\}} x_i \varepsilon_{e_i}$$

whenever $x \in L$. It is immediate that D is a dilation on K_n, and $D(S)$ is a standard measure on K_n. Let $Y := X_{D(S)}$ be the standard experiment corresponding to $D(S)$. Theorem 29.1 together with the equivalence preceding that theorem yields

$$\Delta(Y,X_i) \geq \Delta(X,X_i).$$

3. We shall now show that the supremum of $\Delta(X,X_i)$ will be attained for the experiment X with standard measure $D(S)$ which satisfies the conditions

(1) $D(S)$ is supported by the $(n+1)$-element set

$$\{e_i \in K_n : i \in I\} \cup \{\tfrac{1}{2}(e_k + e_\ell)\}.$$

(2) $\int (x_k \wedge x_\ell) D(S)(dx) = c.$

First of all we note that (1) is satisfied for any standard measure $S \in \mathscr{M}^1(K_n)$. Secondly,

$$\int (x_k \wedge x_\ell) S(dx) = \int (x_k \wedge x_\ell) D(S)(dx),$$

since D is a dilation on K_n and $x \to x_k \wedge x_\ell$ is a continuous affine function on K_n. Therefore

$$\int (x_k \wedge x_\ell) D(S)(dx) = c$$

for all standard measures $S \in \mathscr{M}^1(K_n)$ satisfying

$$\int (x_k \wedge x_\ell) S(dx) = c.$$

4. The only experiment X, whose standard measure T fulfills the conditions (1) and (2) of 3 is of the form $X_b := (I \cup \{I\}, \mathfrak{P}(I \cup \{I\}), (R_i)_{i \in I})$ for $b = (b_1, \ldots, b_n) \in [0,1]^n$ with $b_j := c$ for $j \in \{k, \ell\}$ and $b_j := 0$ for $j \in I \setminus \{k, \ell\}$, where

$$\begin{cases} R_i(\{i\}) := 1 - b_i & \text{for } i \in I \\ R_i(\{I\}) := b_i & \\ R_i(\{j\}) := 0 & \text{for } i, j \in I,\ i \neq j. \end{cases}$$

In fact, from (2) we deduce $D(T)(\{\frac{1}{2}(e_k + e_\ell)\}) = 2 \cdot c$. Let $X := X_{D(T)}$. Then $P_i = X_i \cdot D(T)$ and P_i is supported by the set $\{e_j \in K_n : j \in I\} \cup \{\frac{1}{2}(e_k + e_\ell)\}$ for all $i \in I$. Since for all $i \in I$,

$$P_i(\{\tfrac{1}{2}(e_k + e_\ell)\}) = X_i(\tfrac{1}{2}(e_k + e_\ell)) \cdot 2c = cX_i(e_k + e_\ell),$$

we obtain

$$P_i(\{\tfrac{1}{2}(e_k + e_\ell)\}) = \begin{cases} c & \text{if } i \in \{k, \ell\} \\ 0 & \text{otherwise} \end{cases}.$$

From $P_i(\{e_j \in K_n : j \in I\}) = \sum_{j \in I} X_i(e_j) D(T)(\{e_j\}) = D(T)(\{e_i\})$ we get

$$P_i(\{e_j\}) = \begin{cases} 1-c & \text{if } i = j \text{ and } i \in \{k, \ell\} \\ 1 & \text{if } i = j \text{ and } i \notin \{k, \ell\} \\ 0 & \text{if } i \neq j. \end{cases}$$

and therefore $X_{D(T)} = X_b$. We conclude that

$$\sup \Delta(X,X_i) = \Delta(X_b,X_i),$$

where the supremum is taken over all experiments X satisfying $||P_k \wedge P_\ell|| = c$.

5. Now we shall determine an upper bound for $\Delta(X_b,X_i)$.

<u>1st Case</u>. Let $c \leq 1/(n-1)$. We define a measure $Q \in \mathscr{M}^1(I \cup \{I\}, \mathfrak{P}(I \cup \{I\}))$ by

$$Q(\{j\}) := \begin{cases} \frac{1}{n}(1-(n-1)c) & \text{if } j \in \{k,\ell\} \\ c & \text{if } j = I \\ \frac{1+c}{n} & \text{if } j \in I \setminus \{k,\ell\}. \end{cases}$$

Then $||P_i - Q|| = 2(1 - \frac{1}{n} - \frac{c}{n})$ for all $i \in I$, and by Theorem 33.2,

$$\Delta(X_b,X_i) = \inf_P \sup_{i \in I} ||P_i - P||$$

$$\leq \sup_{i \in I} ||P_i - Q|| = 2(1 - \frac{1}{n} - \frac{c}{n})$$

<u>2nd Case</u>. Let $1 > c > \frac{1}{n-1}$. In this case we consider the experiment X_d for $d := (d_1,\ldots,d_n) \in [0,1]^n$ with

$$d_i := \begin{cases} \frac{1}{n-1} & \text{if } i \in \{k,\ell\} \\ 0 & \text{if } i \in I \setminus \{k,\ell\}. \end{cases}$$

Let $X_d := (I \cup \{I\}, \mathfrak{P}(I \cup \{I\}), (R_i)_{i \in I})$ be defined as in 4. From $b \geq d$ we want to conclude $X_d > X_b$ and hence

$$\Delta(X_b,X_i) \leq \Delta(X_d,X_i).$$

In order to see that $X_d > X_b$ holds we introduce a kernel $N \in \mathrm{Stoch}(I \cup \{I\}, \mathfrak{P}(I \cup \{I\}))$ by

$$\begin{cases} N(I,\{I\}) := 1 & \\ N(i,\{I\}) := \begin{cases} \frac{d_i - b_i}{1 - b_i} & \text{if } b_i < 1 \\ 0 & \text{if } b_i = 1 \end{cases} & \text{for } i \in I \\ N(i,\{i\}) := 1 - N(i,\{I\}) & \text{for } i \in I \\ N(i,\{j\}) := 0 & \text{for } i,j \in I,\ i \neq j. \end{cases}$$

Since

$$N(P_i)(\{i\}) = N(i,\{i\})(1 - b_i) + N(I,\{i\})b_i = 1 - d_i$$

and

$$N(P_i)(\{I\}) = N(i,\{I\})(1 - b_i) + N(I,\{I\}) = d_i,$$

we obtain $N(P_i) = R_i$ for all $i \in I$, which by Theorem 21.5 (or Theorem 33.1) yields the assertion.

We finally apply the 1st case to X_d. This implies

$$\Delta(X_b,X_i) \leq \Delta(X_d,X_i) \leq 2 - \frac{2}{n} - \frac{2}{n(n-1)}$$
$$\leq 2 - \frac{2}{n} - \frac{2c}{n(n-1)} .$$

Altogether we arrived at the inequality

$$\Delta(X_i,X_a) - \Delta(X,X_i) \geq \frac{2c}{n(n-1)} ,$$

which implies the desired estimate for $\Delta(X,X_i)$. □

<u>Corollary 34.4</u>.

$$\frac{1}{n(n-1)^2} \Delta(X,X_a) \leq \Delta(X_i,X_a) - \Delta(X,X_i) \leq \Delta(X,X_a).$$

<u>Proof</u>: From Corollary 32.10 we obtain

$$\frac{1}{n(n-1)^2} \Delta(X,X_a) \leq \frac{1}{n(n-1)} \sup_{\substack{i,j\in I \\ i\neq j}} \Delta(X_{\{i,j\}},X_a).$$

Now, let $i,j \in I$ with $i \neq j$. We apply successively Corollary 32.4, Theorem 31.8, Corollary 33.7 and Theorem 34.3 in order to get

$$\begin{aligned}
\Delta(X_{\{i,j\}},X_a) &= 2 \sup_{\alpha\in[0,1]} r(X_{\{i,j\}}|\alpha\varepsilon_i + (1-\alpha)\varepsilon_j) \\
&= 2 \sup_{\alpha} ||\alpha P_i \wedge (1-\alpha)P_j|| \\
&\leq 2 ||P_i \wedge P_j|| \\
&= 2(1 - \tfrac{1}{2}||P_i - P_j||) \\
&= 2(1 - \Delta(X_{\{i,j\}},X_i)) \\
&\leq n(n-1)(\Delta(X_i,X_a) - \Delta(X,X_i)).
\end{aligned}$$

This implies

$$\frac{1}{n(n-1)^2}\,\Delta(X,X_a) \le \Delta(X_i,X_a) - \Delta(X,X_i),$$

the remaining estimate following from the triangle inequality for Δ. □

Notational Conventions

Most of the basic terminology will be in agreement with the standard literature. As a convenient reference for measure and probability theory we recommend the textbook by H. Bauer [9]. Abstract measure and integration theory is discussed in great detail in Zaanen's book [141]. For concrete measure theory in the sense of Radon the reader is referred to Bourbaki [25]. The relationship between the two approaches can be checked in Parthasarathy [94], §33.

I. MEASURE AND INTEGRATION

Let Ω be any set. We put

$\mathcal{F}(\Omega)$: = set of all real-valued functions on Ω.

$\mathcal{B}(\Omega)$: = subset of $\mathcal{F}(\Omega)$ consisting of the bounded functions.

I.1. If Ω is furnished with a topology $\mathcal{T}$, we denote the corresponding topological space by $(\Omega, \mathcal{T})$.

$\mathcal{C}(\Omega) = \mathcal{C}(\Omega, \mathcal{T})$: = vector space of $\mathcal{T}$-continuous real-valued functions on Ω.

$\mathcal{C}^b(\Omega) = \mathcal{C}^b(\Omega, \mathcal{T})$: = $\mathcal{C}(\Omega) \cap \mathcal{B}(\Omega)$.

For a locally compact space $(\Omega, \mathcal{T})$ we put

$\mathcal{C}^o(\Omega) = \mathcal{C}^o(\Omega, \mathcal{T})$: = subspace of $\mathcal{C}^b(\Omega)$ consisting of the functions that vanish at infinity.

$\mathcal{K}(\Omega) = \mathcal{K}(\Omega, \mathcal{T})$: = subspace of $\mathcal{C}^o(\Omega)$ consisting of the functions having compact support.

Standard examples of topological spaces $(\Omega,\mathcal{T})$ occurring in the text are the spaces $\mathbb{N}$, $\mathbb{Z}$, $\mathbb{Q}$, $\mathbb{R}$ and $\mathbb{C}$ of all natural numbers, integers, rationals, real numbers and complex numbers respectively, together with their natural topologies, the compact space of the extended real line $\overline{\mathbb{R}}$, and the p-dimensional euclidean space $\mathbb{R}^p$ for $p \geq 1$, again with their natural topologies, where by convention $\mathbb{R}^1 = \mathbb{R}$.

I.2. If Ω is furnished with a σ-algebra $\mathfrak{A}$ (of measurable sets), we denote the corresponding measurable space by $(\Omega,\mathfrak{A})$. In the special case of a topological space $(\Omega,\mathcal{T})$, $\mathfrak{A}$ is often chosen to be the Borel-σ-algebra $\mathfrak{B}(\mathcal{T})$ of Ω with respect to the topology $\mathcal{T}$. This σ-algebra will also be abbreviated by $\mathfrak{B}:= \mathfrak{B}(\Omega)$.

If $\Omega:= \mathbb{R}^p$ for $p \geq 1$, we introduce

$\mathfrak{B}^p:= \mathfrak{B}(\mathbb{R}^p)$ with the additional conventions
$\mathfrak{B}:= \mathfrak{B}^1$ if no confusion is to be expected, and
$\overline{\mathfrak{B}}:= \overline{\mathfrak{B}}^1:= \mathfrak{B}(\overline{\mathbb{R}})$.

I.3. Let $(\Omega,\mathfrak{A})$ be a measurable space. We introduce

$\mathfrak{M}(\Omega,\mathfrak{A}):=$ set of all ($\mathfrak{A}$-$\mathfrak{B}$-) measurable real-valued functions on Ω.

In a few cases the symbol $\mathfrak{M}(\Omega,\mathfrak{A})$ will also denote the set of all ($\mathfrak{A}$-$\overline{\mathfrak{B}}$-) measurable extended real (numerical) functions on Ω.

$\mathfrak{M}^b(\Omega,\mathfrak{A}):=$ subset of $\mathfrak{M}(\Omega,\mathfrak{A})$ consisting of bounded functions.

$\mathfrak{M}^{(1)}(\Omega,\mathfrak{A}):=$ subset of $\mathfrak{M}^b(\Omega,\mathfrak{A})$ consisting of all $f \in \mathfrak{M}^b(\Omega,\mathfrak{A})$ satisfying $|f| \leq 1$.

$\mathfrak{M}_{(1)}(\Omega,\mathfrak{A}):= \mathfrak{M}^{(1)}(\Omega,\mathfrak{A}) \cap \mathfrak{M}_+(\Omega,\mathfrak{A})$.

I.4. For any measurable space $(\Omega,\mathfrak{A})$ we consider the vector space $\mathcal{M}(\Omega,\mathfrak{A})$ of all (σ-additive, signed) measures on $(\Omega,\mathfrak{A})$.

$\mathcal{M}^\sigma(\Omega,\mathfrak{A}):=$ subspace of $\mathcal{M}(\Omega,\mathfrak{A})$ consisting of σ-finite measures.

$\mathcal{M}^b(\Omega,\mathfrak{A}):=$ subspace of $\mathcal{M}^\sigma(\Omega,\mathfrak{A})$ consisting of bounded measures

$= \{\mu \in \mathcal{M}^b(\Omega,\mathfrak{A}): |\mu|(\Omega) < \infty\}$.

$\mathcal{M}^{(1)}(\Omega,\mathfrak{A}):= \{\mu \in \mathcal{M}_+^b(\Omega,\mathfrak{A}): \mu(\Omega) \leq 1\}$.

$\mathcal{M}^1(\Omega,\mathfrak{A}):= \{\mu \in \mathcal{M}_+^b(\Omega,\mathfrak{A}): \mu(\Omega) = 1\}$

$=$ set of probability measures on $(\Omega,\mathfrak{A})$.

In the special case of the Borel space $(\Omega, \mathfrak{B}(\mathcal{T}))$ of a locally compact space $(\Omega,\mathcal{T})$ with a countable basis, by the Riesz representation theorem the set $\mathcal{M}_+(\Omega,\mathfrak{B}(\mathcal{T}))$ of positive (inner and outer) regular Borel measures can be identified with the set $\mathcal{M}_+(\Omega)$ of all Radon measures on Ω which are defined as positive linear functionals on $\mathcal{K}(\Omega)$. Correspondingly we use the symbols $\mathcal{M}_+^b(\Omega)$, $\mathcal{M}^{(1)}(\Omega)$ and $\mathcal{M}^1(\Omega)$.

I.5. The vector spaces $\mathfrak{M}^b(\Omega,\mathbf{A})$ and $\mathcal{M}^b(\Omega,\mathbf{A})$ are Banach spaces for the norms

$$f \to ||f||: = \sup_{\omega\in\Omega} |f(\omega)|$$

and

$$\mu \to ||\mu||: = \sup_{f\in\mathfrak{M}^{(1)}(\Omega,\mathbf{A})} \left|\int f\, d\mu\right|$$

respectively. The bilinear functional

$$<\mu,f>: = \int f\, d\mu$$

on $\mathcal{M}^b(\Omega,\mathbf{A}) \times \mathfrak{M}^b(\Omega,\mathbf{A})$ is nondegenerate.

I.6. Let $(\Omega_1,\mathbf{A}_1)$ and $(\Omega_2,\mathbf{A}_2)$ be two measurable spaces. By $\mathbf{A}_1 \otimes \mathbf{A}_2$ we denote the σ-algebra in $\Omega_1 \times \Omega_2$ generated by the set $\{A_1 \times A_2: A_1 \in \mathbf{A}_1, A_2 \in \mathbf{A}_2\}$.

For functions $f_1 \in \mathfrak{M}^b(\Omega_1,\mathbf{A}_1)$ and $f_2 \in \mathfrak{M}^b(\Omega_2,\mathbf{A}_2)$ we define the function $f_1 \otimes f_2 \in \mathfrak{M}^b(\Omega_1 \times \Omega_2, \mathbf{A}_1 \otimes \mathbf{A}_2)$ by

$$f_1 \otimes f_2(\omega_1,\omega_2): = f_1(\omega_1)f_2(\omega_2)$$

for all $\omega_1 \in \Omega_1$, $\omega_2 \in \Omega_2$.

For measures $\mu_1 \in \mathcal{M}^b(\Omega_1,\mathbf{A}_1)$ and $\mu_2 \in \mathcal{M}^b(\Omega_2,\mathbf{A}_2)$ there exists exactly one measure $\mu_1 \otimes \mu_2 \in \mathcal{M}^b(\Omega_1 \times \Omega_2, \mathbf{A}_1 \otimes \mathbf{A}_2)$ given by

$$\mu_1 \otimes \mu_2(A_1 \times A_2): = \mu_1(A_1)\mu_2(A_2)$$

for all $A_1 \in \mathbf{A}_1$, $A_2 \in \mathbf{A}_2$.

I.6.1. $<f_1 \otimes f_2, \mu_1 \otimes \mu_2> = <f_1,\mu_1><f_2,\mu_2>$.

I.6.2. $||f_1 \otimes f_2|| = ||f_1|| \, ||f_2||$.

I.6.3. The set $\{f_1 \otimes f_2: f_1 \in \mathfrak{M}^b(\Omega_1,\mathbf{A}_1), f_2 \in \mathfrak{M}^b(\Omega_2,\mathbf{A}_2)\}$ is total in $\mathfrak{M}^b(\Omega_1 \times \Omega_2, \mathbf{A}_1 \otimes \mathbf{A}_2)$.

In this spirit we accept the notation

$$\mathfrak{M}^b(\Omega_1,\mathbf{A}_1) \otimes \mathfrak{M}^b(\Omega_2,\mathbf{A}_2) = \mathfrak{M}^b(\Omega_1 \times \Omega_2,\mathbf{A}_1 \otimes \mathbf{A}_2)$$

and

$$\mathcal{M}^b(\Omega_1,\mathbf{A}_1) \otimes \mathcal{M}^b(\Omega_2,\mathbf{A}_2) = \mathcal{M}^b(\Omega_1 \times \Omega_2, \mathbf{A}_1 \otimes \mathbf{A}_2).$$

I.7. Appropriate mappings between the spaces $\mathfrak{M}^b(\Omega_1,\mathbf{A}_1)$ and $\mathfrak{M}^b(\Omega_2,\mathbf{A}_2)$ are positive linear operators $T: \mathfrak{M}^b(\Omega_2,\mathbf{A}_2) \to \mathfrak{M}^b(\Omega_1,\mathbf{A}_1)$ such that

(S) T is σ-continuous, i.e., T preserves monotone limits.

(N) $T(1_{\Omega_2}) = 1_{\Omega_1}$.

Let T be such an operator. Then putting

$$N_T(\omega_1,A_2) := [T(1_{A_2})](\omega_1)$$

for all $\omega_1 \in \Omega_1$, $A_2 \in \mathbf{A}_2$ we obtain a mapping $N_T: \Omega_1 \times \mathbf{A}_2 \to \mathbb{R}_+$ which is a stochastic (Markov) kernel from $(\Omega_1,\mathbf{A}_1)$ to $(\Omega_2,\mathbf{A}_2)$.

Here, a mapping $N: \Omega_1 \times \mathbf{A}_2 \to \mathbb{R}_+$ is said to be a *substochastic (stochastic) kernel from* $(\Omega_1,\mathbf{A}_1)$ *to* $(\Omega_2,\mathbf{A}_2)$ if

(i) $\omega_1 \to N(\omega_1,A_2)$ is $\mathbf{A}_1$-measurable for all $A_2 \in \mathbf{A}_2$.

(ii) $A_2 \to N(\omega_1,A_2)$ is a measure in $\mathcal{M}^{(1)}(\Omega_2,\mathbf{A}_2)(\mathcal{M}^1(\Omega_2,\mathbf{A}_2))$ for all $\omega_1 \in \Omega_1$.

If $(\Omega_1,\mathbf{A}_1) = (\Omega_2,\mathbf{A}_2)$, N is called a substochastic (stochastic) kernel on $(\Omega_1,\mathbf{A}_1)$. We introduce

$\mathrm{Stoch}((\Omega_1,\mathbf{A}_1),(\Omega_2,\mathbf{A}_2)) :=$ set of all stochastic kernels from $(\Omega_1,\mathbf{A}_1)$ to $(\Omega_2,\mathbf{A}_2)$

$\mathrm{Stoch}(\Omega_1,\mathbf{A}_1) := \mathrm{Stoch}((\Omega_1,\mathbf{A}_1),(\Omega_1,\mathbf{A}_1))$.

Conversely, for every $N \in \mathrm{Stoch}((\Omega_1,\mathbf{A}_1),(\Omega_2,\mathbf{A}_2))$ there exist positive linear operators

$$T_N: \mathfrak{M}^b(\Omega_2,\mathbf{A}_2) \to \mathfrak{M}^b(\Omega_1,\mathbf{A}_1)$$

and

$$T_N': \mathcal{M}^b(\Omega_1,\mathbf{A}_1) \to \mathcal{M}^b(\Omega_2,\mathbf{A}_2)$$

which satisfy properties (S) and (N), given by

$$[T_N(f_2)](\omega_1) := \int f_2(\omega_2)N(\omega_1,d\omega_2) =: (Nf_2)(\omega_1) =: N(\omega_1,f)$$

for all $f_2 \in \mathfrak{M}^b(\Omega_2,\mathbf{A}_2)$, $\omega_1 \in \Omega_1$, and

$$(T_N'\mu_1)(A_2) := \int N(\omega_1,A_2)\mu_1(d\omega_1) =: (\mu_1 N)(A_2) =: N(\mu_1)(A_2)$$

for all $\mu_1 \in \mathcal{M}^b(\Omega_1,\mathbf{A}_1)$, $A_2 \in \mathbf{A}_2$, respectively.

<u>I.7.2.</u> $T_{N_T} = T$.

<u>I.7.3.</u> $N_{T_N} = N$.

<u>I.8.</u> Let $(\Omega_i,\mathbf{A}_i)$ (for $i = 1,2,3$) and $(\Omega_i',\mathbf{A}_i')$ (for $i = 1,2$) be measurable spaces, and let $N_1 \in \mathrm{Stoch}((\Omega_1,\mathbf{A}_1),(\Omega_2,\mathbf{A}_2))$, $N_2 \in \mathrm{Stoch}((\Omega_2,\mathbf{A}_2),(\Omega_3,\mathbf{A}_3))$ as well as $K_i \in \mathrm{Stoch}((\Omega_i,\mathbf{A}_i),(\Omega_i',\mathbf{A}_i'))$ (for $i = 1,2$). We define the kernels $N_1N_2 \in \mathrm{Stoch}((\Omega_1,\mathbf{A}_1),(\Omega_3,\mathbf{A}_3))$ and $K_1 \otimes K_2 \in \mathrm{Stoch}((\Omega_1 \times \Omega_2, \mathbf{A}_1 \otimes \mathbf{A}_2), (\Omega_1' \times \Omega_2', \mathbf{A}_1' \otimes \mathbf{A}_2'))$ by

$$N_1N_2(\omega_1,A_3) := \int N_1(\omega_1,d\omega_2)N_2(\omega_2,A_3)$$

for all $\omega_1 \in \Omega_1$, $A_3 \in \mathbf{A}_3$, and

$$K_1 \otimes K_2((\omega_1,\omega_2),A_1' \times A_2') := K_1(\omega_1,A_1')\cdot K_2(\omega_2,A_2')$$

for all $\omega_1 \in \Omega_1$, $\omega_2 \in \Omega_2$, $A_1' \in \mathbf{A}_1'$, $A_2' \in \mathbf{A}_2'$ respectively.

In the special case $(\Omega_1,\mathbf{A}_1) = (\Omega_2,\mathbf{A}_2) =: (\Omega,\mathbf{A})$ we obtain the *diagonal tensor product* $K_1 \underset{\sim}{\otimes} K_2$ of the kernels K_1 and K_2, defined as an element of $\mathrm{Stoch}((\Omega,\mathbf{A}),(\Omega_1' \times \Omega_2', \mathbf{A}_1' \otimes \mathbf{A}_2'))$ by

$$K_1 \underset{\sim}{\otimes} K_2(\omega,A_1' \times A_2') := K_1(\omega,A_1')\cdot K_2(\omega,A_2')$$

for all $\omega \in \Omega$, $A_1' \in \mathbf{A}_1'$, $A_2' \in \mathbf{A}_2'$.

Now let $(\Omega,\mathbf{A},\mu)$ be a measure space with $\mu \in \mathcal{M}_+(\Omega,\mathbf{A})$ and $p \in [1,\infty]$.

<u>I.9.</u> We consider functions $N_p: \mathfrak{M}(\Omega,\mathbf{A}) \to \overline{\mathbb{R}}_+$ defined for $p \in [1,\infty[$ by

$$N_p(f) := \left(\int |f|^p d\mu\right)^{1/p}$$

and for $p = \infty$ by

$$N_\infty(f) := \inf_{\substack{A\in\mathbf{A} \\ \mu(\complement A)=0}} \sup_{\omega\in A} |f(\omega)|.$$

Restricted to the vector space

$$\mathscr{L}^p(\Omega,\mathbf{A},\mu) := \{f \in \mathfrak{M}(\Omega,\mathbf{A}) : N_p(f) < \infty\}$$

the functions N_p are seminorms for $p \in [1,\infty]$. The corresponding normed vector spaces will be denoted by $L^p(\Omega,\mathbf{A},\mu)$, their norms by $||\cdot||_p$, for $p \in [1,\infty]$.

<u>I.9.1</u>. $L^p(\Omega,\mathbf{A},\mu)$ is a Banach space for every $p \in [1,\infty]$.

<u>I.9.2</u>. If μ is σ-finite and $p,q \in \,]1,\infty[$ such that $\frac{1}{p} + \frac{1}{q} = 1$, then for all $f \in L^p(\Omega,\mathbf{A},\mu)$, $g \in L^q(\Omega,\mathbf{A},\mu)$ we have Hölder's inequality

$$\int |fg|\,d\mu \leq N_p(f)N_q(g),$$

and the spaces $L^p := L^p(\Omega,\mathbf{A},\mu)$ and $L^q := L^q(\Omega,\mathbf{A},\mu)$ are in duality with respect to the bilinear functional

$$\langle f,g\rangle \to \int fg\,d\mu$$

on $L^p \times L^q$.

<u>I.9.3</u>. The unit ball $\{f \in L^\infty : ||f||_\infty \leq 1\}$ of L^∞ is compact with respect to the topology $\sigma(L^\infty,L^1)$, i.e., the coarsest topology on L^∞ that makes the linear functionals $\langle\cdot,f\rangle$ $(f \in L^1)$ on L^∞ continuous.

In the following we consider measure spaces $(\Omega,\mathbf{A},\mu)$ with $\mu \in \mathscr{M}^b(\Omega,\mathbf{A})$.

<u>I.10</u>. For every $f \in \mathfrak{M}^b(\Omega,\mathbf{A})$ and every $\mu \in \mathscr{M}^b(\Omega,\mathbf{A})$ we define the measure $f\cdot\mu \in \mathscr{M}^b(\Omega,\mathbf{A})$ by

$$f\cdot\mu(A) := \int 1_A f\,d\mu$$

for all $A \in \mathbf{A}$. Under the operation $(f,\mu) \to f\cdot\mu$, $\mathscr{M}^b(\Omega,\mathbf{A})$ becomes an $\mathfrak{M}^b(\Omega,\mathbf{A})$-module.

Since $L^\infty(\Omega,\mathbf{A},\mu)\cdot L^p(\Omega,\mathbf{A},\mu) \subset L^p(\Omega,\mathbf{A},\mu)$, the spaces $L^p(\Omega,\mathbf{A},\mu)$ for $p \in [1,\infty[$ are $L^\infty(\Omega,\mathbf{A},\mu)$-modules.

But $f\cdot\mu$ can be defined also for $f \in L^1(\Omega,\mathbf{A},\mu)$, thus the set $\mathscr{M}_a := \{\nu \in \mathscr{M}^b(\Omega,\mathbf{A}) : \nu << \mu\}$ of all μ-absolutely continuous measures on $(\Omega,\mathbf{A})$ becomes an L^∞-module.

In fact by the Radon-Nikodym theorem the mapping $f \to f\cdot\mu$ provides a norm preserving isomorphism of modules between L^1 and $\mathscr{M}_a$. The inverse image under this isomorphism of a measure $\nu \in \mathscr{M}_a$ (i.e., the Radon-Nikodym derivative of ν with respect to μ) will be denoted by $d\nu/d\mu$.

II. PROBABILITY

For $p \geq 1$ we collect a few types of probability distributions in $\mathcal{M}^1(\mathbb{R}^p, \mathfrak{B}^p)$ in order to fix their notation. Let λ^p denote the Lebesgue measure of $\mathbb{R}^p$.

II.1.1. *Binomial* or Bernoulli *distribution*

$$B(n,q) := \sum_{k=0}^{n} \binom{n}{k} q^k (1-q)^{n-k} \varepsilon_k$$

for $n \geq 1$, $q \in]0,1[$.

II.1.2. *Poisson distribution*

$$\pi(\alpha) := \sum_{k \geq 0} e^{-\alpha} \frac{\alpha^k}{k!} \varepsilon_k$$

for $\alpha \in \mathbb{R}_+^*$.

II.1.3. 1-dimensional normal or *Gauss distribution with mean* $a \in \mathbb{R}$ *and variance* $\sigma^2 \in \mathbb{R}_+^*$

$$\nu_{a,\sigma^2} := n_{a,\sigma^2} \cdot \lambda$$

with

$$n_{a,\sigma^2}(x) := \frac{1}{\sqrt{2\pi}\,\sigma} e^{-\frac{(x-a)^2}{2\sigma^2}}$$

for all $x \in \mathbb{R}$.

II.1.4. *p-dimensional normal distribution with mean vector* $a \in \mathbb{R}^p$ *and covariance matrix* Σ^{-1}, where Σ is a symmetric, positive definite matrix in $\mathbb{M}(p \times p, \mathbb{R})$

$$\nu_{a,\Sigma} := n_{a,\Sigma} \cdot \lambda^p,$$

where

$$n_{a,\Sigma}(x) := \frac{1}{\sqrt{(2\pi)^p \det \Sigma}} \exp[-\tfrac{1}{2}(x-a)^T \Sigma^{-1} (x-a)]$$

for all $x \in \mathbb{R}^p$.

II.1.5. Γ*-distribution with parameters* $\alpha, s \in \mathbb{R}_+$

$$\Gamma_{\alpha,s} := g_{\alpha,s} \cdot \lambda,$$

where

$$g_{\alpha,s}(x) := \frac{\alpha^s}{\Gamma(s)} x^{s-1} e^{-\alpha x} 1_{]0,\infty[}(x)$$

for all $x \in \mathbb{R}$. Special cases are

II.1.5'. *Exponential distribution*

$$\exp(\alpha) := \Gamma_{\alpha,1}$$

II.1.5". χ^2*-distribution with* n *degrees of freedom*

$$\chi_n^2 := \Gamma_{\frac{1}{2},\frac{n}{2}}$$

II.1.6. *Student's* t*-distribution with* n *degrees of freedom*

$$\tau_n := t_n \cdot \lambda$$

with

$$t_n(x) := \frac{1}{\sqrt{\pi n}} \frac{\Gamma(\frac{n+1}{2})}{\Gamma(\frac{n}{2})} (1 + \frac{x^2}{n})^{-\frac{n+1}{2}}$$

for all $x \in \mathbb{R}$.

II.2. For every measure $\mu \in \mathcal{M}^1(\mathbb{R}^p, \mathfrak{B}^p)$ its *distribution function* F_μ is defined as a mapping $\mathbb{R}^p \to \mathbb{R}$ by

$$F_\mu(x) := \mu(]-\infty, x])$$

for all $x := (x_1, \ldots, x_p) \in \mathbb{R}^p$, where

$$]-\infty, x] := \{(y_1, \ldots, y_p) \in \mathbb{R}^p : y_i \le x_i \text{ for all } i = 1, \ldots, p\}.$$

Moreover, for $\mu \in \mathcal{M}^1(\mathbb{R}^p)$ and $\mu \in \mathcal{M}^1(\mathbb{R}_+^p)$ we have the *Fourier transform* and the *Laplace transform* of μ given as mappings $\hat{\mu}: \mathbb{R}^p \to \mathbb{C}$ and $L_\mu: \mathbb{R}_+^p \to \mathbb{R}$ by

$$\hat{\mu}(x) := \int_{\mathbb{R}^p} e^{i\langle x,y\rangle} \mu(dy)$$

and

$$L_\mu(x) := \int_{\mathbb{R}_+^p} e^{-\langle x,y\rangle} \mu(dy)$$

for all $x \in \mathbb{R}^p$ and $x \in \mathbb{R}_+^p$ respectively.

μ is uniquely determined by each of the functions F_μ, $\hat{\mu}$ and L_μ.

II.3. Given a measure $\mu \in \mathcal{M}^b(\Omega, \mathfrak{A})$ and a sub-σ-algebra $\mathfrak{S}$ of $\mathfrak{A}$, we denote by $\mu_{\mathfrak{S}}$ the restriction $\mathrm{Res}_{\mathfrak{S}}\mu$ of μ to $\mathfrak{S}$.

For $P \in \mathcal{M}^1(\Omega,\mathfrak{A})$ we have the *conditional expectation of* P *with respect to a sub-σ-algebra* $\mathfrak{S}$ of $\mathfrak{A}$ as a mapping

$$E_P^{\mathfrak{S}}: L^1(\Omega,\mathfrak{A},P) \to L^1(\Omega,\mathfrak{S},P_{\mathfrak{S}})$$

given by

$$E_P^{\mathfrak{S}}(f) := \frac{d(f\cdot P)_{\mathfrak{S}}}{dP_{\mathfrak{S}}}$$

or equivalently

$$\int_S E_P^{\mathfrak{S}}(f)\,dP = \int_S f\,dP$$

for all $S \in \mathfrak{S}$.

The mapping $P^{\mathfrak{S}}: \mathfrak{A} \to L^1(\Omega,\mathfrak{S},P_{\mathfrak{S}})$ defined by

$$P^{\mathfrak{S}}(A) := E_P^{\mathfrak{S}}(1_A)$$

for all $A \in \mathfrak{A}$ is called the *conditional probability of* P *with respect to* $\mathfrak{S}$.

For a given function $f \in L^1(\Omega,\mathfrak{A},P)$ or a set $A \in \mathfrak{A}$, $E_P^{\mathfrak{S}}(f)$ and $P^{\mathfrak{S}}(A)$ are just the conditional expectation of f and the conditional probability of A (under $\mathfrak{S}$ and with reference to P) respectively.

It should be noted that $E_P^{\mathfrak{S}}$ can also be defined on $\mathfrak{M}_+(\Omega,\mathfrak{A})$ without any integrability conditions.

II.4. Properties of $E_P^{\mathfrak{S}}$ as an operation on $L^1(\Omega,\mathfrak{A},P)$.

II.4.1. $E_P^{\mathfrak{S}}$ is a positive contraction operator.

II.4.2. For any two sub-σ-algebras $\mathfrak{S}_1,\mathfrak{S}_2$ of $\mathfrak{A}$ with $\mathfrak{S}_1 \subset \mathfrak{S}_2$ one has

$$E_P^{\mathfrak{S}_1} = E_P^{\mathfrak{S}_1}E_P^{\mathfrak{S}_2} = E_P^{\mathfrak{S}_2}E_P^{\mathfrak{S}_1}.$$

II.4.3. Let $(f_n)_{n\geq 1}$ be a sequence dominated in modulus by a function in $L^1(\Omega,\mathfrak{A},P)$ and P-a.s. convergent to f. Then

$$\lim_{n\to\infty} E_P^{\mathfrak{S}}(f_n) = E_P^{\mathfrak{S}}(f).$$

II.4.4. For $f \in L^1(\Omega,\mathfrak{A},P)$ and $g \in L^\infty(\Omega,\mathfrak{S},P_{\mathfrak{S}})$ one has

$$E_P^{\mathfrak{S}}(g\cdot f) = gE_P^{\mathfrak{S}}(f).$$

II.4.5. $E_P^{\mathfrak{S}}(L^\infty) \subset L^\infty$ and $E_P^{\mathfrak{S}}(L^2) \subset L^2$.

The latter property yields this.

II.4.6. $\mathrm{Res}_{L^2} E_P^{\mathfrak{S}}$ is an orthogonal projection of L^2. Conversely, every positive orthogonal projection of L^2 leaving 1_Ω invariant is the restriction to L^2 of a conditional expectation (with respect to some sub-σ-algebra of $\mathfrak{A}$).

II.4.7. (Jensen's inequality). Let $f_1,\ldots,f_n \in L^1(\Omega,\mathfrak{A},P)$, and let $C \subset \mathbb{R}^n$ be a convex set in $\mathfrak{B}^n$ such that $(f_1,\ldots,f_n)(\Omega) \subset C$, and let $q: C \to \mathbb{R}$ be measurable and convex such that $q\circ(f_1,\ldots,f_n) \in L^1(\Omega,\mathfrak{A},P)$. Then for every sub-σ-algebra $\mathfrak{S}$ of $\mathfrak{A}$ we have

(i) $q\circ(E_P^{\mathfrak{S}}(f_1),\ldots,E_P^{\mathfrak{S}}(f_n)) \leq E_P^{\mathfrak{S}}\{q\circ(f_1,\ldots,f_n)\}[P]$.

(ii) If q is strictly convex, then for every $i = 1,\ldots,n$, $f_i = E_P^{\mathfrak{S}}(f_i)[P]$ on the set

$$[q\circ(E_P^{\mathfrak{S}}(f_1),\ldots,E_P^{\mathfrak{S}}(f_n)) = E_P^{\mathfrak{S}}\{q\circ(f_1,\ldots,f_n)\}].$$

We note that statement (ii) assures that for strictly convex q, the inequality of (i) is strict with the possible exception of a set on which all of the function $f_1,\ldots,f_n$ are $\mathfrak{S}$-measurable.

For the proof of this form of Jensen's inequality the reader is referred to Pfanzagl [99], or Mussmann [89].

II.5. Let $(\Omega,\mathfrak{A})$ and $(\Omega',\mathfrak{A}')$ be measurable spaces and $T: (\Omega,\mathfrak{A}) \to (\Omega',\mathfrak{A}')$ a measurable mapping. For given $P \in \mathcal{M}^1(\Omega,\mathfrak{A})$ and $f \in L^1(\Omega,\mathfrak{A},P)$ we put

$$E_P^T(f) := \frac{dT(f\cdot P)}{dT(P)}.$$

By definition $E_P^T(f)$ consists of equivalence classes modulo $T(P)$. For any function $g' \in \mathfrak{M}(\Omega',\mathfrak{A}')$ we get $g' = E_P^T(f)[T(P)]$ iff

$$\int_{A'} g'\,dT(P) = \int_{T^{-1}(A')} f\,dP$$

for all $A' \in \mathfrak{A}'$.

Analoguously one defines $E_P^T(f)$ for $f \in \mathfrak{M}_+(\Omega,\mathfrak{A})$ and draws the same conclusion.

II.6. Let $(\Omega,\mathfrak{A},P)$ be a probability space and $\mathfrak{S}$ a sub-σ-algebra of $\mathfrak{A}$. For every $A \in \mathfrak{A}$ the conditional probability $P^{\mathfrak{S}}(A)$ of A

under the hypothesis $\mathfrak{S}$ is an $\mathfrak{S}$-measurable, P-a.s. nonnegative function on Ω for which

$$\int_S P^{\mathfrak{S}}(A)\,dP = P(A \cap S)$$

holds for all $S \in \mathfrak{S}$. By this equality $P^{\mathfrak{S}}(A)$ is P-a.s. uniquely determined.

We have the following properties

II.6.1. $0 \leq P^{\mathfrak{S}}(A) \leq 1[P]$.

II.6.2. $P^{\mathfrak{S}}(\emptyset) = o[P]$ and $P^{\mathfrak{S}}(\Omega) = 1[P]$.

II.6.3. For $A, B \in \mathfrak{A}$ with $A \subset B$,

$$P^{\mathfrak{S}}(A) \leq P^{\mathfrak{S}}(B)[P].$$

II.6.4. For every sequence $(A_n)_{n\geq 1}$ of pairwise disjoint sets in $\mathfrak{A}$,

$$P^{\mathfrak{S}}\left(\bigcup_{n\geq 1} A_n\right) = \sum_{n\geq 1} P^{\mathfrak{S}}(A_n)[P].$$

It is known that the above properties do not imply that the mapping $A \to P^{\mathfrak{S}}(A)(\omega)$ is a probability measure on $(\Omega, \mathfrak{A})$ for P-a.s. $\omega \in \Omega$.

A mapping $P_{\mathfrak{S}}: \Omega \times \mathfrak{A} \to \mathbb{R}$ is said to be an *expectation kernel corresponding to* $\mathfrak{S}$ if

(EK1) For every $A \in \mathfrak{A}$ the mapping $\omega \to P_{\mathfrak{S}}(\omega, A)$ is a version of $P^{\mathfrak{S}}(A)$.

(EK2) For every $\omega \in \Omega$ the mapping $A \to P_{\mathfrak{S}}(\omega, A)$ is a probability measure on $(\Omega, \mathfrak{A})$.

Obviously, any expectation kernel $P_{\mathfrak{S}}$ corresponding to a sub-σ-algebra $\mathfrak{S}$ of $\mathfrak{A}$ is an element of $\mathrm{Stoch}((\Omega,\mathfrak{S}),(\Omega,\mathfrak{A}))$.

II.7. Let $(\Omega, \mathfrak{A}, P)$ be a probability space and $(\mathfrak{A}_n)_{n\in\mathbb{Z}}$ an ascending system of sub-σ-algebras of $\mathfrak{A}$ in the sense that for any $n, m \in \mathbb{Z}$ with $n \leq m$ one has $\mathfrak{A}_n \subset \mathfrak{A}_m$.

We introduce two more σ-algebras by setting

$$\mathfrak{A}_{-\infty} := \bigcap_{n\in\mathbb{Z}} \mathfrak{A}_n$$

and

$$\mathfrak{A}_{\infty} := \bigvee_{n\in\mathbb{Z}} \mathfrak{A}_n.$$

II.7.1. A sequence $(f_n)_{n\in\mathbb{Z}}$ of functions in $\mathscr{L}^1(\Omega, \mathfrak{A}, P)$ is called a *martingale* if for all $n, m \in \mathbb{Z}$ with $n \leq m$,

$$E_P^{\mathfrak{A}_n}(f_m) = f_n.$$

II.7.2. If $(f_n)_{n\in\mathbb{Z}}$ is a nonnegative martingale, then there exist functions $f_{-\infty}, f_{+\infty} \in L^1(\Omega,\mathfrak{A},P)$ satisfying

$$f_{-\infty} = \lim_{n\to-\infty} f_n$$

and

$$f_{+\infty} = \lim_{n\to\infty} f_n$$

P-a.s. and in L^1.

Moreover, for all $n \in \mathbb{Z}$ one has

$$E_P^{\mathfrak{A}_n}(f_{+\infty}) = f_n$$

and

$$E_P^{\mathfrak{A}_{-\infty}}(f_n) = f_{-\infty}.$$

II.7.3. For every $f \in L^1_+(\Omega,\mathfrak{A},P)$ the sequence $(f_n)_{n\in\mathbb{N}}$ with

$$f_n := E_P^{\mathfrak{A}_n}(f)$$

for all $n \in \mathbb{N}$ is a nonnegative martingale satisfying

$$\lim_{n\to\infty} f_n =: f_\infty = E_P^{\mathfrak{A}_\infty}(f)$$

P-a.s. and in L^1.

Moreover, the martingales of the form $(E_P^{\mathfrak{A}_n}(f))_{n\in\mathbb{N}}$ with $f \in L^1_+(\Omega,\mathfrak{A},P)$ are exactly those nonnegative martingales in L^1 which converge in L^1.

For the proofs of the results concerning nonnegative martingales one consults H. Bauer [9], 11.4 and Neveu [91], II-2.

Appendix

I. STANDARD BOREL SPACES

Definition I.1. Two measurable spaces $(\Omega_1,\mathbf{A}_1)$ and $(\Omega_2,\mathbf{A}_2)$ are called *isomorphic* if there exists a bimeasurable bijection (measurable isomorphism) from $(\Omega_1,\mathbf{A}_1)$ onto $(\Omega_2,\mathbf{A}_2)$.

Definition I.2. A measurable space $(\Omega,\mathbf{A})$ is called a *Borel space (standard Borel space)* if there exists a topology (polish topology) $\mathcal{T}$ on Ω such that $\mathbf{A} = \mathbf{B}(\mathcal{T})$.

Theorem I.3. For any measurable space $(\Omega,\mathbf{A})$ with an uncountable set Ω the following statements are equivalent:

(i) $(\Omega,\mathbf{A})$ is a standard Borel space.

(ii) There exists a compact metrizable topology $\mathcal{T}$ on Ω such that $\mathbf{A} = \mathbf{B}(\mathcal{T})$.

(iii) The measurable spaces $(\Omega,\mathbf{A})$ and $([0,1], \mathbf{B}([0,1]))$ are isomorphic.

Remark I.4. If Ω is countable, only the equivalence (i) $\Leftrightarrow$ (ii) remains valid.

Proof: While the equivalence (i) $\Leftrightarrow$ (ii) is well-known from general topology, the equivalence (i) $\Leftrightarrow$ (iii) is proved in Kuratowski [61], p. 227.

See also Christensen [29], pp. 38,43; Parthasarathy [93], Chapter I, Sections 2,3 and Chapter V, Section 2; Parthasarathy [94], Remark 24.27 and Proposition 26.5.

Definition I.5. Two measurable spaces $(\Omega_1,\mathbf{A}_1)$ and $(\Omega_2,\mathbf{A}_2)$ are called *σ-isomorphic* if there exists a bijective σ-homomorphism (σ-isomorphism) from $\mathbf{A}_1$ onto $\mathbf{A}_2$.

Definition I.6. A measurable space $(\Omega,\mathbf{A})$ is called a *weakly standard Borel space* if there exists a standard Borel space $(\Omega',\mathbf{A}')$ such that $(\Omega,\mathbf{A})$ and $(\Omega',\mathbf{A}')$ are σ-isomorphic.

Theorem I.7. (O. J. Björnsson). For any measurable space $(\Omega,\mathbf{A})$ the following statements are equivalent:

(i) $(\Omega,\mathbf{A})$ is a weakly standard Borel space.

(ii) There exists a countable algebra $\mathfrak{C}$ in Ω with $\mathbf{A} = \mathbf{A}(\mathfrak{C})$ such that every finite (nonnegative) content on $\mathfrak{C}$ is a premeasure on $\mathfrak{C}$.

Proof: Björnsson [16].

Standard Borel spaces $(\Omega,\mathbf{A})$ possess the important property that for every measure $\nu \in \mathcal{M}^1(\Omega,\mathbf{A})$ and every sub-σ-algebra $\mathbf{S}$ of $\mathbf{A}$ there exists an expectation kernel $N \in \mathrm{Stoch}((\Omega,\mathbf{S}),(\Omega,\mathbf{A}))$ satisfying

$$\int_S N(\omega,A)\nu(d\omega) = \nu(A \cap S)$$

for all $A \in \mathbf{A}$, $S \in \mathbf{S}$.

More generally, we have the following

Theorem I.8. Let $(\Omega_1,\mathbf{A}_1)$ and $(\Omega_2,\mathbf{A}_2)$ be two standard Borel spaces, ϕ a measurable mapping from Ω_1 onto Ω_2, and $\nu \in \mathcal{M}_+^\sigma(\Omega_1,\mathbf{A}_1)$. Then there exists a *conditional measure* $\nu_{\phi|\mathbf{A}_2}$ *of* ϕ *under* $\mathbf{A}_2$ defined as a kernel N from $(\Omega_2,\mathbf{A}_2)$ to $(\Omega_1,\mathbf{A}_1)$ satisfying the conditions

(a) There exists a set $C \in \mathbf{A}_2$ such that $\phi(\nu)(C) = 0$ and $N(\omega_2,\Omega\setminus\Omega_{\omega_2}) = 0$ for all $\omega_2 \in \Omega_2 \setminus C$, where $\Omega_{\omega_2} := \{\omega_1 \in \Omega_1 : \phi(\omega_1) = \omega_2\}$.

(b) $\displaystyle\int_{\Omega_2} N(\omega_2,A)\phi(\nu)(d\omega_2) = \nu(A)$ for all $A \in \mathbf{A}_1$.

Remark I.9. We note that for the conditional measure N of the theorem and sets $A \in \mathbf{A}_1$, $B \in \mathbf{A}_2$ we have

$$\int_{\phi^{-1}(B)} N(\phi(\omega_1),A)\nu(d\omega_1) = \nu(A \cap \phi^{-1}(B)).$$

Proof: For weakly standard Borel spaces $(\Omega_1,\mathbf{A}_1)$ and $(\Omega_2,\mathbf{A}_2)$ and probability measures $\nu \in \mathscr{M}^1(\Omega_1,\mathbf{A}_1)$ the theorem is proved in Parthasarathy [93], as Theorem 8.1 of Chapter V. The extension to arbitrary measures $\nu \in \mathscr{M}_+^\sigma(\Omega_1,\mathbf{A}_1)$ can be performed with the help of Theorem I.7.

Let $(\Omega,\mathbf{A},\mu)$ be a σ-finite measure space and K a linear subspace of $L^\infty(\Omega,\mathbf{A},\mu)$ with $\mathbb{1}: = [1_\Omega]_\mu \in K$.

Definition I.10. A linear mapping $L: K \to \mathbb{M}^b(\Omega,\mathbf{A})$ is said to be a (linear) *lifting* on K if

(L1) $[L(f)]_\mu = f$ for all $f \in K$.

(L2) $L(f) \geq 0$ for all $f \in K$ with $f \geq 0$.

(L3) $L(\mathbb{1}) = 1_\Omega$.

Here $[g]_\mu$ denotes the μ-equivalence class of $g \in \mathbb{M}^b(\Omega,\mathbf{A})$.

Theorem I.11. Let K be a separable linear subspace of $L^\infty(\Omega,\mathbf{A},\mu)$ with $\mathbb{1} \in K$. Then there exists a lifting on K.

Theorem I.12. If $(\Omega,\mathbf{A})$ is a standard Borel space, then there exists a lifting on the entire space $L^\infty(\Omega,\mathbf{A},\mu)$ and consequently on any of its linear subspaces K with $\mathbb{1} \in K$.

Proofs of both theorems within the framework of concrete measure theory can be found in Edwards [39], pp. 579-581, where Ω is a locally compact space and μ a positive Radon measure on Ω, with Ω admitting a countable basis of its topology in the case of Theorem I.12. A proof of Theorem I.12 for complete probability spaces is contained in Meyer [82], p. 154 or Schwartz [117], p. 130ff. See also Schwartz [118]. As a general reference for lifting problems the reader might consult the standard monograph by A. and C. Ionescu-Tulcea [57].

II. INVARIANT MEANS

Let G be a semigroup. For any $f \in \mathscr{B}(G)$ and $x \in G$ we denote by ${}_xf$ the *left translate* of f by x, defined by ${}_xf(y): = f(xy)$ for all $y \in G$.

Definition II.1. A (left) *invariant mean* on G is a linear functional m on $\mathscr{B}(G)$ with the following properties:

(IM1) $m(f) \geq 0$ for all $f \in \mathscr{B}_+(G)$.

(IM2) $m(1) = 1$.

(IM3) $m({}_xf) = m(f)$ for all $f \in \mathscr{B}(G)$, $x \in G$.

Definition II.2. A semigroup G is called *amenable* if there exists an invariant mean on G.

Properties II.3 of the class $\mathscr{A}$ of all amenable *groups*.

II.3.1. If $G \in \mathscr{A}$ and π is a homomorphism onto a group H, then $H \in \mathscr{A}$.

II.3.2. If $G \in \mathscr{A}$ and H is a subgroup of G, then $H \in \mathscr{A}$.

II.3.3. If N is a normal subgroup of G and if N and G/N are members of $\mathscr{A}$, then $G \in \mathscr{A}$.

Theorem II.4. For any discrete group G the following statements are equivalent:

(i) $G \in \mathscr{A}$.

(ii) (Følner's condition). Given $\varepsilon > 0$ and a finite subset K of G there exists a nonempty finite subset U of G such that

$$\frac{1}{\text{card}(U)} \text{card}(gU \,\Delta\, U) < \varepsilon$$

for all $g \in K$.

(iii) (Asymptotic left invariance). There exists a net $(\mu_s)_{s \in S}$ in $\mathscr{M}^1(G, \mathfrak{B}(G))$ such that

$$\lim_{s \in S} (\mu_s(B) - \mu_s(Bg^{-1})) = 0$$

for all $B \in \mathfrak{B}(G)$, $g \in G$.

Proof: Greenleaf [45], Section 3.6 for (i) $\Leftrightarrow$ (ii), Sections 3.6 and 2.4 for (ii) $\Leftrightarrow$ (iii).

Theorem II.5. (A. A. Markov, S. Kakutani, M. M. Day). Let K be a nonempty compact convex subset of a locally convex vector space E, and let G be a semigroup of continuous, affine linear mappings from K into K. Suppose that $G \in \mathscr{A}$. Then there exists an $x_0 \in K$ such that $g(x_0) = x_0$ for all $g \in G$.

Proof: Day [32].

III. SUBLINEAR FUNCTIONALS AND CONVEXITY

Let E be a vector space over $\mathbb{R}$.

Definition III.1. A real-valued function ψ on E is called a *sublinear functional* if the following axioms are satisfied:

(SF1) (Subadditivity) $\psi(x+y) \leq \psi(x) + \psi(y)$ for all $x,y \in E$.

(SF2) (Positive homogeneity) $\psi(tx) = t\psi(x)$ for all $x \in E$, $t \in \mathbb{R}_+$.

Theorem III.2. (S. Banach, H. Hahn). Let F a linear subspace of E and ψ a sublinear functional on E. We consider a linear functional $L \in F^*$ such that $L \leq \mathrm{Res}_F \psi$. Then there exists a linear functional $\overline{L} \in E^*$ satisfying the following properties:

(i) $\mathrm{Res}_F \overline{L} = L$.

(ii) $\overline{L} \leq \psi$.

Proof: Bourbaki [23], Chapitre 2, p. 65.

Now let E be a locally convex Hausdorff space with dual E'. By K we denote a nonempty compact, convex subset of E. Let $C: = \mathbb{R}_+ K = \{cx: c \in \mathbb{R}_+, x \in K\}$. C is a so-called pointed convex cone having K as its basis.

For any measure $\mu \in \mathcal{M}_+(K)$ there exists exactly one point $x_\mu \in E$ such that

$$\ell(x_\mu) = \int_K \ell(x)\mu(dx)$$

for all $\ell \in E'$. We have $x_\mu \in C$, and if $||\mu|| = 1$, then $x_\mu \in K$.

Definition III.3. The point $x_\mu \in C$ is called the *resultant* of μ and will be denoted by $r(\mu)$. If $\mu \in \mathcal{M}^1(K)$, then $b(\mu): = r(\mu)$ is said to be the *barycentre* of μ.

Remark III.4. For arbitrary $\mu \in \mathcal{M}_+(K)$ we have

$$r(\mu) = \begin{cases} ||\mu|| b(\frac{\mu}{||\mu||}) & \text{if } \mu \neq 0 \\ 0 & \text{if } \mu = 0. \end{cases}$$

Let $S(K)$ denote the convex cone consisting of all continuous concave functions on K. Then $A(K): = S(K) \cap (-S(K))$ is the vector space of all continuous affine-linear functions on K. The vector space $S(K) - S(K)$ generated by $S(K)$ is closed under the operations $\vee$ and $\wedge$, contains the constants and separates the points of K. It follows by the Stone-Weierstrass theorem that $S(K) - S(K)$ is dense in the Banach space $\mathcal{C}(K)$.

For every $\mu \in \mathcal{M}^1(K)$ we obtain that

$$k(b(\mu)) \leq \mu(k)$$

for all lower semicontinuous, convex real-valued functions k on K, in particular for all $k \in -S(K)$.

Theorem III.5. (G. Choquet, P. A. Meyer). Let K be metrizable, and let K_e denote the set of extreme points of K (which in this case is a Borel subset of K). Then for every $x \in K$ there exists a measure $\mu_x \in \mathcal{M}^1(K)$ with $\mu_x(CK_e) = 0$ such that

$$b(\mu_x) = x$$

or equivalently,

$$\int_{K_e} \ell d\mu_x = \ell(x)$$

for all $\ell \in A(K)$.

If, moreover, K is a simplex, then the *representing measure* μ_x is unique.

In this statement the *simplex* is defined as a convex compact subset K of E with the property that any pointed convex cone corresponding to a natural embedding of K is a lattice.

Proof: Meyer [82], Chapter XI, Section 2. See also Phelps [100].

For the rest of this section we restrict ourselves to the special case $E := \mathbb{R}^p$ for $p \geq 1$.

Let $\Psi(\mathbb{R}^p)$ denote the totality of sublinear functionals on $\mathbb{R}^p$.

Properties III.6.

III.6.1. If $\psi_1, \psi_2 \in \Psi(\mathbb{R}^p)$ and $c \in \mathbb{R}_+$, then $\psi_1 \vee \psi_2$, $\psi_1 + \psi_2$ and $c\psi_1$ belong to $\Psi(\mathbb{R}^p)$.

In particular,

III.6.2. For $\ell_1, \ldots, \ell_r \in (\mathbb{R}^p)^*$ the function $\psi := \bigvee_{i=1}^{r} \ell_i \in \Psi(\mathbb{R}^p)$.

For every $r \geq 1$ let

$$\Psi_r(\mathbb{R}^p) := \{\psi \in \Psi(\mathbb{R}^p) : \psi = \bigvee_{i=1}^{r} \ell_i,\ \ell_i \in (\mathbb{R}^p)^* \text{ for } i = 1,\ldots,r\}.$$

Then

III.6.3. $\Psi_1(\mathbb{R}^p) \subset \Psi_2(\mathbb{R}^p) \subset \ldots \subset \Psi(\mathbb{R}^p)$.

III.6.4. Every $\psi \in \Psi(\mathbb{R}^p)$ is convex and uniformly continuous on $\mathbb{R}^p$.

III.6.5. A real-valued function ψ on $\mathbb{R}^p$ is a sublinear functional iff ψ is positive homogeneous and convex.

Theorem III.7. Let $\psi \in \Psi(\mathbb{R}^p)$ and $y \in \mathbb{R}^p$. Then there exists a $c \in \mathbb{R}^p$ satisfying

$$\psi(x) \geq \langle c,x\rangle$$

for all $x \in \mathbb{R}^p$, and

$$\psi(x) = \langle c,x\rangle$$

for $x = y$.

Theorem III.8. For every $\psi \in \Psi(\mathbb{R}^p)$ there exists a sequence $(\psi_r)_{r\geq 1}$ with $\psi_r \in \Psi_r(\mathbb{R}^p)$ $(r \geq 1)$ such that

$$\psi = \lim_{r\to\infty} \psi_r.$$

Given a compact convex subset K of $\mathbb{R}^p$, the function ψ_K on $\mathbb{R}^p$ defined by

$$\psi_K(x) := \sup_{y\in K} \langle x,y\rangle$$

for all $x \in \mathbb{R}^p$ is an element of $\Psi(\mathbb{R}^p)$. It is called the *support functional* of K.

Theorem III.9. Let $a_1,\ldots,a_r \in \mathbb{R}^p$ and $\langle a_1,\ldots,a_r\rangle := \mathrm{conv}(\{a_1,\ldots,a_r\})$. Then

$$\psi_{\langle a_1,\ldots,a_r\rangle}(x) = \bigvee_{i=1}^{r} \langle x,a_i\rangle$$

for all $x \in \mathbb{R}^p$ and $\psi_{\langle a_1,\ldots,a_r\rangle} \in \Psi_r(\mathbb{R}^p)$. Moreover, we have the equivalence of the following three statements:

(i) $\psi \in \Psi_r(\mathbb{R}^p)$.

(ii) $\psi = \psi_K$ for the convex hull K of at most r points of $\mathbb{R}^p$.

(iii) $\psi = \psi_K$ for some compact, convex subset K of $\mathbb{R}^p$ having at most r extreme points.

Theorem III.10. Every $\psi \in \Psi(\mathbb{R}^p)$ is the support functional ψ_K of some compact, convex subset K of $\mathbb{R}^p$.

Properties III.11. Let K_1,K_2 be two compact, convex subsets of $\mathbb{R}^p$, $c \in \mathbb{R}_+$.

III.11.1. $K_1 \subset K_2 \Leftrightarrow \psi_{K_1} \leq \psi_{K_2}$, thus

III.11.2. $K_1 = K_2 \Leftrightarrow \psi_{K_1} = \psi_{K_2}$.

That is to say: There exists a one-to-one correspondence between compact, convex subsets of $\mathbb{R}^p$ and sublinear functionals on $\mathbb{R}^p$.

III.11.3. $\psi_{K_1+K_2} = \psi_{K_1} + \psi_{K_2}$.

III.11.4. $\psi_{cK_1} = c\psi_{K_1}$.

For the proofs of the properties and theorems concerning the set $\Psi(\mathbb{R}^p)$ the reader is referred to Valentine [136], but also to Blackwell-Girshick [19], Chapter 2, Section 2.

IV. WEAK COMPACTNESS LEMMA AND ERGODIC THEOREM

Let $(\Omega,\mathbf{A},\mu)$ be a (positive) measure space, and let M be a subset of $\mathfrak{M}(\Omega,\mathbf{A})$.

Definition IV.1. M is said to be *equiintegrable* if for every $\varepsilon > 0$ there exists a function $g \in L^1(\Omega,\mathbf{A},\mu)$ with $g \geq 0$ such that

$$\int_{[|f|\geq g]} |f| \, d\mu < \varepsilon$$

holds for every $f \in M$.

Theorem IV.2. (Weak Compactness Lemma). Let $\mu \in \mathscr{M}_+^b(\Omega,\mathbf{A})$. Then for every subset M of $L^1(\Omega,\mathbf{A},\mu)$ the following statements are equivalent:

(i) M is equiintegrable.

(ii) M is $||\cdot||_1$-bounded, and for every $\varepsilon > 0$ there exists a function $h \in L^1(\Omega,\mathbf{A},\mu)$ with $h \geq 0$ as well as a $\delta > 0$ such that for any $A \in \mathbf{A}$

$$\int_A h d\mu < \delta \quad \text{implies} \quad \int_A |f| d\mu < \varepsilon$$

for all $f \in M$.

(iii) M is relatively compact in $L^1(\Omega,\mathbf{A},\mu)$ with respect to the topology $\sigma(L^1,L^\infty)$.

(iv) M is sequentially relatively compact in $L^1(\Omega,\mathbf{A},\mu)$ with respect to $\sigma(L^1,L^\infty)$.

Proof: 1. (i) $\Leftrightarrow$ (ii) (valid for an arbitrary positive measure μ). Meyer [82], p. 17.

2. (ii) $\Leftrightarrow$ (iii) $\Rightarrow$ (iv). Meyer [82], p. 20.

3. (iv) $\Leftrightarrow$ (ii). Dunford-Schwartz [38], p. 294.

4. (iv) $\Leftrightarrow$ (iii). Dunford-Schwartz [38], p. 430. (Eberlein-Smulian theorem).

Theorem IV.3. (Dunford-Schwartz Ergodic Theorem). Let T be a linear operator on $L^1(\Omega,\mathbf{A},\mu)$ satisfying on $L^1(\Omega,\mathbf{A},\mu) \cap L^\infty(\Omega,\mathbf{A},\mu)$ the inequalities $||T||_1 \leq 1$ and $||T||_\infty \leq 1$ (which by the Riesz convexity theorem imply $||T||_p \leq 1$ for all $p \in [1,\infty[$). Then for every $p \in [1,\infty[$ and every $f \in L^p(\Omega,\mathbf{A},\mu)$ there exists a function $f^* \in L^p(\Omega,\mathbf{A},\mu)$ satisfying $Tf^* = f^*$ such that

$$\lim_{n\to\infty} \frac{1}{n} \sum_{k=0}^{n-1} T^k f = f^* [\mu].$$

Proof: Dunford-Schwartz [38], p. 675 ff.

References

[1] ALFSEN, E. M.: Compact Convex Sets and Boundary Integrals. Springer (1971).

[2] BAHADUR, R. R.: Sufficiency and Statistical Decision Functions. Ann. Math. Statist. 25(1954), 432-462.

[3] BAHADUR, R. R.: A Characterization of Sufficiency. Ann. Math. Statist. 26(1955), 286-293.

[4] BAHADUR, R. R.: Statistics and Subfields. Ann. Math. Statist. 26(1955), 490-497.

[5] BAHADUR, R. R.: On Unbiased Estimates of Uniformly Minimum Variance. Sankhyā 18(1957), 211-224.

[6] BARNDORFF-NIELSEN, O.: Information and Exponential Families in Statistical Theory. John Wiley & Sons (1978).

[7] BARRA, J. -R.: Notions Fondamentales de Statistique Mathématique. Dunod (1971).

[8] BARTENSCHLAGER, H.: Charakterisierung universell zulässiger Entscheidungsverfahren. Z. Wahrscheinlichkeitstheorie verw. Gebiete 33(1975), 187-194.

[9] BAUER, H.: Probability Theory and Elements of Measure Theory. Second English Edition. Academic Press (1981).

[10] BAUMANN, V.: Eine parameterfreie Theorie der ungünstigsten Verteilungen für das Testen von Hypothesen. Z. Wahrscheinlichkeitstheorie verw. Gebiete 11(1968), 41-60.

[11] BEDNARSKI, T.: Binary Experiments, Minimax Tests and 2-Alternating Capacities. Ann. Statist. 10(1982), 226-232.

[12] BELL, C. B., BLACKWELL, D., BREIMAN, L.: On the Completeness of Order Statistics. Ann. Math. Statist. 31(1960), 794-797.

[13] BERGER, J. O.: Statistical Decision Theory. Springer (1980).

[14] BIRNBAUM, A.: On the Foundations of Statistical Inference: Binary Experiments. Ann. Math. Statist. 32(1961), 414-435.

[15] BIRNBAUM, A.: On the Foundations of Statistical Inference II. Institute of Math. Sciences, New York University, 275(1960).

[16] BJÖRNSSON, O. J.: A Note on the Characterization of Standard Borel Spaces. Math. Scand. 47(1980), 135-136.

[17] BLACKWELL, D.: Comparison of Experiments, Proc. 2nd Berkeley Symp. Math. Stat. Prob. (1951), 93-102.

[18] BLACKWELL, D.: Equivalent Comparison of Experiments. Ann. Math. Statist. 24(1953), 265-272.

[19] BLACKWELL, D., GIRSCHICK, M. A.: Theory of Games and Statistical Decisions. Dover Publ. (1979).

[20] BOLL, C. H.: Comparison of Experiments in the Infinite Case. Ph.D. Thesis, Stanford University (1955).

[21] BONNESEN, T. FENCHEL, W.: Theorie der konvexen Körper. Chelsea Publ. Comp. (1971).

[22] BORGES, R., PFANZAGL, J.: A Characterization of the One-Parameter Exponential Family of Distributions by Monotonicity of Likelihood Ratios. Z. Wahrscheinlichkeitstheorie verw. Gebiete 2(1963), 111-117.

[23] BOURBAKI, N.: Espaces Vectoriels Topologiques, Chapitres I, II, 2e Edition. Hermann (1966).

[24] BOURBAKI, N.: Espaces Vectoriels Topologiques, Chapitres III, IV. Hermann (1964).

[25] BOURBAKI, N.: Intégration, Chapitres 1-4, 2e Edition. Hermann (1965).

[26] BURKHOLDER, D. L.: Sufficiency in the Undominated Case. Ann. Math. Statist. 32(1961), 1191-1200.

[27] CARTIER, P., FELL, J. M. G., MEYER, P. A.: Comparaison des Mesures Portées par un Ensemble Convexe Compact. Bull. Soc. Math. France 29(1964), 435-445.

[28] CHERNOFF, H.: A Measure of Asymptotic Efficiency for Tests of a Hypothesis Based on the Sum of Observations. Ann. Math. Statist. 23(1952), 493-507.

[29] CHRISTENSEN, J. -P.R.: Topology and Borel Structure. North-Holland Publ. Comp. (1974).

[30] CSISZAR, I.: Eine informationstheoretische Ungleichung und ihre Anwendung auf den Beweis der Ergodizität von Markoffschen Ketten. Publ. Math. Inst. Hung. Acad. Sci., Ser. A, 8(1963), 85-108.

[31] CSISZAR, I.: Information-Type Measures of Difference of Probability Distributions and Indirect Observations. Studia Sci. Math. Hung. 2(1967), 299-318.

[32] DAY, M. M.: Fixed Point Theorems for Compact Convex Sets. Ill. J. Math. 5(1961), 585-589. Correction. Ill. J. Math. 8(1964), 713.

[33] DE GROOT, M. H.: Uncertainty, Information, and Sequential Experiments. Ann. Math. Statist. 33(1962), 404-419.

[34] DE GROOT, M. H.: Optimal Allocation of Observations. Ann. Inst. Statist. Math. 18(1966), 13-28.

[35] DENNY, J. L.: Sufficient Conditions for a Family of Probabilities to be Exponential. Proc. Nat. Acad. Sci. (USA) 57(1967), 1184-1187.

[36] DETTWEILER, E.: Über die Existenz überall trennscharfer Tests im nicht-dominierten Fall. Metrika 25(1978), 247-254.

[37] DEVILLE, J. C.: Information et Exhaustivité Relative dans Certaines Structures Statistiques. Proc. 10th Session, ISI Warsaw 1975, Vol. 3, Bull. Inst. Internat. Stat. 46(1976), 217-223.

[38] DUNFORD, N., SCHWARTZ, J. T.: Linear Operators. Part I: General Theory. Interscience (1958).

[39] EDWARDS, R. E.: Functional Analysis: Theory and Applications. Holt-Rinehart-Winston (1965).

[40] FELLER, W.: Diffusion Processes in Genetics. Proc. 2nd Berkeley Symp. Math. Stat. Prob. (1951), 227-246.

[41] FERGUSON, Th. S.: Mathematical Statistics: A Decision Theoretic Approach. Academic Press (1967).

[42] FRASER, D. S.: Nonparametric Methods in Statistics. John Wiley & Sons (1967).

[43] GOEL, P. K., DE GROOT, M. H.: Comparison of Experiments and Information Measures. Ann. Statist. 7(1979), 1066-1077.

[44] GOSH, J. K., MORIMOTO, H., YAMADA, S.: Neyman Factorization and Minimality of Pairwise Sufficient Subfields. Ann. Statist. 9(1981), 514-530.

[45] GREENLEAF, F. P.: Invariant Means on Topological Groups. Van Nostrand-Reinhold (1969).

[46] GRETTENBERG, Th. L.: The Ordering of Finite Experiments. Trans. 3rd Prague Conference on Information Theory, Statistical Decision Functions, Random Processes. Prague (1964).

[47] HALMOS, P. R., SAVAGE, L. J.: Applications of the Radon-Nikodym Theorem to the Theory of Sufficient Statistics. Ann. Math. Statist. 20(1949), 225-241.

[48] HANSEN, O. H., TORGERSEN, E. N.: Comparison of Linear Normal Experiments. Ann. Math. Statist. 2(1974), 367-373.

[49] HARDY, G. H., LITTLEWOOD, J. E., POLYA, G: Inequalities. Cambridge University Press (1934).

[50] HASEGAWA, M., PERLMAN, M. D.: On the Existence of a Minimal Sufficient Subfield. Ann. Statist. 2(1974), 1049-1055. Correction. Ann. Statist. 3(1975), 1371-1372.

[51] HEYER, H.: Erschöpftheit und Invarianz beim Vergleich von Experimenten. Z. Wahrscheinlichkeitstheorie verw. Gebiete 12(1969), 21-55.

[52] HEYER, H.: Zum Erschöpftheitsbergriff von D. Blackwell. Metrika 19(1972), 54-67.

[53] HEYER, H.: Invariante Markoff-Kerne und der Vergleich von Translationsexperimenten. Mh. Math. 88(1979), 123-135.

[54] HEYER, H.: Information-Type Measures and Sufficiency. Symposia Mathematica XXV (1981), 25-54.

[55] HEYER, H., TORTRAT, A.: Sur la Divisibilité des Probabilités dans un Groupe Topologique. Z. Wahrscheinlichkeitstheorie verw. Gebiete 16(1970), 307-320.

[56] HOFFMANN-JØRGENSEN, J.: The Theory of Analytic Sets. Lecture Notes, Aarhus University (1970).

[57] IONESCU-TULCEA, A., IONESCU-TULCEA, C.: Topics in the Theory of Lifting. Springer (1969).

[58] KRAFFT, O., WITTING, H.: Optimale Tests und ungünstigste Verteilungen. Z. Wahrscheinlichkeitstheorie verw. Gebiete 7(1967), 289-302.

[59] KULLBACK, S.: Information Theory and Statistics. John Wiley & Sons (1959).

[60] KULLBACK, S., LEIBLER, R.: On Information and Sufficiency. Ann. Math. Statist. 22(1951), 79-86.

[61] KURATOWSKI, C.: Topology I. Mathematical Monographs, Warszawa-Lwów (1933).

[62] KUSAMA, T., YAMADA, S.: On Compactness of the Statistical Structure and Sufficiency. Osaka J. Math. 9(1972), 11-18.

[63] LANDERS, D.: Sufficient and Minimal Sufficient σ-Fields. Z. Wahrscheinlichkeitstheorie verw. Gebiete 23(1972), 197-207.

[64] LANDERS, D., ROGGE, L.: Existence of Most Powerful Tests for Composite Hypotheses. Z. Wahrscheinlichkeitstheorie verw. Gebiete 24(1972), 339-340.

[65] LAURANT, F., OHEIX, M., RAOULT, J. -P.: Tests d'Hypothèses. Ann. Inst. Henri Poincaré 5(1969), 385-414.

[66] LE BIHAN, M. -F., LITTAYE-PETIT, M., PETIT, J. -L.: Exhaustivité par Paire. C. R. Acad. Sci. Paris Sér. A 270(1970), 1753-1756.

[67] LECAM, L.: An Extension of Wald's Theory of Statistical Decision Functions. Ann. Math. Statist. 26(1955), 69-81.

[68] LECAM, L.: Sufficiency and Approximate Sufficiency. Ann. Math. Statist. 35(1964), 1419-1455.

[69] LECAM, L.: Limits of Experiments. Proc. 6th Berkeley Symp. Math. Stat. Prob., Vol. 1 (1972), 245-261.

[70] LECAM, L.: Notes on Asymptotic Methods in Statistical Decision Theory. Centre de Recherches Mathématiques, Université de Montréal (1974).

[71] LECAM, L.: Distances between Experiments. In: Survey of Statistical Designs and Linear Models. Edited by J. N. Srivastava. North Holland Publ. Comp. (1975), 383-395.

[72] LEHMANN, E. L.: Notes on the Theory of Estimation. Associated Student's Store, University of California, Berkeley (1950).

[73] LEHMANN, E. L.: Testing Statistical Hypotheses. John Wiley & Sons (1959).

[74] LINDLEY, D. V.: On a Measure of the Information Provided by an Experiment. Ann. Math. Statist. 27(1956), 986-1005.

[75] LINDQVIST, Bo: How Fast Does a Markov Chain Forget the Initial State? A Decision Theoretic Approach. Scand. J. Statist. 4(1977), 145-152.

[76] LINDQVIST, Bo: A Decision Theoretic Characterization of Weak Ergodicity. Z. Wahrscheinlichkeitstheorie verw. Gebiete 44(1978), 155-158.

[77] LINDQVIST, Bo: On the Loss of Information Incurred by Lumping States of a Markov Chain. Scand. J. Statist. 5(1978), 92-98.

[78] LITTAYE-PETIT, M., PIEDNOIR, J. -L., VAN CUTSEM, B.: Exhaustivité. Ann. Inst. Henri Poincaré 5(1969), 289-322.

[79] LUKACS, E.: Characteristic Functions. 2nd Edition. Griffin (1970).

[80] LUSCHGY, H.: Sur l'Existence d'une Plus Petite Sous-Tribu Exhaustive par Paire. Ann. Inst. Henri Poincaré 14(1978), 391-398.

[81] MARTIN, F., PETIT, J.-L., LITTAYE-PETIT, M.: Comparaison des Expériences. Ann. Inst. Henri Poincaré 7(1971), 145-176.

[82] MEYER, P. A.: Probability and Potentials. Blaisdell Publ. Comp. (1966).

[83] MORIMOTO, H.: Statistical Structure of the Problem of Sampling From Finite Populations. Ann. Math. Statist. 43(1972), 490-497.

[84] MORSE, N., SACKSTEDER, R.: Statistical Isomorphisms. Ann. Math. Statist. 37(1966), 203-214.

[85] MÜLLER, D. W.: Statistische Entscheidungstheorie. Lecture Notes, University of Göttingen (1971).

[86] MUSSMANN, D.: Vergleich von Experimenten im schwach dominierten Fall. Z. Wahrscheinlichkeitstheorie verw. Gebiete. 24(1972), 295-308.

[87] MUSSMANN, D.: Suffiziente Vergröberungen im schwach dominierten Fall. Metrika 20(1973), 219-229.

[88] MUSSMANN, D.: Equivalent Statistical Experiments. Trans. 8th Prague Conference on Information Theory, Statistical Decision Functions, Random Processes (1978), 51-58.

[89] MUSSMANN, D.: Sufficiency and f-Divergence. Studia Sci. Math. Hung. 14(1979), 37-41.

[90] NEMETZ, D.: Information-Type Measures and Their Applications to Finite Decision Problems. Lecture Notes No. 17, Carleton University (1977).

[91] NEVEU, J.: Discrete-Parameter Martingales. North Holland Publ. Comp. (1975).

[92] OWEN, G.: Game Theory. W. B. Saunders (1968).

[93] PARTHASARATHY, K. R.: Probability Measures on Metric Spaces. Academic Press (1967).

[94] PARTHASARATHY, K. R.: Introduction to Probability and Measure. MacMillan India (1977).

[95] PETIT, J.-L.: Exhaustivité, Ancillarité et Invariance. Ann. Inst. Henri Poincaré 6(1970), 327-334.

[96] PFANZAGL, J.: Über die Existenz überall trennscharfer Tests. Metrika 3(1960), 169-176. Eine ergänzende Bemerkung hierzu. Metrika 4(1961), 105-106.

[97] PFANZAGL, J.: Überall trennscharfe Tests und monotone Dichtequotienten. Z. Wahrscheinlichkeitstheorie verw. Gebiete 1(1963), 109-115.

[98] PFANZAGL, J.: A Characterization of Sufficiency by Power Functions. Metrika 21(1974), 197-199.

[99] PFANZAGL, J.: Convexity and Conditional Expectations. Ann. Prob. 2(1974), 490-494.

[100] PHELPS, R. R.: Lectures on Choquet's Theorem. D. Van Nostrand (1966).

[101] PITCHER, T. S.: Sets of Measures not Admitting Necessary and Sufficient Statistics or Subfields. Ann. Math. Statist. 28(1957), 267-268.

[102] PITCHER, T. S.: A More General Property than Domination for Sets of Probability Measures. Pacific J. Math. 15(1965), 597-611.

[103] RAOULT, J. -P.: Structures Statistiques. Presses Universitaires de France (1975).

[104] RAUHUT, B., SCHMITZ, N., ZACHOW, E. -W.: Spieltheorie. Teubner (1979).

[105] ROCKAFELLAR, R. T.: Convex Analysis. Princeton University Press (1970).

[106] ROGGE, L.: The Relation between Sufficient Statistics and Minimal Sufficient σ-Fields. Z. Wahrscheinlichkeitstheorie verw. Gebiete 23(1972), 208-215.

[107] ROGGE, L.: Compactness and Domination. Manuscripta Math. 7(1972), 299-306.

[108] ROMIER, G.: Modèle d'Expérimentation Statistique. Ann. Inst. Henri Poincaré 5(1969), 275-288.

[109] ROMIER, G.: Decision Statistique. Ann. Inst. Henri Poincaré 5(1969), 323-355.

[110] ROY, K. K., RAMAMOORTHI, R. V.: Relationship between Bayes, Classical and Decision Theoretic Sufficiency. Tech. Report No. 30, ISI Calcutta (1978).

[111] SACKSTEDER, R.: A Note on Statistical Equivalence. Ann. Math. Statist. 38(1967), 784-794.

[112] SCHAEFER, H. H.: Topological Vector Spaces. Springer (1970).

[113] SCHMETTERER, L.: On Unbiased Estimation. Ann. Math. Statist. 31(1960), 1154-1163.

[114] SCHMETTERER, L.: Über eine allgemeine Theorie der erwartungstreuen Schätzungen. Publ. Math. Inst. Hung. Acad. Sci. Ser. A, 6(1961), 295-

[115] SCHMETTERER, L.: Quelques Problèmes Mathématiques de la Statistique. Université de Clermont, Faculté des Sciences (1967).

[116] SCHMETTERER, L.: Introduction to Mathematical Statistics. Springer (1974).

[117] SCHWARTZ, L.: Radon Measures on Arbitrary Topological Spaces and Cylindrical Measures. Oxford University Press (1973).

[118] SCHWARTZ, L.: Lectures on Desintegration of Measures. Tata Inst. Fund. Research, Bombay (1976).

[119] SHERMAN, S.: On a Theorem of Hardy, Littlewood, Polya, and Blackwell. Proc. Nat. Acad. Sci. (USA) 37(1951), 826-831.

[120] SIEBERT, E.: Pairwise Sufficiency. Z. Wahrscheinlichkeitstheorie verw. Gebiete 46(1979), 237-246.

[121] SPEED, T.P.: A Note on Pairwise Sufficiency and Completions. Sankhyā 38(1976), 194-196.

[122] SPEED, T. P.: A Review of Some Results Concerning the Completion of Sub-σ-Fields. Preprint (1976).

[123] STEIN, C.: Notes on the Comparison of Experiments. University of Chicago (1951).

[124] STEPNIAK, C., TORGERSEN, E. N.: Comparison of Linear Models with Partially Known Covariances with Respect to Unbiased Estimators. Scand. J. Statist. 8(1981), 183-184.

[125] STONE, M.: Non-Equivalent Comparisons of Experiments and Their Use for Experiments Involving Location Parameters. Ann. Math. Statist. 32(1961), 326-332.

[126] SWENSEN, A. R.: Deficiencies between Linear Normal Experiments. Ann. Statist. 8(1980), 1142-1155.

[127] TORGERSEN, E. N.: Comparison of Experiments when the Parameter Space is Finite. Z. Wahrscheinlichkeitstheorie verw. Gebiete 16(1970), 219-249.

[128] TORGERSEN, E. N.: Comparison of Translation Experiments. Ann. Math. Statist. 43(1972), 1383-1399.

[129] TORGERSEN, E. N.: Notes on Comparison of Statistical Experiments, Chapters 0-8. University of Oslo (1973/74).

[130] TORGERSEN, E. N.: Comparison of Statistical Experiments. Scand. J. Statist. 3(1976), 186-208.

[131] TORGERSEN, E. N.: Deviations from Total Information and from Total Ignorance as Measures of Information. Statistical Research Report, Institute of Mathematics, University of Oslo (1976).

[132] TORGERSEN, E. N.: Mixtures and Products of Dominated Experiments. Ann. Statist. 5(1977), 44-64.

[133] TORGERSEN, E. N.: Deviations from Total Information and from Total Ignorance as Measures of Information. Math. Statistics, Banach Center Publications, Volume 6, PWN-Polish Scientific Publishers, Warsaw (1980), 315-322.

[134] TORGERSEN, E. N.: Measures of Information Based on Comparison with Total Information and with Total Ignorance. Ann. Statist. 9(1981), 638-657.

[135] TORGERSEN, E. N.: On Complete Sufficient Statistics and Uniformly Minimum Variance Unbiased Estimators. Symposia Mathematica XXV (1981), 137-153.

[136] VALENTINE, F. A.: Convex Sets. McGraw-Hill (1964).

[137] WALD, A.: Statistical Decision Functions. Chelsea Publ. Comp. (1971).

[138] WALD, A., WOLFOWITZ, J.: Two Methods of Randomization in Statistics and the Theory of Games. Ann. Math. 53(1951), 581-586.

[139] VON WEIZSÄCKER, H.: Zur Gleichwertigkeit zweier Arten der Randomisierung. Manuscripta Math. 11(1974), 91-94.

[140] WITTING, H.: Mathematische Statistik. B. G. Teubner (1966).

[141] ZAANEN, A.C.: Integration, 2nd Edition. North-Holland Publ. Comp. (1967).

Symbol Index

Subject Index